T0324940

Genetic Suspects

Global Governance of Forensic DNA Profiling and Databasing

As DNA forensic profiling and databasing become established as key technologies in the toolbox of the forensic sciences, their expanding use raises important issues that promise to touch everyone's lives. In an authoritative global investigation of a diversity of countries, including those at the forefront of these technologies' development and use, this book identifies and provides critical reflection upon the many issues of privacy; distributive justice; who shapes and governs DNA information systems; biosurveillance; function creep; the reliability of collection, storage and analysis of DNA profiles; the possibility of transferring medical DNA information to forensics databases; and democratic involvement and transparency in governance, an emergent key issue. This book is timely and significant in providing the essential background and discussion of the ethical, legal and societal dimensions for academics, practitioners, public interest and criminal justice organisations, and students of the life sciences, law, politics and sociology.

RICHARD HINDMARSH is Associate Professor at Griffith School of Environment, and Centre for Governance and Public Policy, Griffith University, Australia. He specialises in co-produced sociotechnical systems analysis informed by science, technology and society (STS) studies; governance and regulation studies; environmental policy; and the politics and sociology of green biotechnology and forensic DNA technologies. Professor Hindmarsh is also an international expert reviewer for both the Australian Research Council and the UK Economic and Social Research Council and invited International Consultative Group member of the (US) Council for Responsible Genetics. Currently, as its co-founder, he is further establishing the Asia–Pacific STS Network, a new regional research community spanning Australasia, East and Southeast Asia and Oceania, as its convenor for 2010–2011.

BARBARA PRAINSACK is Reader at the Centre for Biomedicine & Society (CBAS) at King's College London, UK. A political scientist by training, her research focuses on how politics, bioscience, religion and 'culture' mutually shape each other, and how they interact with how we understand ourselves as human beings, individuals and citizens. Her research on regulatory and societal aspects of human cloning, stem cell research and DNA testing (both medical and forensic) has featured in national and international media such as BBC News, ABC National Radio (Australia), and *Die Zeit*. She is a member of the Editorial Advisory Boards of *Science as Culture* and *Personalized Medicine*, and a member of the National Bioethics Commission in Austria.

Genetic Suspects

Global Governance of
Forensic DNA Profiling
and Databasing

Edited by
**RICHARD
HINDMARSH**
Griffith University, Australia

**BARBARA
PRAINSACK**
King's College London

CAMBRIDGE
UNIVERSITY PRESS

University Printing House, Cambridge CB2 8BS, United Kingdom

One Liberty Plaza, 20th Floor, New York, NY 10006, USA

477 Williamstown Road, Port Melbourne, VIC 3207, Australia

314-321, 3rd Floor, Plot 3, Splendor Forum, Jasola District Centre, New Delhi - 110025, India

79 Anson Road, #06-04/06, Singapore 079906

Cambridge University Press is part of the University of Cambridge.

It furthers the University's mission by disseminating knowledge in the pursuit of education, learning and research at the highest international levels of excellence.

www.cambridge.org
Information on this title: www.cambridge.org/9781108829076

© Cambridge University Press 2010

First published 2010
First paperback edition 2020

A catalogue record for this publication is available from the British Library

Library of Congress Cataloging in Publication data
Genetic suspects : global governance of forensic DNA profiling and databasing / edited by Richard Hindmarsh, Barbara Prainsack.
p. cm.
Summary: "The introduction of DNA profiling and databasing into the criminal justice system, which began in 1988, when English baker Colin Pitchfork was the first person convicted through the use of DNA evidence (Sanders 2000C001-001)" – Provided by publisher.
Includes bibliographical references and index.
ISBN 978-0-521-51943-4
1. DNA fingerprinting. I. Hindmarsh, R. A. (Richard A.) II. Prainsack, Barbara. III. Title.
RA1057.55.G46 2010
614´.1-dc22
2010011237

ISBN 978-0-521-51943-4 Hardback
ISBN 978-1-108-82907-6 Paperback

Contents

Contributors

Jay D. Aronson
Department of History, Carnegie Mellon University, Pittsburgh, PA, USA

Gali Ben-Or
Legal Counsel and Legislation Department, Israeli Ministry of Justice, Israel

Simon A. Cole
Department of Criminology, Law & Society, University of California, Irvine, CA, USA

Johanne Yttri Dahl
Department of Sociology and Political Science, NTNU Social Research, Norwegian University of Science and Technology, Trondheim, Norway

Maria Corazon A. De Ungria
DNA Analysis Laboratory, Natural Sciences Research Institute, University of the Philippines, Quezon City, Philippines

Gabriela Fisman
Legal Counsel and Legislation Department, Israeli Ministry of Justice, Israel

Richard Hindmarsh
Griffith School of Environment and Centre for Governance and Public Policy, Griffith University, Brisbane, Australia

Sheila Jasanoff
John F. Kennedy School of Government, Harvard University, Cambridge, MA, USA

Jose Manguera Jose
Legal Consultant, DNA Analysis Laboratory, Natural Sciences Research Institute, University of the Philippines, Diliman, Quezon City, Philippines

Mairi Levitt
Department of Philosophy, Lancaster University, Lancaster, UK

Michael Lynch
Department of Science & Technology Studies, Cornell University, Ithaca, NY, USA

Helena Machado
Department of Sociology, University of Minho, Braga, Portugal

Gerald Midgley
Institute of Environmental Science & Research, Christchurch, New Zealand

Barbara Prainsack
Centre for Biomedicine & Society, King's College London, London, UK

Susana Silva
Department of Hygiene and Epidemiology, University of Porto Medical School, Porto, Portugal

Victor Toom
Amsterdam School for Social Science Research, University of Amsterdam, Amsterdam, the Netherlands

Richard Tutton
ESRC Centre for the Economic and Social Aspects of Genomics (Cesagen), Lancaster University, Lancaster, UK

Johanna S. Veth
Institute of Environmental Science & Research, Christchurch, New Zealand

Harriet A. Washington
Independent researcher, USA

Robin Williams
School of Applied Social Sciences, University of Durham, Durham, UK

Elazar Zadok
Independent researcher and consultant in forensic science and management (former Director of the Division of Identification and Forensic Science (DIFS) of the Israel Police)

About the contributors

Jay D. Aronson is an Associate Professor of Science, Technology, and Society at Carnegie Mellon University. His research and teaching focus is on the interactions of science, technology, law and human rights in a variety of contexts. His first book, *Genetic Witness: Science, Law, and Controversy in the Making of DNA Profiling* (Rutgers University Press, 2007), examines the development of forensic DNA analysis in the American legal system. He is currently engaged in a long-term study of the ethical, political and social dimensions of post-conflict and post-disaster DNA identification of the missing and disappeared. He received his PhD in History of Science and Technology from the University of Minnesota and was both a pre- and postdoctoral fellow at Harvard University's John F. Kennedy School of Government.

Gali Ben-Or did her law degree at the Hebrew University in Jerusalem in 1992, and was admitted to the Israeli Bar in 1993. Since 1995, she has worked in the Legal Counsel and Legislation Department of the Israeli Ministry of Justice. She coordinated the work of the governmental team that prepared the National Forensic Databases Bill, which established the DNA database in Israel. She is also in charge of legal advice and legislation in the fields of judicial review, administrative courts, surrogacy, cloning, genetics and bioethics, and she participates and represents the government during the legislation process in the Israeli Parliament committees. Gali also lectures in different forums in Israel regarding issues relating to her work.

Simon A. Cole is an Associate Professor and Chair of the Department of Criminology, Law and Society at the University of California, Irvine. He took his first degree in History at

Princeton University and a PhD in Science and Technology Studies at Cornell University. His most recent book is *Truth Machine: The Contentious History of DNA Fingerprinting* (with Michael Lynch, Ruth McNally & Kathleen Jordan; University of Chicago Press, 2008). He is a member of the American Judicature Society Commission on Forensic Science & Public Policy. His current interests are the sociology of forensic science and the development of criminal identification databases and biometric technologies.

Johanne Yttri Dahl is a Lecturer at the Norwegian Police University College. She has submitted her doctoral thesis in sociology about the use of DNA evidence in courts and forensic DNA databases at the Norwegian University of Science and Technology. Previously Johanne worked as a research assistant on the project UrbanEye: On the Threshold to Urban Panopticon. Her research interests lie within the fields of security, surveillance and gender.

Maria Corazon A. De Ungria is the Head of the DNA Analysis Laboratory, Natural Sciences Research Institute at the University of the Philippines. She obtained her PhD in Microbiology at the University of New South Wales in Australia. She has written numerous papers on the forensic use of DNA and has been recognised as an expert witness by the Philippine Supreme Court. She was awarded as one of the 2007 Ten Outstanding Women in the Nations Service (TOWNS) and one of the 2003 Ten Outstanding Young Scientists by the National Academy of Science and Technology. In 2007, she was appointed as one of five regional Affiliate Fellows for East and Southeast Asia by the Academy of Sciences for the Developing World (TWAS).

Gabriela Fisman obtained her LLB (2000) and MA in political Science (2002) at Tel Aviv University, Israel. Gabriela was admitted to the Israeli Bar in 2001. Between 1995 and 2001, Gabriela held different positions as an academic assistant at Tel-Aviv University and the Israeli branch of Manchester University. In 2002, Gabriela joined the Legal Counsel and Legislation Department of the Israeli Ministry of Justice and has since represented the government in the legislating processes relating to criminal law, with emphasis on inspection and enforcement powers (investigation, detention and arrest, search and body

cavity search), suspects' rights, criminal record and police operation. Gabriela participated in the legislation process regulating
the establishment and management of Israel's National Forensic
DNA Database.

Richard Hindmarsh is an Associate Professor at the Centre for
Governance and Public Policy, Griffith University, and the
Griffith School of Environment, Brisbane, Australia. In playing
a major role in the rise of Australasian biotechnology and society
studies, he has drawn together cutting-edge anthologies and
journal issues in new genetics, including *New Genetics and Society*
and *Science as Culture*. His monograph *Edging Towards BioUtopia:
A New Politics of Life and the Democratic Challenge* (University of
Western Australia Press, 2008) is a foundational social history
of green biotechnology in Australia, highlighting particularly the
systemic failure of regulatory frameworks either to deal with
genetically modified organisms released into the environment
or to adequately reflect public concern. His engagement in forensic DNA technologies was especially stimulated through the
international collaborative research project Genes Without
Borders – Towards Global Genomic Governance (2006–2008). He
is currently establishing a new regional research community
called the Asia–Pacific Science, Technology and Society (STS)
Network.

Sheila Jasanoff is Pforzheimer Professor of Science and
Technology Studies at Harvard University's John F. Kennedy
School of Government. She is affiliated with the Department of
the History of Science and the program in Environmental Science
and Public Policy; she holds a visiting appointment at Harvard Law
School. She has held academic appointments at Cornell, Yale,
Cambridge, Oxford, MIT, and Kyoto. At Cornell, she founded and
chaired the Department of Science and Technology Studies.
She has been Leverhulme Visiting Professor at Cambridge,
Karl Deutsch Professor at the Science Center Berlin and Fellow at
the Berlin Institute for Advanced Study (*Wissenschaftskolleg*). She has
done pioneering research on the relationship of science and technology with law, politics and policy in modern democratic societies, looking particularly at the role of science in cultures of public
participation and public reasoning. She has written and lectured
widely on environmental regulation, risk management, the politics of the life sciences and the governance of science and

technology in the USA, Europe and India. Her books on these topics include *Controlling Chemicals* (with Ronald Brickman and Thomas Ilgen; Cornell University Press, 1985), *The Fifth Branch* (Harvard University Press, 1990), *Science at the Bar* (Harvard University Press, 1995) and *Designs on Nature* (Princeton University Press, 2005).

Jose Manguera Jose is the Secretary of the Integrated Bar of the Philippines (IBP) Makati Chapter, a professional organisation of lawyers, and is active in its legal aid programme. In 1999 and 2001, he was named the Most Outstanding Legal Aide Lawyer by the IBP. He obtained his law degree from the University of the Philippines College of Law in 1988 and was the 15th placer in that year's bar examination. He has given lectures on expert witness testimony, collection of crime scene DNA evidence and the rule on DNA evidence promulgated by the Philippine Supreme Court. He is a practising litigation lawyer and is currently working on a book on the legal issues concerning the use of DNA evidence.

Mairi Levitt is a Senior Lecturer and Head of the Philosophy Department at Lancaster University. She has a social science background and a PhD; since 1993, she has engaged in multidisciplinary research on the ethical and social implications of genetics and biotechnology. Her research projects in different substantive areas have involved engagement work with the general public, young people and stakeholders. Her research on genetic databases includes exploring issues of trust in relation to the UK Biobank, a project on children's and parent's attitudes to the National DNA database and the Criminal Genes and Public Policy project, which examined the possible legal, social and political issues raised by current research in behavioural genetics through discussions with relevant researchers and professional groups.

Michael Lynch is a Professor in the Department of Science & Technology Studies at Cornell University. His research is on discourse, visual representation and practical action in research laboratories, clinical settings and legal tribunals. His most recent book, *Truth Machine: The Contentious History of DNA Fingerprinting* (with Simon Cole, Ruth McNally and Kathleen Jordan; University of Chicago Press, 2008) examines the interplay between law and science in criminal cases involving DNA evidence. He is Editor of the journal *Social Studies of Science* and served as President of the Society for Social Studies of Science from 2007 to 2009.

Helena Machado is an Associate Professor in the Department of Sociology and Senior Researcher at the Research Centre for the Social Sciences, University of Minho, Portugal. She has a Ph.D. in sociology and has written books and papers in a wide range of topics related to justice system and uses of DNA technology. Her research interests are primarily in the fields of forensic genetics, press representations of genetics and interfaces between the criminal justice system and the mass media. She has recently published in *Crime, Media and Culture* and *Public Understanding of Science*.

Gerald Midgley is a Senior Science Leader in the Institute of Environmental Science and Research, New Zealand. He has Visiting Professorships at the University of Hull (UK), the University of Queensland (Australia), the University of Canterbury (New Zealand) and Victoria University of Wellington (New Zealand). He is the author of *Systemic Intervention: Philosophy, Methodology, and Practice* (Kluwer/Plenum, 2000); the editor of *Systems Thinking* (Sage, 2003); and the co-editor of *Community Operational Research: OR and Systems Thinking for Community Development* (with Alejandro Ochoa-Arias; Kluwer/Plenum, 2004).

Barbara Prainsack is Reader at the Centre for Biomedicine & Society (CBAS) at King's College London, UK. Prior to joining King's in 2007, she worked at the Department of Political Science and the Life Science Governance Platform at the University of Vienna, Austria, where she led the international collaborative project Genes Without Borders – Towards Global Genomic Governance (2006–2008), out of which this volume developed. Her research focuses on how politics, bioscience, religion and 'culture' mutually constitute each other, and how they interact with how we understand ourselves as human beings, individuals and citizens. She is a member of the Editorial Advisory Boards of *Science as Culture* and *Personalized Medicine*, and of the National Bioethics Commission in Austria.

Susana Silva has a PhD in sociology and worked as a postdoctoral researcher and lecturer at the Research Centre for the Social Sciences, Department of Sociology, University of Minho, before moving to Cardiovascular Research and Development Unit, Department of Hygiene and Epidemiology, University of Porto Medical School, Portugal, where she works as a senior researcher.

Her main research interests are the processes of the mutual shaping of science, technology, law and gender within the medically assisted reproduction in Portugal, with a primary focus on interactions between expert and lay knowledge, informed consent and rhetoric related to donation of embryos, eggs and sperm. She has recently published in *Health, Risk and Society*, and *New Genetics and Society*.

Victor Toom has submitted his PhD on forensic genetic practices in the Netherlands. He considers himself a scholar of science and technology studies and is a member of the Amsterdam School for Social Science Analysis, a research institute at the University of Amsterdam. He teaches several courses at the university, among them 'Our Genetic Identity' at the Institute of Interdisciplinary Studies, and 'Policy, Ethics and Media' for a masters of science course. His research interests include genetics and society, (international) law and biomedical practices.

Richard Tutton is a Senior Lecturer at the Centre for Economic and Social Aspects of Genomics (Cesagen) at Lancaster University, UK. He works at the intersections of the social studies of science and the sociology of health and illness. His research interests are in the social and technical aspects of biobanking for biomedical research and the implications of developments in science, technology and medicine for cultural and social identities. He co-edited *Genetic Databases: Socio-ethical Issues in the Collection and Use of DNA* (with Oonagh Corrigan; Routledge, 2004).

Johanna S. Veth is a forensic scientist at the Institute of Environmental Science and Research in New Zealand. She is responsible for the analysis and interpretation of biological evidence recovered from crime scenes and regularly provides expert witness testimony. Johanna is also a doctoral candidate at the University of Canterbury and her current research interests include investigating the extent and nature of the gap between lay and professional understandings of forensic DNA technologies and identifying potential consequences for the criminal justice system.

Harriet A. Washington spent 2002–2005 as a Research Fellow in Medical Ethics at Harvard Medical School after a John S. Knight Fellowship at Stanford University and a Harvard Fellowship for Advanced Studies in Public Health. Her books include *Medical*

Apartheid: The Dark History of Medical Experimentation on Black Americans from Colonial Times to the Present, which won the 2007 National Book Critics Circle Award. She has been a visiting faculty member at DePaul University, and the University of Chicago. Her work has appeared in *Nature*, the *American Journal of Public Health*, the *New England Journal of Medicine*, the *New York Times*, and *Harper's*. She co-authored a 2008 *Journal of the American Medical Association* paper that provided the basis for the association's apology to US black physicians.

Robin Williams is Professor Emeritus in the School of Applied Social Sciences at Durham University and a Professor in the School of Applied Sciences at Northumbria University. He co-authored a book on the growth of forensic DNA databases, *Genetic Policing: The Use of DNA in Criminal Investigations* (with Paul Johnson; Willan, 2008) and has recently co-edited a *Handbook of Forensic Science* (with Jim Fraser; Willan, 2009). He was a member of the Nuffield Council on Bioethics Working Party on the Forensic Uses of Bioinformation and is currently completing a Wellcome Trust-funded study of the growth of forensic DNA databases across the European Union. He has recently begun work on the first UK study of the use of forensic science in homicide investigations.

Elazar (Azi) Zadok obtained his PhD in organic chemistry at the Weizman Institute of Science, Rehovot, Israel in 1983. He served in different senior scientific positions in the Israeli Defense Forces and the Israeli chemical industries. During 2000–2007, he was the Director of the Division of Identification and Forensic Science of the Israel National Police, ranking Brigadier General. He was deeply involved in the legislation and the establishment processes of the Israeli National Forensic DNA Database, becoming operational early in 2006. He also led the process of accreditation of the National Forensic Laboratories. He is now retired, serving as consultant for forensic services construction in developing countries, and is still active in lecturing on legislative and ethical perspectives of forensic databases.

Foreword

Without science and its muscular twin technology, contemporary societies would be reduced to chaos. We would lose much of our ability to read, write, communicate, travel, grow crops, raise animals, cook food or find clean water to sustain our lives. Commercial transaction would stop; financial institutions be crippled; emergency services incapacitated, and hospitals no longer able to provide essential treatment. In that devastated, dying world, law and order would break down, and violence would flourish. Not insignificantly, we would lose the capacity to track and prosecute lawbreakers and criminals. Today, even law enforcement has become a 'high-tech' business, and DNA profiling, the subject of this book, is the most highly valued recent addition to the toolkit of the forensic sciences. For law enforcement agencies, it is hard to imagine life before or without it.

Technology's benefits for social order are obvious, ubiquitous and unquestionable. Yet, since long before the scientific revolution, human beings have looked upon the unchecked thirst for knowledge and its applications as dangerous things. Humanity's Faustian bargain with science set us on a path of discovering more and more about the way the world works and accomplishing more impressive feats with the results of that knowledge. But around the bends of the brightly lit corridors of enlightenment lurked unintended consequences that threatened to usurp our humanity and even annihilate us physically. Advances in the life sciences and technologies have proved particularly alarming because they destabilize the worth of life itself. Biological inventiveness calls for heightened attention to ensure that important human values are not lost because no one is watching. How to cultivate that sense of social alertness – an instinct as valuable to advanced industrial civilizations as keenness of sight, smell and hearing were to our prehistoric ancestors – remains one of modernity's most pressing problems.

This volume provides one appealing answer: a multi-sited case study of a powerful, emerging sociotechnical system that invites readers to address its benefits, its risks and its governance. The authors follow a single new technology – DNA profiling – from the whiff of promise in a UK laboratory to the reality of institutionalized law enforcement practices around the world. It is a relatively short history, but revolutionary in its implications. In 1985, the British scientist Alec Jeffreys discovered almost by chance that random variations in the structure of genes could be used as a technique for identifying individuals from samples of their DNA. Within a bare quarter century, DNA profiling became the best known and most celebrated instrument of forensic science, a virtually failsafe tool, if properly used, for linking violent crimes to the persons who perpetrated them. To prosecutors, this was fingerprinting on steroids: a technique based on the soundest of basic science and seeming to eliminate virtually all possibility of false identifications.

But DNA profiling turned out to be less than failsafe in practice and to have troubling uses beyond the identification of guilty and violent people. Compiled into databases, DNA identifiers offer a highly efficient means of storage and retrieval of personal information that can be used to track, group and classify people with or without their acquiescence. Once a person's biological identity gets locked up in a DNA database, the profiled individual has very little say in how that information will be used and managed. If mistakes were made, or inappropriate data collected, those facts remain largely outside the individual's capacity to detect or correct. Instead, the technology seems ideally suited to feeding the appetites of the all-seeing state as conceived by Michel Foucault and other students of late modernity. In the aggregate, the DNA profiles of all those 'genetic suspects' who give this book its title provide the raw material for constructing innumerable knowable and manageable populations, whose identities and traits the state can call up whenever such groupings serve its purposes of surveillance and control.

The precision of DNA profiling, the very feature that makes it miraculous in the hands of responsible law enforcement authorities, also enables its potentially grave misuse. Consider the following description by Jeffrey Rosen (2009)

> In March 2003, a drunk in southern England threw a brick off a bridge late at night, striking and killing a truck driver traveling along the freeway below. Armed with DNA from the blood on the brick, the British

police searched the United Kingdom's national DNA database, which includes convicted felons and people who have been arrested, but failed to get a direct match. They then conducted a DNA dragnet, asking hundreds of young men in the area to donate a sample voluntarily, but still came up short. Without any other leads, the police decided to conduct what's called a 'familial search' of the national DNA database.

That search, when trimmed to two counties near the crime scene, produced 25 partial matches that were deemed close enough to warrant further follow-through. Police investigators interviewed the person with the largest number of shared alleles and discovered he had a brother, who was then asked to provide a sample of his DNA. This time, the match was perfect; confronted with the evidence, the brother confessed and was subsequently convicted of manslaughter.

The case presents all the features that have won the allegiance of police departments the world over. A seemingly random act that in the past might have entered the annals of unsolved tragedy now proves traceable. The proverbially long arm of the law catches up with a criminally irresponsible man who killed and ran. The accused is brought to justice; the victim is avenged. Neat, orderly, satisfying. Case closed. And yet the story has resonances that are not altogether pretty. There is first of all the 'voluntary' dragnet – one of those 'offers' an invitee cannot refuse. There is the question of what happens to all those samples collected from persons who had no business being swept into a suspect DNA database other than their random association with the time and place of a crime. Then there are issues of distributive justice. In chronically unequal societies, what will prevent the state's suspicious eye from landing more frequently on the poor and under-represented, thereby distorting and, almost inevitably, making mistakes in the delivery of justice? What new uses will be found as the technologies of extracting information from DNA mature, and who will participate in the design of ever-expanding information systems? Overarching all, are questions deriving from the special attributes of the DNA profile, a far more comprehensive record of a person's biological characteristics, and of familial and racial relationships, than the varied footprints, fingerprints, voiceprints, teeth marks, signatures, video images, eye-witness recollections and even blood samples left behind as possible identifiers by earlier generations of suspect individuals.

Part historical, part contemporary, part case study and part comparison, this collection of essays by a distinguished array of international experts presents DNA profiling as a technology manifestly in need of better governance. The editors, Richard Hindmarsh and Barbara Prainsack, have divided the collection into two richly informative sections, each of which strengthens and underlines the significance of the other. The first part introduces DNA profiling as a complex technological system, composed of heterogeneous social and material elements and practices that give rise to problems of governance at the same time that they promise to solve the 'who dunnit' question that is one of any law-abiding society's prime concerns. For example, the practice of the DNA dragnet, so effectively deployed in the UK case of the bloody brick, may through its focus on specific geographical locations lead to increased surveillance of ethnic and racial minorities. Elazar Zadok, Gali Ben-Or and Gabriela Fisman explicitly note that possibility in their chapter on forensic DNA practices in Israel – a nation in which geography and ethnicity reinforce each other with toxic effects. But concern may be warranted in much tamer political settings, as Harriet Washington and Jay Aronson suggest in their studies of the entanglement of race and policing in forensic DNA practices in the USA.

Forensic DNA profiling, moreover, sits in both historical and contemporary proximity to other techniques that illuminate and help to reinforce the governance challenges that this technology presents. As Richard Tutton and Mairi Levitt point out, forensic DNA databases coexist with medical ones, and yet the implications of the two compilations differ from the standpoint of the persons profiled in each storage system. The 'genetic suspect' who falls within the state's dragnet loses privacy, freedom and autonomy; for suspects, there is little prospect of the self-fashioning and group affiliation, or the exercise of active citizenship discussed in biomedical contexts by authors such as Shobita Parthasarathy (2007), Nikolas Rose (2006) and Paul Rabinow (1992). Mistakes, too, are graver when a suspect's life and liberty may be at stake. Here, the historical gaze imparted by Simon Cole and Michael Lynch, who draw analogies between DNA profiling and other flawed biometric technologies of the past, offers a healthy antidote to unbridled enthusiasm. The juxtaposition of historical and contemporary practices adds depth to this collection, as is well illustrated by the pairing of the Cole and Lynch chapter with the chapters by Robin Williams and Jay Aronson on the origins of forensic DNA profiling in England and Wales and in the USA, respectively.

The book's second part reviews the introduction of DNA profiling into national law enforcement systems in three largely white, industrially advanced regions of the world, the European Union, the USA and

Australasia. These studies offer valuable insights into the conditioning of this technology by specific features of national legal culture. The contrasts reach deep down, not only into divergent histories of judging and prosecuting but also into different national expectations of how to run a responsible law enforcement system. As Victor Toom recounts in the Dutch case, for instance, a rapid and relatively uncontroversial expansion of forensic DNA tests from only violent crimes, such as murder, to routine high-volume ones, such as burglary, reflected traditions of legal centralization and the belief in the impartiality of judicially managed fact finding in the Netherlands. More generally, all of the country case authors show that prior histories of public experiences with government and law enforcement – whether relatively trusting as in Johanne Yttri Dahl's Norway or mistrustful as in the Portugal described by Helena Machado and Susana Silva – influence policies for managing DNA databases, including rules of access and accountability.

What emerges from these individual accounts is a technology in flux. DNA profiling is everywhere founded on the same secure core of biological knowledge, and it raises similar questions about human rights and liberties; however, national practices and debates differ, reflecting political tensions and concerns that vary substantially from state to state. These socially conditioned differences, emphasized in each chapter, add up to a strong argument for more explicit forms of citizen engagement in the governance of forensic DNA profiling; indeed, calls for more transparency form the book's most compelling *leitmotiv*. We are left wondering, however, about the global implications of these analyses as the technology spreads through regions with far less experience of democratic involvement.

Apart from the chapter on the Philippines by Maria Corazon De Ungria and Jose M. Jose and that on Portugal by Helena Machado and Susana Silva, describing countries that did not have operational national forensic databases as of this writing, the distribution of the country cases reflects the typical diffusion pattern of 'high-tech' innovation. Novel technologies originate where wealth and knowledge are most concentrated and are then exported to other societies and cultures with considerably different histories of technology and governance. We know from decades of work on the co-production of natural knowledge and social order that technologies are never ethically or politically neutral: they carry within them, particular, culturally conditioned imaginaries of good and evil, what (and, in the case of DNA profiles, *who*) should be encouraged and what (or who) should be suppressed (Jasanoff 2004, 2005). What moves in transfers of technology is not simply expertise or technique, it is an entire mode of knowledge and organization, in this case of the human subject in relation to institutions of power. The very fact of technologies of social control imagined and produced in the West for use in the rest of the world raises political and ethical dilemmas that the editors and authors invite readers to ponder for themselves.

These observations point to one caveat concerning this timely and thought-provoking compendium. Empirical richness, local specificity and breadth of coverage are gained to some extent at the expense of a unifying theoretical vision and deep, cross-national comparative insight. The authors offer a hugely intelligent examination of a technology of control that is still in the making, when handholds still exist in many places for intervening in its modes of governance. The book addresses only briefly in its conclusion how DNA profiling will fit into emerging structures of global governance, and it leaves unaddressed how the transitions and frictions discussed here relate to the broader repositioning of the human in the post-DNA analytics of law and political theory. Necessarily, then, this book will have to sit beside works that attempt a more ambitious social theorization of the genetic revolution in all its complexity. As a study of one key, technologically mediated component of that transformation, however, this book sets a high standard of scholarship and insight that will be hard to beat.

Sheila Jasanoff
Harvard University

REFERENCES

Jasanoff, S. (ed.) (2004). *States of Knowledge: The Co-Production of Science and Social Order*. London: Routledge.

Jasanoff, S. (2005). *Designs on Nature: Science and Democracy in Europe and the United States*. Princeton, NJ: Princeton University Press.

Parthasarathy, S. (2007). *Building Genetic Medicine: Breast Cancer, Technology, and the Comparative Politics of Health Care*. Cambridge, MA: MIT Press.

Rabinow, P. (1992). Artificiality and enlightenment: from sociobiology to biosociality. In *Incorporations*, eds. J. Crary and S. Kwinter. New York: Bradbury Tamblyn and Boorne, pp. 234–252.

Rose, N. (2006). *The Politics of Life Itself: Biomedicine, Power, and Subjectivity in the Twenty-First Century*. Princeton, NJ: Princeton University Press.

Rosen, J. (2009). Genetic surveillance for all. *Slate*, March 17 http://www.slate.com/id/2213958/pagenum/all/

Acknowledgements

We conceived the idea of *Genetic Suspects* at the conference *The Global Governance of Genomics: Testing Genes, Profiling DNA: Medicine, Forensics, Ethics* (Mendel Museum, Masaryk University, Brno, Czech Republic, 1–3 November 2007), from which a number of the contributors were drawn. In that respect we gratefully acknowledge the support for the conference of Genomeresearch in Austria (www.gen-au.at) programme of the Federal Austrian Ministry of Science and Research, the Austrian Science and Research Liaison Office Brno (ASO) and the Centre for Governance and Public Policy, Griffith University, Brisbane, Australia.

We thank our contributors, who represent an important and international group of authors concerned with profound issues relating to contemporary DNA databasing and profiling. We are grateful to our language editor, Jane Neuda, for her critical reading of the manuscript, which went far beyond the call of duty. We also thank all those who supported and accompanied the genesis of this volume in different ways including David Gurwitz, Reinhard Schmid, Hendrik Wagenaar, Nikolas Rose and Herbert Gottweis.

Last, but not least, we express our appreciation to Dominic Lewis and Katrina Halliday at Cambridge University Press for their invaluable help at every stage in the process; they have well contributed to the production of this book as a very enjoyable experience.

Richard Hindmarsh and Barbara Prainsack

RICHARD HINDMARSH AND BARBARA PRAINSACK

1

Introducing *Genetic Suspects*

This book investigates the impacts and implications for governance of one of the most successful and yet controversial developments in recent science and technology history: the introduction of DNA profiling and databasing into the criminal justice system, which began in 1988, when English baker Colin Pitchfork was the first person convicted through the use of DNA evidence (Sanders 2000). The increasing use of DNA evidence in criminal investigations and in court soon assumed the role of a new 'language of truth', following on from traditional fingerprinting. While DNA profiling, on a case-by-case basis, had been used since the late 1980s, it was the establishment of centralised national registries of DNA profiles for police and forensic use 5 to 10 years later which made possible the wider and systematic use of DNA technologies in criminal investigation. Computerised forensic DNA databases enabled authorities to compare profiles from crime scenes and subjects against and between each other on an automated basis and on a large scale. The first national forensic DNA database of this kind was implemented in 1995 in the UK, followed by New Zealand, several European countries, and the USA and Canada (Walsh *et al.* 2004: 36). Australia and many other countries across the globe followed suit.

Despite the many declarations by proponents, especially law enforcement agencies and forensic scientists, that forensic DNA profiling and databasing play an increasingly useful part in curbing crime, and despite popular TV shows like *Crime Scene Investigation* (*CSI*) and *Silent Witness* and other cultural sites that show these elements almost exclusively in a favourable light, many issues have, in fact, steadily emerged and grown in connection with this relatively new practice. Such issues belie the claim that forensic DNA profiling and databasing provide simple solutions to complex problems.

Genetic Suspects: Global Governance of Forensic DNA Profiling and Databasing, ed. Richard Hindmarsh and Barbara Prainsack. Published by Cambridge University Press. Copyright © Cambridge University Press 2010.

In the next decade, the use of forensic DNA databases will increase in breadth and scope at national, transnational and international levels. Moreover, the importance and use of DNA profiling and databasing will increase and broaden with the rise of security, biometrics and anti-terrorist issues on public and political agendas. Parallel to such expansion, which is also occurring in developing countries, will be an increasing need to balance the benefits of the new genetic technologies of identification, surveillance and security against civic concerns, informed by criminal, genealogical and, potentially, medical and health histories, and new shifting definitions and identities and stakes pertaining to the criminal and the suspect. One of the biggest challenges will be building and maintaining public trust, which involves the creation of arenas where multiple viewpoints, interests and values can be articulated and heard.

The many social, ethical and political (less so economic) concerns and issues pertaining to forensic DNA profiling and databasing are situated at the intersection of civil rights, science and governance. They are intimately linked to the constitution of new and wider groups of populations as 'genetic suspects'. Such concerns include, but are not limited to, privacy; (non)standardisation of jurisdictional databases; surveillance; ideological and scientific interpretation of DNA evidence; scientific reliability of DNA testing; potential misuse and abuse of databases; sample collection and analysis, security and/or contamination; database implementation and expansion; function creep; public trust and participation; and mass testing or DNA dragnets.

In addressing this profoundly rich debate, the goal of *Genetic Suspects* is to contribute to the discussion of what 'good governance' of DNA profiling and databasing is, and could be. By 'governance' here we refer to a cluster of rules for, and practices of, conduct and decision making that go beyond parliamentary legislation and governmental measures. Governance includes practices and institutional arrangements of and between both governmental and non-governmental actors. We understand as 'good governance', institutional and practical arrangements and modes of governance that enjoy high levels of public trust. However, questions about 'good governance' cannot be answered in a general way. What 'makes sense' in one national context can seem starkly different in another, as they are marked by differences in informal practices; legal, cultural and political traditions; policy narratives; and, sometimes, religious beliefs.

An international group of contributors such as those gathered in this book is thus uniquely placed to examine the differing, sometimes partly conflicting, meanings and objectives that various stakeholders and interests attribute to the technologies, objectives and infrastructures that constitute forensic DNA profiling and databasing internationally, and their good governance. An investigation of the emergence, implications and governance of forensic DNA profiling and databasing also requires critical and comparative analysis rather than assertions as to their 'negative' or 'positive' consequences.

As such, *Genetic Suspects* explores the scientific, technical, legislative, crime management related and social and cultural contexts of forensic DNA profiling and database governance in a range of countries that display different stages of development of forensic DNA databases. Examples range from the highly developed practices of the UK and the USA to the emergent ones of Portugal and the Philippines. This variety reflects the original contextualisation of the book at a conference[1] organised as part of an international research collaboration on forensic DNA profiling and databasing situated within a broader project called *Genes Without Borders: Towards Global Genomic Governance*.[2]

A number of questions informed the writing of *Genetic Suspects*. How are individuals and populations governed through new DNA technologies of profiling and databasing and their practices? What shapes and patterns of governance and regulation, at both the global and national levels, can be discerned? What new arenas of conflict have emerged? Which actors have been central in shaping the debate, and what has been their role? What are the key ethical, legal, social and cultural implications and challenges? Are there any convergences between forensic and biomedical DNA databases and practices? What impacts did differing regulatory strategies have on scientific–technological development, society, the political system and the criminal justice system? How did they accommodate and/or act upon scientific debate

---

[1] *The Global Governance of Genomics: Testing Genes, Profiling DNA: Medicine, Forensics, Ethics*, Mendel Museum, Masaryk University, Brno, Czech Republic, 1–3 November 2007. Organisers included the editors, Ursula Naue and Anna Durnová (University of Vienna, Austria), Peter Kakuk (Central European University, Budapest, Hungary), Frank Yeruham Leavitt (Ben Gurion University of the Negev, Beer Sheva, Israel) and Josef Kure and Renata Veselská (Masaryk University, Brno, Czech Republic).

[2] The project (2006–2008) was led by Barbara Prainsack and funded by the GEN-AU (Genomeresearch in Austria) program of the Austrian Federal Ministry of Science and Research (www.gen-au.at). Project team members were Herbert Gottweis, Richard Hindmarsh, Ursula Naue, Jenny Reardon, Jeantine E. Lunshof and Nikolas Rose.

and public attitudes? How inclusive has the debate been about forensics profiling and DNA databasing? What overall implications are there for current directions of, and what are the challenges for, governance?

Our contributors' approaches and perspectives are arranged into two sections, reflecting the major areas of analysis and concerns of the contributors, which also mirror the international debate. Section 1 provides a context for examining and understanding key areas in DNA profiling and databasing, while Section 2 addresses how issues related to forensic DNA databasing emerge in national contexts.

THE CONTRIBUTORS' PERSPECTIVES

Section 1

In Chapter 2, Barbara Prainsack provides an overview of the most current issues and trends pertaining to forensic DNA profiling and databasing, and their implications for governance. Prainsack frames the discussion from the beginning with reference to two proactive developments of late. First was the recent decision of the European Court of Human Rights in the case of *S and Marper* v. *the United Kingdom* (2008), which upheld the argument that retention of fingerprints, DNA profiles and DNA samples represents infringements of citizen's rights to private life as provided in Article 8 of the *European Convention on Human Rights* (Council of Europe 1950). Second was the firing of the laboratory director of the Baltimore Police Department as a result of serious contamination of crime scene samples and other 'operational issues'. Key issues highlighted included securing traces at the crime scene, obtaining DNA elimination samples from suspects and volunteers, transport and storage of crime scene traces, laboratory analysis, profile matching, courtroom evidence dilemmas, profile storage in the database, sample storage and function creep. Prainsack concludes with a call for equal attention to be paid to both the potentials and the limits of DNA technologies, and in particular, to what they can and what they cannot prove.

In Chapter 3, Elazar Zadok, Gali Ben-Or and Gabriela Fisman focus on the forensic use of voluntarily collected DNA samples, which they contend disadvantages individuals while pursing maximising law enforcement for the benefit of society. Three topics are examined. First, operational considerations in carrying out DNA dragnets or mass screenings; second, ethical and legislative considerations encountered in sample collection; and, third, ethical and legislative considerations related to the fate of these samples and the profiles derived from them.

These topics are applied to a murder case in Israel and the possible dilemmas, particularly regarding the considerations of Israeli courts addressing claims about the presumably illegal use of such samples, which leads the authors to make recommendations for better governance of voluntarily collected DNA samples.

In Chapter 4, Harriet Washington explores how ethnic issues, and in particular how the 'tangled calculus' of race, inform the debate on DNA use and governance for the purposes of law enforcement. She asks: 'Are such questions particularly relevant for the US context, or do other nations share these challenges?' In discussing such concerns in the context of racialised DNA sweeps, Washington contributes to the important discussion about how genetic databases should be designed and governed to maximise citizens' security while protecting privacy, autonomy and social justice. The author concludes, 'the challenge for good governance lies in determining how best to exploit genetic power without abusing it. One place to start is to abandon racialised DNA sweeps as inefficient, expensive, scientifically inaccurate, and most of all, as dramatic violations of social justice'.

In Chapter 5, Richard Tutton and Mairi Levitt address the 'trade' or flow of DNA samples across the boundaries between the medical and the forensic, and set out not simply to enumerate the similarities and differences between the way police databases and medical research databases have developed but also to reflect on how we conceptualise their different sociotechnical configurations. In other words, how do we analyse the simultaneous development of forensic DNA databases and the establishment of biomedical biobanks? In addressing this question in the UK context, the authors refer to notions of 'biolegality' but with specific reference to the forensic as opposed to the biomedical context in order to illuminate issues posed by the parallel development of genetic databases for policing purposes.

Our final chapter in Section 1, Chapter 6, by Simon Cole and Michael Lynch, examines how other forensic systems have provided models for the organisation of DNA databases. In that context, the authors find that many, but by no means all, aspects of DNA profiling follow patterns established by earlier techniques, most prominently fingerprint identification – the technique most closely analogous to DNA profiling on several levels. Both fingerprinting and DNA profiling are used to identify particular bodies as sources of crime scene evidence, and both involve comparisons between traces found at crime scenes and reference samples taken from persons in police custody. However, even though both techniques proved useful enough from a social control perspective – warranting large government investments and enjoying primacy as 'gold standards' in an imagined hierarchy of forensic techniques – the utility of both diverged from many of the hopes put forward about the 'sciences' of individual identification. Both techniques require practical judgment and are subject to error and abuse when caught up in the bureaucratic prerogatives of criminal justice administration. In sum, Cole and Lynch make a powerful argument that any hopes invested in DNA databases as a means of governance needs to take account of the historical lessons afforded by examining the biometric technologies that have preceded this latest forensic technology.

Section 2

Contributors in this second section of *Genetic Suspects* discuss the introduction of forensic DNA technologies into their countries and identify key issues that appear central to these developments and that need raising and addressing in a critical context for good governance. The contributions are organised into three geographical areas at the forefront of development and critical discourse: first, the European Union countries and Norway; second, the USA; and third, Australasia (Australia and New Zealand). In turn, we look at frontline countries adopting the technologies and others planning to do so within the above areas or close to them. This provides an interesting and representative analysis of the issues in relation to global trends in a range of countries that reflect diverse cultures and development. What are the convergences and divergences in the issues, concerns and developments of forensic DNA profiling and databasing we ask.

In Chapter 7, Robin Williams discusses the development of forensic DNA profiling as a particularly conspicuous instance of a method for observing, knowing and recording 'individuality' to achieve the individuation of bodies though specific techniques of bio-identification, such as those of anthropometry or fingerprinting. With a focus on England and Wales as the first criminal jurisdiction in which a national DNA database was established, Williams first outlines the historical sources of the contemporary enthusiasm for the use of DNA profiling to support the control of crime, and then the ways in which scientific advances were harnessed to legislative changes and financial support to enable the creation of a DNA database that currently holds more than five million profiles. This sets the scene for considering a range of issues as the enthusiasm for forensic DNA profiling and databasing as resources for governing individuals and populations has been supplemented by how such technologies should themselves be governed. In acknowledging historical trajectories taken by these innovations (similar to the discussions in Chapter 6), he focuses on significant operational, legal and ethical challenges, as well as the forms of regulation, oversight and accountability that may have to be put in place to meet these challenges.

In Chapter 8, Barbara Prainsack outlines the relevant legal provisions as well as actual practices pertaining to forensic DNA profiling and databasing in Austria, and discusses the circumstances that led to the establishment of Austria's forensic DNA database. Drawing upon empirical research carried out in 2006 and 2007, Prainsack then provides an overview of understandings and practices of forensic DNA databasing on the part of individuals in law enforcement, on the one hand, and the understandings and responses of convicted criminal offenders on the other. A final section reflects on the impact that DNA profiling and databasing has had on crime commission and crime investigation. This discussion is particularly relevant to one of the core issues of this volume, namely how new technologies reinforce and/or alter prevailing configurations of power.

In Chapter 9, Victor Toom describes how Dutch DNA profiling became governed through legal measures and the inquisitorial orientation of the Dutch legal system. Second, he describes the trajectory – the lines of development – of Dutch DNA profiling practices, outlining who and what has been involved in DNA profiling. This account provides insight into the strategies employed by various stakeholders to deploy DNA profiling extensively and routinely in volume crimes and to apply DNA profiling in the process of crime investigations. Toom's analysis contributes to the understanding of how current DNA profiling practices were realised in a country – the Netherlands – with what he refers to as an 'inquisitorial legal orientation', where judges and other involved jurists in legal cases act impartially. Finally, he highlights some implications for current directions in the governance of Dutch forensic DNA profiling practices, especially the view that broad and informed public debates need to better address and resolve the many issues arising with regard to forensic genetic bodies and the civic protection of genetic suspects.

In Chapter 10, Johanne Yttri Dahl explores the case of Norway with regard to the use of DNA databases in criminal law administration, and the issues raised by the expansion the national database, which occurred in September 2008, for example issues of sample storage and deletion. Such issues Dahl asserts require further attention and critical discussion as they pose important challenges for responsible and transparent governance of Norway's expanding forensic DNA database. Special attention in her analysis is paid to the fact that one governmental DNA laboratory has a monopoly on DNA analysis in Norway. This issue was prominently debated in connection with the expansion of the DNA database, which was reinforced by respondents in in-depth interviews of key stakeholders that Dahl conducted on the use of DNA evidence in courts and the expansion of the database. In this chapter, she presents her highly interesting thematic analysis and findings. Like Toom in Chapter 9, her key finding is that 'Increased transparency, use of second opinion, improved governance and increased knowledge is ... essential.'

In Chapter 11, Helena Machado and Susana Silva write in the context of the Portugal forensic DNA database that will become operational during 2010. The authors describe the legislative basis for the Portuguese forensic DNA database and address the social, ethical and practical implications of the legal framework by focusing on important concerns and issues prominent in Portugal's public debate, further informed by insights obtained from interviews carried out with law and forensic science experts on the committee drafting the law for the DNA database. As such, the authors' aim is to contribute to the debate on emerging practices related to DNA databases for crime investigation and prevention. They discuss important governance and regulatory challenges, especially in exploring loopholes in the existing legislation, amidst discussing issues around the likely public acceptance of future governmental efforts to expand the database in a society where there are low levels of trust in the criminal justice system, commonly regarded as corrupt, discriminating, slow, inefficient and protective of the more powerful social groups. They finish by arguing for appropriate participatory avenues for decision making in the interest of public confidence and good governance of the Portuguese forensic DNA database.

In Chapter 12, Jay Aronson charts the history of DNA profiling and the national DNA database within the context of the US legal system and society. The author pays particular attention to the initial introduction of DNA profiling from the late 1980s through to the mid 1990s because it was in this period that most major patterns regarding its governance and oversight were established, particularly the centrality of the Federal Bureau of Investigation. Towards the end of the chapter, Aronson then turns to recent issues that have emerged in the context of the expansion of the DNA database and the creation of ever more powerful technologies for recovering and analysing DNA from potential criminal suspects. An aspect that Aronson finds unique in the US context is that the introduction of the technique into the courtroom was much more contentious than in most other countries, while the expansion of the national DNA database and other associated ethical and legal issues have been significantly less controversial compared with the rest of the world. Further, the involvement of private companies in the US DNA profiling market has created significant challenges for the legal and scientific communities. In the end, the author finds the omission of voices at the policy-making table that are critical of both the presumed infallibility of DNA profiling and the unambiguous social utility of the expansion of DNA databases to be a troublesome reality of the American context. He argues that this situation is one that ought to be avoided for good governance of forensic DNA technologies in other countries.

In Chapter 13, Richard Hindmarsh finds the Australian development of forensic DNA technologies raises many 'biocivic' concerns, not least policy formation by a narrow policy network of proponents who present DNA profiling and databasing as a simple solution to a complex problem. Hindmarsh addresses these concerns in the context of what many have posed as an increase of biosurveillance through extended forensic DNA technologies, and the associated socio-political implications of that for good governance in a democratic society characterised by the integrity of and public trust in governance. The author explores the terrain through policy and media analysis of developments and the public debate and concludes by suggesting a transition from the notion of 'DNA as a language of truth' to one of 'DNA as a language of trust' as part of new governance strategies for addressing the tensions arising between forensic DNA-driven law enforcement and biocivic concerns.

In Chapter 14, Johanna Veth and Gerald Midgley trace the evolution of DNA profiling and review the legislative aspects of governance in New Zealand. The authors first direct their attention to forensic DNA analysis prior to legislation and consider two key judicial decisions that illustrate the legal debate surrounding this technology. Focusing on whose samples may be obtained and under what circumstances, Veth and Midgley summarise the main themes and concerns in submissions made to the Parliamentary Select Committee considering the 1994 Criminal Investigations (Blood Samples) Bill and the subsequent 2002 Amendment Bill. In their summary, the authors emphasise cultural

and ethical concerns, as these signal possible unintended consequences that a purely technical evaluation of the technology's
effectiveness would not anticipate. They find areas of both contention and acceptance. Finally, the authors review present-day operations of New Zealand's DNA Profile Databank, but, overall, find
that wide public support exists for forensic DNA technologies and
practices owing to successes in detecting criminals and securing
convictions.

In the last chapter of this section (Chapter 15), Maria Corazon De
Ungria and Jose Manguera Jose note that the Philippines does not yet
have a national DNA database or any legislation to facilitate its establishment, but DNA evidence is admissible in courts and judicial guidelines exist for the collection, handling and storage of biological
samples. Forensic DNA testing in criminal investigations is performed
by the National Bureau of Investigation and the Philippine National
Police, institutions that exist under different governmental departments. The authors provide an overview of policy developments and
discuss legislative issues in the mooted establishment of a national
DNA database. They discuss the implications for governance, especially
resource allocations for a developing country as well as potential
misuses of such a database, and they conclude with suggestions to aid
in effective policing that include the minimisation of infringements of
individual rights.

CONCLUDING REMARKS

The chapters in this book provide readers with a variety of
approaches and perspectives to the important topic of DNA profiling
and databasing, with, as Aronson emphasises in the closure of his
chapter, 'DNA playing an increasingly important role in law enforcement around the world'. Our broad aim again is to contribute to
discussions about 'good governance' pertaining to some of the most
profound issues of DNA profiling and databasing yet facing societies
and cultures around the globe. Perhaps most provocative are current
initiatives to further expand the scopes and uses of forensic DNA
databases, which stimulates the need for enhanced public engagement on the many issues raised more generally by DNA databasing
and profiling, as many of our contributors highlight. We hope that it
will be helpful to a range of readers, both professionals and non-
professionals, to whom these issues are of interest, in theory and
practice.

REFERENCES

Council of Europe (1950). *Convention for the Protection of Human Rights and Fundamental Freedoms*. Strasbourg: Council of Europe http://conventions. coe.int/Treaty/Commun/QueVoulezVous.asp?NT=005&CL=ENG (accessed February 2010).

Sanders, J. (2000). *Forensic Casebook of Crime*. London: True Crime Library/Forum Press.

Walsh, S., Ribaux, O., Buckleton, J. *et al.* (2004). DNA profiling and criminal justice: a contribution to a changing debate. *Australian Journal of Forensic Sciences*, 36, 34–43.

CASE

S and Marper v. *the United Kingdom* (2008). A summary of the judgment is available from http://cmiskp.echr.coe.int/tkp197/view.asp?action=html& documentId=843937&portal=hbkm&source=externalbydocnumber&table= F69A27FD8FB86142BF01C1166DEA398649 (accessed January 2009).

Section 1   Key areas in DNA profiling
and databasing

BARBARA PRAINSACK

2

Key issues in DNA profiling and databasing: implications for governance

INTRODUCTION

On 4 December 2008, the European Court of Human Rights (ECHR) delivered its Grand Chamber Judgment in the case of *S and Marper v. the United Kingdom* (2008). The judgment dealt a major blow to the national forensic DNA database (United Kingdom National DNA Database (NDNAD)) in England and Wales, the largest forensic DNA database in Europe.[1] That judgment informs the purpose of this chapter, which is to review key issues of DNA profiling and databasing, many of which contribute to contesting forensic DNA technologies as an infallible means to truth finding in the criminal justice system (e.g. Lynch *et al.* 2008). The chapter concludes with a discussion of the implications that these issues have for the governance of DNA forensic profiling and databasing.

The case of *S and Marper v. the United Kingdom* (2008) was brought to the ECHR by Michael Marper and another young man known only as 'S', whose fingerprints and DNA profiles were stored in the English police database following their arrest in 2001. Charges had been dropped in both the case of Marper (arrested on harassment charges) and of S (at age 11 arrested for attempted robbery). In the absence of convictions, both men demanded their fingerprints and DNA data be removed from the database. In both cases, the Court of Appeal ruled against this.

Having exhausted all legal remedies in the UK, appealing to the ECHR offered the last resort. There the two men argued that retention of their fingerprints, DNA profiles and DNA samples represented

---

[1] Scotland maintains a separate database. In addition, it exports profiles (but not the samples from which they originated) to the NDNAD (Williams and Johnson 2008).

Genetic Suspects: Global Governance of Forensic DNA Profiling and Databasing, ed. Richard Hindmarsh and Barbara Prainsack. Published by Cambridge University Press. Copyright © Cambridge University Press 2010.

infringements of their rights to private life as provided in Article 8 of the *European Convention on Human Rights* (Council of Europe 1950). In its judgment, the Court acknowledged a violation of Article 8 (Council of Europe 2008). Noting that England, Wales and Northern Ireland were the only countries within the Council of Europe to allow indefinite retention of fingerprint and DNA material for those arrested for any recordable offence, the Court in *S and Marper* v. *the United Kingdom* argued that

> the protection afforded by Article 8 of the Convention would be unacceptably weakened if the use of modern scientific techniques in the criminal-justice system were allowed at any cost and without carefully balancing the potential benefits of the extensive use of such techniques against important private-life interests. Any State claiming a pioneer role in the development of new technologies bears special responsibility for striking the right balance in this regard.

The Court also expressed a particular concern for possible risks of stigmatisation, arguing that the presumption of innocence of the applicants had been infringed by the retention of their samples and data in the absence of conviction. This concern was deemed particularly relevant with regard to minors.

A few months later, the UK Home Office reacted to the judgment with a consultation paper, proposing a change of practices pertaining to the indefinite and indiscriminate retention of DNA profiles, samples and fingerprints from individuals who were not convicted (Home Office 2009). Although some stakeholders were disappointed that the government had, in fact, not decided to remove all DNA material and fingerprints of non-convicted individuals from operational police data-bases (BBC News online 2009a), the policy shift represented a turning point in the rather rocky history of forensic DNA profiling and data-basing after the equivalent of a hero's welcome some two decades previously when DNA evidence arrived as a new 'language of truth' for law enforcement (*New York* v. *Wesley* 1988: 644; see Chapter 12). Despite ongoing discussions about the correct interpretation of the use of forensic DNA data, it has since come to be seen or claimed as being more reliable than many, if not all, other kinds of evidence (Neufeld and Coleman 1990; Cole 2001). Public representations and popularisation of forensic policing activities in TV series such as *Crime Scene Investigation (CSI)*, *Cold Case* and *Silent Witness*, reinforce the perception of contemporary police work as a morally neutral, objective activity of 'following the evidence' to its conclusive end (Kruse 2010).

In recent years, increasing public attention has been drawn to the flaws and 'dangers' of DNA profiling (e.g. Dolan and Felch 2008; Gilbert 2010). In 2007, for example, Erin Murphy argued that forensic DNA profiling exacerbated the problems and biases of earlier generations of forensic evidence (see also Thompson *et al.* 2003). She contended that the so-called 'cold hit' case scenario[2] (where an unsolved case is reinvestigated using genetic material which at the time of the crime could not be processed owing to insufficient scientific and technical knowledge and means) increased the risk of wrongful convictions because some suspects were being convicted *solely* on the grounds of DNA evidence (Murphy 2007: 743).

Murphy also noted that some requirements of scientific rigor are routinely not met in forensic DNA profiling. In contrast to the medical field, where every new method and technique is, at least in theory, subject to academic peer review and scrutiny, the scope of academic scrutiny and peer review regarding DNA technologies in the field of forensics has been arguably more narrow, because of the prominent position of individuals affiliated with, or working for, law enforcement authorities in this process.[3] Furthermore, independent methodological research and the number of non-government experts working on the development of forensic DNA technologies issues are notoriously limited (Murphy 2007: 753). As Murphy (2007: 749) noted, the companies developing the kits used for forensic DNA profiling, 'vigorously guard the methods and validation studies underlying their technological as intellectual property ... [T]o suggest that the geneticist's broader interest in genomics validates DNA typing for forensic purposes is like suggesting that the widespread market for electricity somehow ensures the proper functioning of an electric chair.' In addition, the 'law-like treatment of forensic methodologies' by courts, which regularly presume the scientific soundness of the DNA evidence, further discourages independent or peer scrutiny of forensic DNA methodologies (Murphy 2007: 765).

Murphy's arguments are also a reaction to recent reports of high numbers of so-called adventitious 'matches' ('false positive' matches; see below), which question the assumptions of statistical independence that

---

[2] Chakraborty and Ge (2009) are critical of this concept.
[3] See Aronson (2007). However, as Lynch *et al.* (2008, Chapter 3) argue, the observation that requirements of scientific rigor remain routinely unmet in the development of forensic 'science' does not mean that scientific research and practice in other contexts *does* always meet these requirements of rigor.

underlie traditional methods of calculating random match probabilities (as discussed by Dan Krane, William Thompson and others; see Krane *et al.* 2004, Felch 2008).[4] They also call for caution in assessing DNA evidence in court. Interestingly, when research in Arizona showed that profiles of supposedly unrelated individuals matched at 9 out of 13 loci (a much higher number of locus matches than statistically predicted),[5] the US Federal Bureau of Investigation (FBI) sought to suppressed further research into this issue (Jefferson 2008; see also Chapter 12). Nevertheless, other states have since followed suit and carried out 'Arizona searches' with similar results (Felch 2008).

In August 2008, news broke about another issue that had been brewing. A news item reading like an episode of the celebrated TV series *The Wire* reported the firing of the laboratory director of the Baltimore Police Department as a result of serious contamination and other 'operational issues' (Bykowicz and Fenton 2008). The scandal was discovered when DNA profiles of laboratory workers entered into the database by a laboratory supervisor resulted in 2500 instant matches to unknown crime scene profiles. The apparent scale of contamination of crime scene samples by laboratory employees' DNA seemed so large that defence attorneys advanced that 'flaws in the city's handling of DNA could raise broader questions about evidence that is generally considered infallible' (Bykowicz and Fenton 2008). Indeed, scandals such as the story of the 'phantom of Heilbronn' – which was a

---

[4] The 'random match probability' is the frequency of a DNA profile in a reference population. A low random match probability – such as one in a million – would mean that the chances that a randomly drawn DNA sample from a given population 'coincidentally' matches the DNA sample from the crime scene is one in a million (see also Jha 2004). For further information on the issue of adventitious matches, see European Network of Forensic Science Institutes (2009: Chapter 6).

[5] According to FBI estimates, the odds of profiles of unrelated individuals matching in nine genetic loci are 1 in 113 billion (Felch 2008). However, within the 65 493 DNA profiles on the database, 122 pairs of profiles matched at nine loci; 20 pairs matched at 10 loci; and one pair each matched at 10, 11 and 12 loci (the latter two pairs were siblings). Some experts state that this unexpectedly high number of matches can (at least partly) be explained by the fact that people who are deemed unrelated are indeed biological relatives, perhaps without being aware of it. In addition, comparing every single profile in the database with every other profile generates such a high number of comparisons that 'coincidental matches' are to be expected; this is very different from a 'normal' profile search in the course of a criminal investigation, when one crime scene profile is compared with potential *full* matches (that is, matches that are identical in all loci; Felch 2008, Jefferson 2008). Chakraborty and Ge (2009) discuss the concept of 'database match probability').

Table 2.1. *Potential issues in forensic DNA profiling*

Phase	Potential issues
Securing traces at crime scenes	Contamination, mistakes in labelling, overlooking potentially incriminating or exonerating material
Obtaining DNA elimination samples from suspects and volunteers	Contamination, coercion
Transport of crime scene traces	Contamination. loss of material
Storage of crime scene traces	Contamination, loss of material
Analysis at crime lab	Contamination, loss of material
Profile matching process	Searches in 'forbidden' databases, issues concerning calculation of random match probabilities
Courtroom	Use of flawed random match probabilities, 'prosecutor's fallacy', *CSI* effect'
Profile storage in the database	Backlog in deletion of profiles that may no longer be used, privacy risks
Sample storage	See function creep, below
Function creep	Extension of inclusion criteria in DNA database, retention criteria, phenotypic profiling, familial searching

large-scale multinational police search for an unknown female offender,[6] until it finally turned out that the DNA found at the crime scenes had come from a woman packaging police cotton swabs – highlight how widespread the problem of contamination is (BBC News online 2009b).

But contamination at crime scenes and crime laboratories are not the only potential issues for forensic DNA profiling and databasing (Dolan and Felch 2008). As Table 2.1 illustrates, issues capable of compromising the quality and reliability of DNA evidence affect every stage in the process, from securing the crime scene to the use of DNA evidence in court. (Issues pertaining to the transnational exchange of DNA profiles are referred to in Chapter 16.)

---

[6] In most of the criminal cases in question, the perpetrators were identified and the cases could be closed. The puzzle that the phantom of Heilbronn posed to the police was that there seemed to be an additional, female perpetrator implicated in the crime, so investigators kept trying to track down the mysterious woman.

The next section discusses some of the most current of these issues. It also addresses some overarching trends (subsumed under the heading of 'function creep', see Dierickx 2008) that do not pertain to a particular phase or stage of DNA profiling and databasing; instead, these trends represent a widening of the scope of purposes for which DNA profiling and databasing is used. In other words, the notion of 'function creep' signifies the assumption that once a technology has been adopted for a particular purpose, and infrastructures have been built, it is almost impossible to prevent the extension of the scope of purposes for which the technology is used. This is deemed problematic by some commentators not only because it potentially increases surveillance but also because it exacerbates existing issues pertaining to the practice and governance of forensic DNA databasing (Dahl and Rudinow 2009; see also Chapter 7).

ISSUES IN DNA PROFILING AND DATABASING

Securing traces at the crime scene

Problems stemming from scarce resources, such as understaffing, and lack of adequate training for crime scene investigators renders the process of securing evidence at crime scenes vulnerable to mistakes. These mistakes include overlooking potentially useful evidence, the use of less-than-optimal techniques to secure or store the evidence, and possibly contamination. These risks can be reduced but not completely eliminated. Even the most well-trained crime scene investigators are not immune to contaminating or overlooking evidence (e.g. Lounsbury and Thompson 2006, Proff *et al.* 2006).

Obtaining DNA elimination samples from suspects and volunteers

The key issues in the process of obtaining DNA samples for elimination purposes are twofold: first, it is difficult to assess in which cases infringements of bodily integrity are justified when DNA samples are taken from suspects by force. Second, the possibility of ever obtaining fully informed consent from volunteers is questionable.[7] 'Volunteers'

---

[7] As recent work at the interface of ethics and the bioscience has shown, 'fully' informed consent is difficult if not impossible to achieve, because nobody can anticipate all possible future uses of donated bodily material, including those that will only become conceivable when research has advanced further (e.g. Felt *et al.*

typically comprise two groups. The first group comprises those who, in connection with a dragnet (aka intelligence-led mass DNA screening), are asked to volunteer a DNA sample for purposes of comparing their DNA with traces from the crime scene. The second group subsumed under the heading of 'volunteers' are those who may have left their DNA at a crime scene for legitimate reasons (e.g. when visiting a friend, or if the crime scene is their workplace); in some investigations, they are requested to volunteer a DNA sample to eliminate their DNA profile from the crime scene stains. In some countries, also victims are subsumed under the category of 'volunteers'.

With regard to suspects, it could be argued that obtaining profiles by force is legitimate especially if the crime is serious, and if the sample and profile are later destroyed when no conviction follows (Kaye 2006). With regard to obtaining samples from volunteers, however, coercion is not commonly seen as acceptable, and consent should be as fully informed as possible. In practice, the situation is complicated further by the high stakes of a criminal investigation, which can blur the meaning and the boundaries of coercion. For example, if a neighbour of a murder victim is being asked to 'volunteer' a DNA profile for elimination purposes, and is told that if s/he refuses to provide a sample voluntarily s/he will be considered a suspect, how 'voluntary' are we to consider cooperation to be? More explicit discussion is needed on the meaning of consent regarding the provision of a DNA sample, and of how feasible protocols for this procedure can be enacted practically.

Transport and storage of crime scene traces

The main problems in transport and storage of samples pertain to contamination risks or loss of trace evidence. Besides simple human errors of swapping or misplacing samples, which are assumed to occur in cases where DNA samples are not adequately sealed and labelled at the crime scene, contamination could occur either deliberately or accidentally while traces are being transported to the laboratory.

2008; Hoeyer 2008; Taylor 2008). This problem is even more pressing in the context of criminal justice, where the use of a volunteer's DNA sample that he or she has not anticipated could have immediate consequences on their personal life (such as rendering the person a suspect, or even lead to a conviction).

Before and during the analysis at the crime laboratory

The level of risk of contamination and loss of evidence at the analysis stage depends on the safety and quality standards of a particular laboratory (it is the case that not all laboratories carrying out DNA analysis for the police are accredited or certified).[8] In the European Union (EU), plans exist to make laboratory accreditation mandatory. Precautionary measures range from simply requesting laboratory personnel to be 'careful' to videotaping samples when they come in. The risk of biased analysis of DNA samples – both from crime scenes as well as subject samples – can be reduced further by not allowing laboratory workers access to case details (Dror 2009). The varying policies and practices that abound need standardising. For example, whereas workers in forensics laboratories in the UK or Austria receive solely the biological material and a bar code, Swedish laboratory workers, in some cases know more about the context of the case (Corinna Kruse, Linköping University, Sweden, personal communication 18 January 2009). Furthermore, chronically understaffed crime laboratories can occasion backlogs in the analysis of DNA traces, which can increase risk of contamination and, not unimportantly, lower detection rates based on DNA evidence (Raymond et al. 2008)

On a technical level, the analysis of mixed samples (samples containing the DNA of more than one person, often the victim and the perpetrator, such as in rape cases) or otherwise contaminated samples is technically more challenging than analysis of 'clean' DNA from the cheeks of subjects (buccal swabs), and is, therefore, more prone to error (Whitall 2008). The development of algorithms 'to accurately and robustly determine whether individuals are in a complex genomic DNA mixture' can be expected to alleviate some difficulties relating to the analysis of mixed samples (Homer et al. 2008: 1).

Profile matching

For a long time, a key problem in the process of profile matching used to lie in the calculation of so-called random match probabilities with regard to the use of reference populations (Chakraborty and Kidd 1991;

---

[8] For more information on ISO standardisation for laboratories, see http://www.iso. org/iso/home.htm.

Lewontin and Hartl 1991). The random match probability, as mentioned above, is meant to indicate the likelihood with which a DNA profile drawn at random from a particular population matches a particular DNA profile. More specifically, it should indicate to a judge or jury the likelihood with which the DNA 'match' between a person's (say Paula's) profile and the DNA trace found at the crime scene could be an adventitious match; this means that the crime scene trace does not really match Paula's DNA but instead coincidentally matches somebody else's who happens to have the same genetic characteristics as Paula at the relevant genetic loci. Assuming that individuals who are genetically more closely related – that is, who are in the same 'ethnic' group – on average share more genetic characteristics with one another than they do with individuals outside their ethnic group, the calculated random match probability can change, depending on the reference population used – particularly when profiles are only partial. For example, compared with the average inhabitant of any central European city, the DNA profile of an immigrant from South America is relatively rare. It would therefore be *less* likely for the South American DNA profile to 'randomly' match somebody else's profile if the reference population were from central Europe than if the reference population were of South American origin. As a result, the lower random match probability, if presented at court, would portray the suspect as much *more* likely to have committed the crime (because it is seen as *less* likely that the match between his DNA and the crime scene profile is a coincidence). For this reason, some commentators advocated the use of reference populations from similar geographical/ethnic backgrounds for suspects who are members of an ethnic minority (M'Charek 2005), suggested ways of calculating random match probabilities very conservatively – that is, to 'err' on the side of caution – to reduce the risk of bias disadvantaging ethnic minorities (see Chapter 12).[9] Others, however, argue that ethnic criteria in the stratification of population in genetics are always problematic and misleading and should therefore be abolished altogether (e.g. Kahn 2006, 2008). It should also be noted that today, with advanced technological means and especially when *full* profiles are being compared, random match

---

[9] This problem is different from the issue of ethnic discrimination, or ethnic bias, with respect to the overrepresentation of members of ethnic minorities both in terms or convictions and entries in police databases. For a discussion of this latter form of ethnic discrimination, see Ossorio and Duster (2005), Duster (2006) and Chapter 4.

probabilities are often in the range of one in hundreds of billions; in that range, 'ethnic' differences are less problematic (Lander and Budowle 1994; Kahn 2008).

Indeed, as mentioned above, more recent accounts of a high number of 'coincidental' matches of nine or more loci of profiles within DNA databases have posed new questions regarding the calculation of random match probabilities; in particular, the assumed independence of genetic markers (Murphy 2007; Felch 2008).

In the courtroom

Biases arising from problematic calculations of random match probabilities have found their way into the courtroom. Because many judges and jurors tend to regard DNA science as infallible (Briody 2004; Nance and Morris 2005), numerical interpretations of DNA evidence (random match probabilities, and also the likelihood ratios[10] – see Williams and Johnson 2008: 52–55) are likely to be taken as facts. This renders it difficult to create room for discussion of the statistical and conceptual issues inherent in the calculation of both (see Chapter 10).

Perceptions of forensic DNA technologies as objective and scientific have been attributed by some to the so-called 'CSI effect'. *Crime Scene Investigation* is a heroic police drama about a group of detectives solving tricky cases with the help of cutting-edge forensic technology (Gilbert 2006; Houck 2006). As Kruse (2010) argues, the real hero of the show is not the detectives but the technology itself. Whereas the human mind is portrayed as fallible, forensic science and DNA profiling is depicted as a 'language of truth'. With the focus on securing DNA evidence, however, other, potentially much more useful, evidence is neglected or downplayed (Prainsack 2007). This articulates itself in jurors or judges who decline to convict a suspect if no DNA evidence is presented; as some would argue, it leads to judges and jurors taking DNA evidence more seriously than other forms of evidence, arguably often more than warranted (Briody 2004; Nance and Morris 2005).

Another problem pertaining to court trials is the so-called 'prosecutor's fallacy' (Thompson and Schumann 1987; Leung 2002; Williams and Johnson 2008: 54). This term signifies a variety of possible fallacies related to assessing the probabilities of guilt and innocence of

---

[10] The likelihood ratio is the ratio of probability that a crime scene profile and a subject profile stem from the same person divided by the probability that they do *not* stem from the same person but randomly match at all loci.

a suspect based on the presentation of forensic evidence. In its most prominent form, the prosecutor's fallacy happens when, for example, the expert witness in a person's (say Paula's) case asserts that the chance that the DNA obtained from the crime scene randomly matches somebody other than Paula in the same reference population is one in one billion. It would be a fallacy if this was understood to mean that the chances that Paula *is innocent* are one in one billion. This, of course, is incorrect, because even if it could be established with certainty that the DNA from the crime scene is indeed Paula's, that does not prove she has committed the crime. The DNA could be there for legitimate reasons (Paula had been at the crime scene a few weeks earlier to visit a friend), or it could have been 'planted' or the sample could have been contaminated at some stage with Paula's DNA. Obviously, such 'misunderstandings' could have disastrous consequences, especially for Paula (Garret and Neufeld 2009). For a discussion of a wider range of variants of the prosecutor's fallacy see Lynch *et al.* (2008).

The phenomenon of the 'defence attorneys' fallacy' pertains to the reverse scenario: when confronted with a random match probability of, for example, one in one million, then some defence attorneys argue that in a city of eight million inhabitants (half of whom can be assumed to be of the same sex as the defendant), there are four other people who could be the originator of the DNA stain. Therefore, so the argument continues, the DNA evidence cannot be seen as incriminating the defendant. This, of course, is misleading as it ignores all other circumstances of the crime: not all of these four people who could 'randomly' match the DNA profile obtained from the crime scene could have had access to the crime scene; some could even be children, or they could have been out of town. This only shows again that DNA evidence should not be treated in isolation but always in the context of other kinds of evidence and known circumstances of the crime.

Profile storage in the database

While privacy risks related to profile storage in DNA databases have received attention in the field of medical DNA databasing (Hindmarsh and Abu Bakar 2007; Homer *et al.* 2008), this topic has been little discussed in the context of forensic DNA databases. This is partly explained by the fact that forensic DNA databases are not publicly accessible, are typically not linked with many other computers (and, therefore, less 'hackable') and that only a limited number of people are able to access them directly. Nevertheless, based on the evaluation of several forensic databases, Stahlberg *et al.* (2007) argue that even deleted DNA data leave unintended traces and can, therefore, pose a privacy risk.

Besides privacy risks, the most obvious disadvantage for individuals whose profiles are stored in a forensic DNA database is that they can be linked to crimes when a sample found at a crime scene matches their profile. For example, if Paula and Paul commit a crime together, and Paula's profile is in a forensic DNA database while Paul's is not, then Paula will likely become a suspect immediately if DNA from the crime scene matches her profile (Prainsack and Gurwitz 2007). Paul, whose profile is not in the database, could be linked to the crime only by other means (such as eyewitnesses or finger or shoe prints). Thus, Paul has a 'competitive advantage' over Paula, which might be justified if Paula's profile is on the database because she was earlier convicted of a crime. However, if Paula's profile is there because she had submitted one voluntarily,[11] or even if Paula had been the subject of an investigation and later acquitted, then the situation could be quite different. Without prior convictions, why then should Paula be more vulnerable to being identified and linked to a crime on the basis of DNA than Paul (Nuffield Council 2007; see Chapter 5)?

Moreover, a higher vulnerability to being linked to a crime also means that one is more vulnerable to being wrongfully convicted. This means that, also from a human and civil rights perspective, such a 'disadvantage' for people whose profile is on the DNA database compared with those whose profile is not is problematic. Some commentators argue that such eventualities constitute infringements of the presumption of innocence (Guillén *et al.* 2000; Kaye and Smith 2004; GeneWatch 2005; van Camp *et al.* 2007; Innocence Project 2009). While this issue cannot be discussed in detail here, it is important to be aware of the fact that rules and practices of profile inclusion and profile retention always have confounding effects.

Sample storage

Issues related to the storage of DNA *profiles* are different from those related to the storage of the material from which the DNA profile was derived. With regard to the latter, it should be noted first that there is

---

[11] Many countries, however, keep volunteers' databases separate from databases of convicts (and suspects). Neither in Austria nor in Israel, for example, are profiles of volunteers used for routine speculative searches (see Chapters 3 and 8). England and Wales intends to adopt the same policy (Home Office 2009).

an essential difference between DNA samples from crime scenes and DNA samples from subjects (convicts, arrestees, suspects, volunteers). Whereas the former are typically contaminated with other materials present at the crime scene (e.g. a bloodstain obtained from the floor that could contain dust, fibres or even blood from additional people, victims or perpetrators), samples from subjects are regularly obtained from the inside of the mouth (cheek or buccal swab). Also, with DNA profiling techniques and technologies constantly improving, DNA evidence of too low quality for analysis in the past or at present can potentially be analysed in the future.

Retention of subject samples (convicts, suspects, arrestees and, in some countries, also volunteers and even victims) is often seen as ethically more problematic than the storage of crime scene samples. As Dierickx (2008) points out, the regulation of subject sample storage and retention varies greatly from country to country. First, there are countries where samples are destroyed immediately after a DNA profile has been derived; second, some countries distinguish between DNA profiles from suspects and convicts and treat their storage differently; third, there are countries that retain samples from both suspects and convicted offenders for a particular period of time; and fourth, some countries have no set term for the destruction of samples.

The risks and considerations emerging from DNA subject sample retention relate to DNA samples being potentially available for more wide-ranging analysis, for example, of the coding parts of the DNA. These additional analyses could reveal information relating to genetic diseases and the likely phenotype of the originator of the DNA. None of this kind of information can be obtained from a 'traditional' DNA profile, which consists of a string of numbers and digits derived solely from genetic regions that do not code for proteins (we cannot, at present, infer any information on disease risks and traits from non-coding regions in the genome). Therefore, many commentators see a conflict with privacy rights when such phenotypic information is derived from samples (see Phenotypic profiling, below). Others contend that most national laws prohibit the use of DNA samples obtained in the course of criminal investigation for any purpose other than forensic purposes. In addition, the retention of samples also serves a practical purpose, especially in the context of the EU Prüm Decision (see below), where certain kinds of preliminary 'match' between two countries always need to be confirmed by an additional DNA analysis. If the sample has not been retained, then police would need to find the originator of the DNA profile and request another sample, both of which might not always be feasible. In that case, the effort of finding a preliminary match was in vain, and suspects might remain inaccessible for further investigations (see also European Network of Forensic Science Institutes (2009), recommendation 27).

Function creep

Proposals to carry out such additional analyses with stored or available DNA samples are often discussed as a form of *function creep*. This term, as mentioned above, signifies the widening of the scope of uses of a particular device or technology. In other words, function creep occurs when a device or technology is used for purposes other than those for which it was designed or designated (Dahl and Rudinow 2009). In addition, function creep can also be seen as manifesting itself in an ever-wider inclusion of individuals from whom DNA is taken and of types of crime or offence for which DNA data is seen as appropriate for storage in the database (e.g. Chapters 7 and 13). Function creep has taken place since the mid 1990s with respect to the widening scope for groups of people whose DNA profiles have been included in forensic DNA databases in some countries. At present, we witness the occurrence of function creep also in the form of so-called familial searching (or genetic proximity testing), and phenotypic profiling, which we will focus on in this subsection.

Familial searching can be performed if a DNA profile derived from a crime scene stain does not produce a 'full' match with a subject profile but only a so-called partial match. This is the case when the two profiles do not match in all of the analysed genetic loci but only in most of them (e.g. in 11 of the 13 loci used in the US Federal CODIS database). Such a relatively high level of genetic concordance can indicate genetic relatedness. The 'hit' in the database could have identified a genetic relative of the person who actually left the DNA at the crime scene. For example, in the US CODIS system, siblings share an average of 16.7 of the total 26 alleles (alleles are genetic variants; there are two alleles at each of the 13 loci) (Greely *et al.* 2006: 253; Fimmers *et al.* 2008; Reid *et al.* 2008;). Police could approach the person identified and investigate whether s/he has any biological relatives who could be suspects for the crime. In the UK – where the method has been used since 2002 (Haimes 2006) – a famous case of manslaughter was solved by this method (the so-called Craig Harman case: Greely *et al.* 2006; McCartney 2006). In the USA, too, several cases have been solved by means of familial searching – arguably more accurately called genetic proximity testing (because not all genetic relatives consider each other 'family' in the social sense of the word). In many other countries, however, familial searching is explicitly or implicitly forbidden (Lazer and Meyer 2004; Curran and Buckleton 2008).

Why do some authors see genetic proximity testing as problematic? The ECHR (2008) addressed the practice critically. The Court argued that because of the 'capacity of DNA profiles to provide a means of identifying genetic relationships between individuals', the retention of such profiles interfered with the right to the private life of those individuals. While this does not necessarily mean that such interference is unjustified, the Court's line of reasoning highlighted ethical complexities inherent in genetic proximity testing. Arguably, the very act of opposing genetic proximity testing endorses a variant of genetic exceptionalism (namely of treating information as 'special', as different from other kinds of information), because it is a common occurrence that relatives of suspects are approached by investigating law enforcement officers; this scenario is not limited to those cases where clues to relatedness come from genetic analyses (Greely *et al.* 2006: 257). Such clues could come from hints from neighbours or acquaintances, who, for example, might suggest that the suspect's brother could have been involved in the crime.

It is also argued that genetic proximity testing could violate the privacy of the person whose profile has been identified on the database through the matching attempt, as well as the genetic privacy of the (genetic) relatives of the person whose DNA profile has been identified (Willing 2005; Greely *et al.* 2006). Furthermore, genetic proximity testing could 'reinforce views about the alleged prevalence of criminality in certain families', reveal to relatives that a genetic relative has a profile on the database or even reveal a genetic link (or lack thereof) between individuals unaware of it. For example, this might reveal paternity information that the parties involved had not asked for and which potentially could disrupt social and familial structures (Haimes 2006; see also Greely *et al.* 2006).

Moreover, some commentators argue that the use of genetic proximity testing repeats existing demographic disparities in the criminal justice system, in which arrests and convictions differ widely based on race, ethnicity, geographic location and social class (Bieber and Lazer 2004; Bieber *et al.* 2006: 1316; Greely *et al.* 2006). Thus, genetic proximity testing is seen as an instance of 'function creep', mainly because it adds an objectionable feature to the existing search functions in forensic DNA databases and consequently broadens the scope of individuals who can be affected by police investigations.

Another instance of function creep is phenotypic profiling. It increases the kinds of information inferred from forensic DNA material. 'Traditional' forensic DNA profiling looks only at genetic characteristics in non-coding regions of the genome. But in most countries, DNA samples – containing a person's full genome, including those regions that can be analysed to learn about susceptibilities for diseases and other traits (Prainsack et al. 2008) – are being retained by officialdom. Through phenotypic profiling, the authorities could access these stored samples to examine coding regions in order to gain probabilistic information about a person's hair colour and ethnic background, very much as genetic ancestry companies do (Graham 2008; Prainsack et al. 2008).

Because of the particular means of analysis, and because of its probabilistic nature, phenotypic profiling is especially prone to error (Lowe et al. 2001; Cho and Sankar 2004). Perhaps also because it is expensive and labour intensive, it is not widely used (Koops and Schellekens 2008; M'charek 2008a, M'charek 2008b: 522–523). In the UK, however, phenotypic profiling is used in connection with dragnets, where the phenotypic analysis of the perpetrator's DNA can provide information about his or her likely genetic ancestry, from which skin tone and other phenotypic traits are inferred. In addition, it is used to analyse the DNA of murder victims in cases where their bodies cannot be identified by other means.

However, function creep also occurs when the scale of inter-linked databases increases. In the European context, the Prüm Decision is the most important example. This Decision is named after the German town where it was originally signed by Belgium, Germany, Spain, France, Luxembourg, the Netherlands and Austria, on 27 May 2005. Two years later, the EU adopted the Treaty (with minor changes) into EU law, the formal adoption taking place in June and the Decision becoming effective 20 days later in August 2008 (Council of the European Union 2008). Member States are obliged to have amended or changed national law by June 2011 to comply with the Decision.

The Prüm Decision has met with notable criticism, as increasing links between national databases are associated with increasing sur-veillance and, by some, with a 'police society'.[12] The EU Member States

---

[12] Besides DNA, the Prüm Decision also includes information about vehicles, fin-gerprints and collaboration in deportation measures (e.g. cooperation in deport-ing asylum seekers by means of air transportation).

are obliged to establish national DNA databases and to grant access to their databases to other Member States. However, this does not mean that police officers in country A have immediate access to the entire DNA database in country B. For example, while officers at the 'national contact point' in Belgium automatically see whether another Member State holds profiles that match Belgian data, they are not able to retrieve any identifying personal details. Only when a match is obtained are further details exchanged through existing channels (depending on whether DNA information is owned by the police or the judiciary, which varies from one country to another). Indicating the scope of this exchange, by the end of 2008, Austria and Germany alone had obtained more than 5000 DNA profile matches from each other.

Of course, the idea of increasing mutual exchange of data stored on forensic DNA databases predates the Prüm Decision. Interpol has been providing an infrastructure for transnational forensic DNA data exchange since 2003, when its international DNA Gateway became operational. Exchange of DNA profiles between countries usually occurs on an individual case-by-case basis. At every national Interpol contact point, a special Interpol DNA matching system is installed on a separate computer. The data in that computer do not include any nominal data (names, addresses, and so on of existing persons) nor are they connected to any other forensic information network. Member countries can define which countries they do not want to compare their data with; such exclusions can stem from concerns regarding policies pertaining to the DNA police database in particular countries, or they can derive from purely practical considerations (e.g. European countries that have already ratified the Prüm Decision compare their data only via the Prüm network and thus 'exclude' each other in the list of countries with which they exchange data via the Interpol DNA Gateway). The DNA profile queries that member countries upload to the Interpol Gateway 'pop up' as a question on the computer screens at other national contact points. Whether individual member countries respond to the query by checking whether the profile matches any profiles in their own database is up to them, and anecdotal evidence suggests that response rates vary considerably across countries.[13]

---

[13] The country that enters the query receives an automated answer. If there is a potential hit, all affected countries will receive an electronic notification and can then exchange data and information through traditional channels (bilateral agreements etc.) in order to confirm the match before they consider further steps. Only at that point may identifying data be exchanged.

The Interpol DNA Gateway was never meant to assume the role of a 'super-database', or be a replacement for national databanks. In this regard, the Prüm system is different from the Interpol Gateway. Whereas Interpol runs a database that stores DNA profiles centrally, the Prüm network does not consist of a central database but instead facilitates mutual access to the databases of EU Member States. This is why critics remain sceptical of the Prüm system, which they see as dangerous function creep that will edge Europe towards a police state (e.g. Balzacq 2006).

Closely related to the issue of function creep are the issues of extension of the criteria for inclusion in forensic DNA databases (i.e. whose profiles and samples should be included), and the length of time they are to be retained. The issue of inclusion and retention criteria has been subject to debate for nearly a decade (Cronan 2000; Kaye and Smith 2003; Rosen 2003; Duster 2004; Simoncelli 2006; Levitt 2007; Prainsack 2007). For example, in the ethical and regulatory literature, it is relatively uncontested that profiles of volunteers – who have never been suspected of involvement in any crime – should not be used for routine speculative searches (see Chapter 3). The judgment of the ECHR of December 2008 set a political precedent in the sense that the indefinite retention of profiles and samples from individuals who were not subsequently convicted was seen as ethically and politically problematic.

CONCLUSIONS

In reviewing the key issues related to the collection, analysis, storage, computations and evidentiary use of DNA material in the criminal justice system, it is obvious that the 'story' of police uses of DNA technologies follows the path of other forensic technologies, especially fingerprinting, which was also bestowed with high hopes and equally high levels of trust when it emerged into wide use in the criminal justice system. However, forensic DNA technologies are particularly prone to be seen as a promissory technology because of their focus on what is often seen as the 'essence' of the body, namely, DNA.[14] Even many who do not subscribe to such genetic essentialism are, it seems, inclined to trust genetic evidence more than they trust evidence obtained from observing the morphology of the body.

---

[14] For excellent reflections on the 'philosophy of depth' in the context of genetics, see Chapter 4 in Rose (2006).

However, it is also notable that the image of forensic DNA profiling as an infallible technology is now being increasingly questioned. Despite the undoubtedly impressive successes of forensic DNA technologies in helping to solve crimes – particularly 'cold' cases – and in exonerating the wrongfully convicted, limits and problems are being acknowledged to the extent that some see DNA profiling to be as fallible and error prone as any other technology. Therefore, it is advanced that DNA profiling cannot – and must not – supersede or render obsolete human experience, observation and judgment. This observation and argument has implications for DNA forensic governance and practice at every stage of its process from securing forensic evidence to the assessment of evidence in court (Table 2.1).

To secure traces at crime scenes (Table 2.1), this means that DNA evidence should not be treated as the most valuable evidence a priori. Instead, it should be used in conjunction with or as a possible complement to other evidence-finding methods if relevant. Anecdotal evidence from law enforcement practice already suggests that crime scene investigators are in danger of overlooking other potentially valuable evidence in some situations because they focus on securing DNA traces. This problem is currently exacerbated in situations where the quality of crime scene work is also measured according to the number of DNA traces officers manage to secure (Prainsack 2007). This, however, applies particularly to scenes of violent and severe crimes. With regard to volume crimes, on the contrary, DNA evidence might deserve more rather than less attention in many countries (see Chapter 9).

With respect to obtaining DNA profiles for elimination purposes from suspects and volunteers (Table 2.1), a strong case can be mounted for addressing more explicitly the meaning and limits of coercion and informed consent in regulatory and public debates. As argued earlier in this chapter, the boundaries of these avenues of criminal investigation are highly problematic. Even if a person states that s/he is providing a DNA sample voluntarily, an element or suggestion of coercion can be present, for example when the volunteer knows that s/he would be treated as a suspect if s/he did not volunteer DNA (see Chapters 3 and 8). Moreover, to ensure an equitable and transparent procedure in any situation in which informed consent is required from a suspect or a volunteer in the process of sample provision, all possible future uses of their DNA profile should be disclosed to the individual. Similarly, the individual should receive information about the duration of storage of both the biological sample and the DNA profile derived from it.

The transport and storage of crime scene traces (Table 2.1), as well as the analysis of the sample in crime laboratories (Table 2.1), should also be governed by clear and explicit quality standards, which should be transparent to all parties involved. With regard to profile matching (Table 2.1), general information about what kinds of profile are routinely compared with profiles in which databases should be made publicly available (e.g. on governmental websites), so that anybody providing a DNA sample – whether voluntary or not – can obtain information on the possible scope and level of the search. Furthermore, pertaining to the courtroom (Table 2.1), it seems a miscarriage of justice that anybody should be convicted on the basis of DNA evidence alone.

Transparency is also likely to alleviate fears of function creep, as stakeholders could make the undesired uses of DNA profiles a topic of public knowledge and discussion (Table 2.1). Similarly, with regard to the storage of profiles in forensic databases (Table 2.1) where backlogs in deleting 'old' profiles exist, there should be explicit discussions about how this problem could be addressed. For example, what happens if a crime scene profile matches a DNA profile of a suspect or convicted offender who should legally have been deleted from the database? Should such a match then be admissible as evidence in court (Chapter 3)? Such questions should be regarded not only as legal issues but also as arenas for societal negotiations of the boundary between individual and collective interests and rights (redefined as 'biocivic' issues in Chapter 13).

On the issue of sample storage (Table 2.1), however, it would be desirable to move the focus of public discussion from the supposed danger of sample retention, which would in principle enable authorities to obtain information on disease risks and traits of the originator of the sample, to a more sober weighing of risks and benefits inherent in this practice. Sample retention has clear benefits with respect to confirming a suspected 'match' between a crime scene profile and a subject profile, for example. The UK Home Office's (2009) proposal to destroy samples as soon as the derived profile has been successfully uploaded to the database seems driven by a desire to anticipate public concerns and appease them preemptively, perhaps to dampen potential resistance to more problematic aspects of proposed policy changes.

Such issues and their implications for governance build on the many others raised about the pitfalls of excessive trust in DNA technologies (e.g. Aronson 2007; Lynch et al. 2008). Clearly, there is a need for increased public awareness of both the potentials and the limits of DNA technologies, in particular what they can and what they cannot prove. The hope remains that the longer these technologies are with us, the more sober our expectations will be.

ACKNOWLEDGEMENTS

I am grateful to Simon Cole, Corinna Kruse, Reinhard Schmid and Kees van der Beek for helpful comments on the draft of this chapter, and to Troy Duster for providing me with relevant materials. To my 'partner in crime' in the production of this book, Richard Hindmarsh, I owe extensive comments on the manuscript.

REFERENCES

Aronson, J. (2007). *Genetic Witness: Science, Law, and Controversy in the Making of DNA Profiling*. New Brunswick, NJ: Rutgers University Press.

Balzacq, T. (2006). *The treaty of Prüm and the principle of loyalty*. http://www.libertysecurity.org/article1186.html (accessed March 2009).

BBC News online (2009a). DNA data plan comes under fire. BBC News online, 7 May http://news.bbc.co.uk/1/hi/uk/8038090.stm (accessed May 2009).

BBC News online (2009b). 'DNA bungle' haunts German police. *BBC News* online, 28 March http://news.bbc.co.uk/1/hi/world/europe/7966641.stm (accessed April 2009).

Bieber, F. and Lazer, D. (2004). Guilt by association. *New Scientist*, 184, 20.

Bieber, D., Brenner, C. and Lazer, D. (2006). Finding criminals through DNA of their relatives. *Science*, 312, 1315–1316.

Briody, M. (2004). The effects of DNA evidence on homicide cases in court. *Australian and New Zealand Journal of Criminology*, 37, 231–252.

Bykowicz, J. and Fenton, J. (2008). City crime lab director fired: database update reveals employees' DNA tainted evidence, throwing lab's reliability into question. *Baltimore Sun*, 21 August http://www.baltimoresun.com/news/local/baltimore_city/bal-te.md.lab21aug21,0,1849069.story (accessed April 2009).

Chakraborty, R. and Ge, J. (2009). Statistical weight of a DNA match in cold-hit cases. *Forensic Science Communications*, 22(3) http://www.fbi.gov/hq/lab/fsc/current/undermicroscope/2009_07_micro01.htm (accessed March 2010).

Chakraborty, R. and Kidd, K. (1991). The utility of DNA typing in forensic work. *Science*, 254, 1735–1739.

Cho, M. and Sankar, P. (2004). Forensic genetics and ethical, legal and social implications beyond the clinic. *Nature Genetics Supplement*, 36, S8–S12.

Cole, S. A. (2001). *Suspect Identities: A History of Fingerprinting and Criminal Identification*. Cambridge, MA: Harvard University Press.

Council of Europe (1950). *Convention for the Protection of Human Rights and Fundamental Freedoms*. Strasbourg: Council of Europe http://conventions.coe.int/Treaty/Commun/QueVoulezVous.asp?NT=005&CL=ENG (accessed February 2010). For a list of signatory states see http://conventions.coe.int/Treaty/Commun/ChercheSig.asp?NT=005&CM=&DF=&CL=ENG (accessed January 2009).

Council of the European Union (2008). *Decision 2008/615/JHA: Decision on the stepping up of cross-border cooperation, particularly in combating terrorism and* cross-border crime (Prüm Decision). Decision 2008/616/JHA: implementation of Decision 2008/615/JHA. Brussels: Council of the European Union.

Cronan, J. (2000). The next frontier of law enforcement: A proposal for complete DNA databanks. American Journal for Criminal Law, 28, 134.

Curran, J. and Buckleton, J. S. (2008). Effectiveness of familial searches. Science and Justice, 48, 164–167.

Dahl, J. Y. and Rudinow Sætnan, A. (2009). 'It all happened so slowly': on controlling function creep in forensic DNA databases. International Journal of Law, Crime and Justice, 37, 83–103.

Dierickx, K. (2008). The retention of forensic DNA samples: a socio-ethical evaluation of current practices in the EU. Journal of Medical Ethics, 34, 606–610.

Dolan, M. and Felch, J. (2008). The danger of DNA: It isn't perfect. Los Angeles Times, 26 December http://www.latimes.com/news/local/la-me-dna26-2008dec26,0,1922163.story (accessed February 2009).

Dror, I. (2009). On proper research and understanding of the interplay between bias and decision outcomes. Forensic Science International, 191, e17–e18.

Duster, T. (2004). Selective arrests, an ever-expanding DNA forensic database, and the specter of an early-twenty-first-century equivalent of phrenology. In DNA and the Criminal Justice System: The Technology of Justice, ed. D. Lazer. Cambridge, MA: MIT Press, pp. 315–334.

Duster, T. (2006). The molecular reinscription of race: unanticipated issues in biotechnology and forensic science. Patterns of Prejudice, 40, 427–441.

European Court of Human Rights (2008). Grand Chamber Judgement S and Marper v. United Kingdom. Press release issued by the Registrar,4 December http://cmiskp.echr.coe.int/tkp197/viewhbkm.asp?sessionId=16773269&skin=hudoc-pr-en&action=html&table=F69A27FD8FB86142BF01C1166DEA398649&key=74844 (accessed February 2009).

European Network of Forensic Science Institutes (2009). DNA-database management: review and recommendations http://kclmail.kcl.ac.uk/OWA/redir.aspx?C=3a3df141f6834951be272a7dcc47a210&URL=http%3a%2f%2fwww.enfsi.eu%2fget_doc.php%3fuid%3d345 (accessed July 2009).

Felch, J. (2008). How reliable is DNA in identifying suspects? Los Angeles Times, 20 July http://articles.latimes.com/2008/jul/20/local/me-dna20 (accessed August 2009).

Felt, U., Bister, M., Strassnig, M. et al. (2008). Refusing the information paradigm: informed consent, medical research, and patient participation. Health, 13, 87–106.

Fimmers, R., Baur, M., Rabold, U. et al. (2008). STR-profiling for the differentiation between related and unrelated individuals in cases of citizen rights. Forensic Science International: Genetics Supplement Series, 1, 510–513.

Garret, B. and Neufeld, P. (2009). Invalid forensic science testimony and wrongful convictions. Virgina Law Review, 95, 1–97.

GeneWatch (2005). The Police National DNA Database: Balancing Crime Detection, Human Rights, and Privacy. Bixton: GeneWatch.

Gilbert, G. (2006). CSI: The cop show that conquered the world. Independent, 19 December, 2–5 http://www.independent.co.uk/news/media/csi-the-cop-show-that-conquered-the-world-429262.html (accessed August 2009).

Gilbert, N. (2010). Science in court: DNA's identity crisis. Nature, 464, 347–348.

Graham, E. (2008). DNA reviews: predicting phenotype. Forensic Science, Medicine, and Pathology, 4, 196–199.

Greely, H., Riordan, D., Garrison, N. et al. (2006). Family ties: the use of DNA offender databases to catch offenders' kin. Journal of Law, Medicine and Ethics, 34(2), 248–262.

Guillén, M., Lareu, M., Pestoni, C. *et al.* (2000). Ethical–legal problems of DNA databases in criminal investigation. *Journal of Medical Ethics*, 26, 266–271.

Haimes, E. (2006). Social and ethical issues in the use of familial searching in forensic investigations: insights from family and kinship studies. *Journal of Law, Medicine and Ethics*, 34, 63–276.

Hindmarsh, R. and Abu-Bakar, A. (2007). Balancing benefits of human genetic research against civic concerns: *Essentially Yours* and beyond – the case of Australia, *Personalized Medicine*, 4, 497–505.

Hoeyer, K. (2008). The ethics of research biobanking: a critical review of the literature. *Biotechnology and Genetic Engineering Reviews*, 25, 429–452.

Houck, M. (2006). *CSI*: reality. *Scientific American*, 295, 85–89.

Home Office (2009). *Consultation paper: Keeping the Right People on the Database. Science and Public Protection.* London: The Stationery Office http://www.homeoffice.gov.uk/documents/cons-2009-dna-database/dna-consultation?view=Binary (accessed May 2009).

Homer, N., Szelinger, S., Redman, M. *et al.* (2008). Resolving individuals contributing trace amounts of DNA to highly complex mixtures using high-density SNP genotyping microarrays. *PLoS Genetics*, 4, e1000167.

Innocence Project (2009). *Website.* www.innocenceproject.org/ (accessed 8 January 2009).

Jefferson, J. (2008). Cold hits meet cold facts: are DNA matches infallible? *Transcript*, 40, 29–33.

Jha, A. (2004). DNA fingerprinting 'no longer foolproof': pioneer of process calls for upgrade'. *Guardian*, 9 September, p. 5 http://www.guardian.co.uk/science/2004/sep/09/sciencenews.crime (accessed January 2009).

Kahn, J. (2006). Genes, race, and population: avoiding a collision of categories. *American Journal of Public Health*, 96, 1965–1970.

Kahn, J. (2008). Race, genes, and justice: a call to reform the presentation of forensic DNA evidence in criminal trials. *ExpressO* http://works.bepress.com/jonathan_kahn/1 (accessed August 2009).

Kaye, D. and Smith, M. (2003). DNA identification databases: legality, legitimacy, and the case for population-wide coverage. *Wisconsin Law Review*, 3, 413–459.

Kaye, D. and Smith, M. (2004). DNA databases for law enforcement: the coverage question and the case for a population-wide database. In *DNA and the Criminal Justice System: The Technology of Justice*, ed. D. Lazer. Cambridge, MA: MIT Press, pp. 247–284.

Kaye, D. (2006). Who needs special needs? On the constitutionality of collecting DNA and other biometric data from arrestees. *Journal of Law and Medical Ethics*, 34, 188–189.

Koops, B.-J. and Schellekens, M. (2008). Forensic DNA phenotyping: regulatory issues. *Columbia Science and Technology Law Review*, 9, 158 http://www.stlr.org/volumes/volume-ix-2007-2008/koops/ (accessed January 2009).

Krane, D., Doom, T., Mueller, L. *et al.* (2004). Commentary on Budowle, B., Shea, B., Niezgoda, S., Chakraborty, R. CODIS STR loci data from 41 sample populations. *Journal of Forensic Sciences* (2001). 46, 453–489 (multiple letters). *Journal of Forensic Sciences*, **49**, 1388–1393.

Kruse, C. (2010). Producing absolute truth: *CSI* science as wishful thinking. *American Anthropologist*, 112, 79–91.

Lander, E. and Budowle, B. (1994). DNA fingerprinting dispute laid to rest. *Nature*, 371, 735–738.

Lazer, D. and Meyer, M. N. (2004). DNA and the criminal justice system: consensus and debate. In *DNA and the Criminal Justice System: The Technology of Justice*, ed. D. Lazer. Cambridge, MA: MIT Press, pp. 357–390.

Levitt, M. (2007). Forensic databases: benefits and ethical and social costs. *British Medical Bulletin*, 83, 235–248.

Leung, W.-C. (2002). The prosecutor's fallacy: a pitfall in interpreting probabilities in forensic evidence. *Medicine, Science and the Law*, 42, 44–50.

Lewontin, R. and Hartl, D. (1991). Population genetics in forensic DNA typing. *Science*, 254, 1745–1750.

Lounsbury, D. and Thompson, L. (2006). Concerns when using examination gloves at the crime scene. *Journal of Forensic Identification*, 56, 179–185.

Lowe, A., Urquhart, A., Foreman, L. *et al.* (2001). Inferring ethnic origin by means of an STR profile. *Forensic Science International*, 119, 17–22.

Lynch, M., Cole, S. A., McNally, R. *et al.* (2008). *Truth Machine. The Contentious History of DNA Fingerprinting*. Chicago, IL: University of Chicago Press.

McCartney, C. (2006). The DNA expansion programme and criminal investigation. *British Journal of Criminology*, 46, 175–192.

M'charek, A. (2005). *The Human Genome Diversity Project: An Ethnography of Scientific Practice*. Cambridge, UK: Cambridge University Press.

M'charek, A. (2008a). Contrasts and comparisons: three practices of forensic investigation. *Comparative Sociology*, 7, 384–412.

M'charek, A. (2008b). Silent witness, articulate collective: DNA evidence and the inference of visible traits. *Bioethics*, 22, 519–528.

Murphy, E. (2007). The new forensics: criminal justice, false certainty, and the second generation of scientific evidence. *California Law Review*, 95, 721–797.

Nance, D. A. and Morris, S. (2005). Juror understanding of DNA evidence: an empirical assessment of presentation formats for trace evidence with a relatively small random-match probability. *Journal of Legal Studies*, 34, 395–444.

Neufeld, P. and Coleman, N. (1990). When science takes the witness stand. *Scientific American*, 262, 46–53.

Nuffield Council on Bioethics (2007). *The Forensic Use of Bioinformation: Ethical Issues*. London: Nuffield Council on Bioethics http://www.nuffieldbioethics. org/go/ourwork/bioinformationuse/publication_441.html (accessed January 2009).

Ossorio, P. and Duster, T. (2005). Race and genetics: controversies in biomedical, behavioral, and forensic sciences. *American Psychologist*, 60, 115–128.

Prainsack, B. (2007). Forum on the Nuffield Report: an Austrian perspective. *BioSocieties*, 3, 92–97.

Prainsack, B. and Gurwitz, D. (2007). 'Private fears in public places?' Ethical and regulatory concerns regarding human genomic databases [editorial]. *Special Focus Issue of Personalized Medicine*, 4, 447–452.

Prainsack, B., Reardon, J., Hindmarsh, R. *et al.* (2008). Misdirected precaution. *Nature*, 456, 34–35.

Proff, C., Schmitt, C., Schneider, P. *et al.* (2006). Experiments on the DNA contamination risk via latent fingerprint brushes. *International Congress Series*, 1288, 601–603.

Raymond, J., van Oorschot, R., Walsh, S. *et al.* (2008). Trace DNA analysis: do you know what your neighbour is doing? A multi-jurisdictional survey. *Forensic Science International: Genetics*, 2, 9–28.

Reid, T., Baird, M., Reid, J. P. *et al.* (2008). Use of sibling pairs to determine the familial searching efficiency of forensic databases. *Forensic Science International: Genetics*, 2, 340–342.

Rose, N. (2006). *The Politics of Life Itself: Biomedicine, Power, and Subjectivity in the Twenty-First Century*. Princeton, MA: Princeton University Press.

Rosen, C. (2003). Liberty, privacy, and DNA databases. *New Atlantis*, 1, 37–52.

Simoncelli, T. (2006). Dangerous excursions: the case against expanding forensic DNA databases to innocent persons. *Journal of Law, Medicine and Ethics*, 34, 390–397.

Stahlberg, P., Miklau, G. and Levine, B. (2007). Threats to privacy in the forensic analysis of database systems. In *Proceedings of the ACM SIGMOD International Conference on Management of Data*, pp. 91–102.

Taylor, P. (2008). When consent gets in the way. *Nature*, 456, 32–33.

Thompson, W. and Schumann, E. (1987). Interpretation of statistical evidence in criminal trials: the prosecutor's fallacy and the defense attorney's fallacy. *Law and Human Behaviour*, 11, 167–187.

Thompson, W., Taroni, F. and Aitken, C. (2003). How the probability of a false positive affects the value of DNA evidence. *Journal of Forensic Sciences*, 48, 47–54.

Van Camp, N., Dierickx, K. and Leuven, K. (2007). The expansion of forensic DNA databases and police sampling powers in the post-9/11 era: ethical considerations on genetic privacy. *Ethical Perspectives*, 14, 237–268.

Whitall, H. (2008). The forensic use of DNA: scientific success story, ethical minefield. *Biotechnology Journal*, 3, 303–305.

Williams, R. and Johnson, P. (2008). *Genetic Policing: The Use of DNA in Criminal Investigations*. Cullompton, UK: Willan.

Willing, R. (2005). Suspects get snared by a relative's DNA. *USA Today*, 7 June, 1A.

CASES

New York v. *Wesley*, 533 N.Y.S.2d 643 (S. Ct. 1988).

S and Marper v. *the United Kingdom* (2008). A summary of the judgment is available at http://cmiskp.echr.coe.int/tkp197/view.asp?action=html&docu mentId=843937&portal=hbkm&source=externalbydocnumber&table=F69A 27FD8FB86142BF01C1166DEA398649 (accessed January 2009).

ELAZAR ZADOK, GALI BEN-OR AND GABRIELA FISMAN

3

Forensic utilization of voluntarily collected DNA samples: law enforcement versus human rights

INTRODUCTION

Forensic DNA profiling is now an indispensable tool used by law enforcement agencies worldwide. Since its introduction by Sir Alec Jeffries and colleagues (Gill et al. 1985), forensic DNA has caused a revolution in crime scene investigation, similar to that brought about by fingerprint identification capabilities a century ago (see Chapter 6). The annual number of cases solved by means of DNA profiling in the UK, for example, is now approaching the number of those solved with the use of fingerprints. Nevertheless, the DNA profiling 'revolution' would not have occurred without the emergence of computerised forensic databases. Among those databases that focus on individual characteristics, the automated fingerprint identification systems and DNA databases are most valuable for law enforcement authorities. They allow the generation of 'cold hits', namely the identification of a suspect without a classical criminal investigation. The combination of two parallel processes – scientific innovations in the field of molecular genetics and the emergence of computerised databases – underpins contemporary methodology of forensic investigation.

Scientifically, forensic DNA profiling is still considered a young technology. It is also a relatively complicated process, prone to contaminations and misinterpretations. Its probabilistic nature leaves much room for debates regarding interpretation (see also Chapters 2, 6 and 7). Other controversies connected to DNA profiling and databasing involve ethical and legislative issues. Privacy and human rights issues are weighed generally against the well-being of society, but no clear lines can be drawn as in many instances they are complementary (Etzioni 2004). The many ethical questions that have been raised about

Genetic Suspects: Global Governance of Forensic DNA Profiling and Databasing, ed. Richard Hindmarsh and Barbara Prainsack. Published by Cambridge University Press. Copyright © Cambridge University Press 2010.

DNA profiling and databasing are well discussed in the literature (Rothstein and Tallbott 2006). This chapter will focus on the forensic use of voluntarily collected DNA samples, a very controversial aspect of forensic DNA profiling and databasing. Its widespread nature and use prompt various governance issues and proposed solutions. It also represents the basis of the fundamental debate about forensic data-basing: maximising law enforcement for the benefit of the society at the presumed expense of individuals. Three topics are examined. First, operational considerations in carrying out DNA dragnets (also referred to as 'intelligence-led mass screenings'), including cost–benefit argu-ments based on real cases; second, ethical and legislative considera-tions encountered in the process of sample collection from so-called volunteers; and third, ethical and legislative considerations related to the fate of these samples, and the profiles derived from them. We also demonstrate, as exemplified by a murder case in Israel, possible dilem-mas arising from the use of voluntarily submitted DNA samples, and in particular we discuss considerations of Israeli courts regarding claims about the presumably illegal use of such samples.

DNA DRAGNETS: OVERVIEW AND OPERATIONAL PERSPECTIVES

Harlan (2004: 187) defines DNA dragnets as '[e]ssentially warrantless searches administered en masse to large numbers of persons whose only known connection with a given crime is that authorities suspect that a particular class of individuals may have had the opportunity to commit it'. The first DNA dragnet, performed in the UK in 1987, became known as the *Colin Pitchfork* case (Sanders 2000). Two teenage girls were sexually assaulted and murdered in a Leicestershire com-munity in 1983 and 1986, respectively. Although some 5000 men from the neighbourhood volunteered DNA samples, the dragnet failed to produce a match. The case was finally solved on the basis of informa-tion disclosed to the police about a man named Colin Pitchfork, who had asked a friend to provide a blood sample in his place because of a portrayed fear of needles. On 22 January 1988, Pitchfork was sen-tenced to life imprisonment on a double count of murder.

This case sheds light on several problems inherent in DNA dragnets. They can be very expensive; they can take a long time without yielding results, and a perpetrator cannot be prevented from using 'tricks' to avoid the provision of a sample that would lead to detection (and possibly conviction). Despite these drawbacks,

DNA dragnets are performed in many countries. In the UK, more than 280 DNA dragnets were undertaken from 1987 to early 2006. They involved over 80 000 volunteers in total and 285 volunteers per screening on average (Mepham 2006). By 2004, more than 40 dragnets had resulted in positive identification of the offender (Williams *et al.* 2004). Burton, in 1999, reported a 39.4% success rate in DNA dragnets in the UK (where they are called intelligence-led DNA screenings). What accounts for some of the relatively high success rate of these screenings is the inclusion of accredited behavioural scientists closely monitoring the process. In contrast, a University of Nebraska study (Walker and Harrington 2005), which looked at 18 dragnets performed in different states in the USA, found that only one dragnet had resulted in positive identification (and subsequent conviction) of the offender. Most European countries are located in between these extremes. In some countries, DNA dragnets are not even legal, or they are not practised (see Chapter 8). The largest DNA dragnet to date was carried out in 1998 in northern Germany, where 16 400 people were sampled, which led to the identification of a rapist and murderer of an 11-year-old girl (Halbfinger 2003).

Two cases from Israel illustrate extreme scenarios in the conducting of DNA dragnets. The first crime, the *Anat Fliner* murder case, occurred in April 2006 (*State of Israel* v. *John Doe* 2008). Fliner, a lawyer and single mother of two, lived in a quiet, wealthy neighbourhood. One day, on answering a knock at the door, she was stabbed twice in the intestines and was later found dead in front of her house. A knife and gloves containing DNA traces from both the victim and the presumed murderer were retrieved from a nearby trashcan. With no clue to the murderer's identity, police conducted a DNA dragnet, which lasted over two years. Police sampled 500 convicted burglars residing in the vicinity of the crime scene without producing a match. The breakthrough came only after a 19-year-old male was caught trying to steal a motorcycle. The DNA taken as part of the arrest procedure matched the profile retrieved from the knife and gloves at the earlier crime scene. The suspect then confessed that while attempting to rob Fliner he had stabbed her when she started to scream. Police later admitted that they had suspected the murderer to be an inexperienced youngster from the very beginning but this line of investigation had not been thoroughly pursued, with a DNA dragnet resorted to instead.

In the second case, in January 2007, a teenage girl named *Maayan Ben-Horin* was murdered during a rape attempt (*State of Israel* v. *John Doe*

2007). The crime occurred in a remote area, and the chances that the murderer had come from outside the area were considered minimal. Twenty-one men residing in the vicinity were sampled and the case was quickly solved. Unlike the first case, this case exemplifies a situation where a DNA dragnet offered an efficient and even inexpensive tool when applied within a small and well-defined population.

However, another case, that of *Antoni Imiela* (Forensic Science Service 2004), again pointed up the problematic nature of dragnets. In different parts of the UK between November 2001 and October 2002, six rapes occurred. The first one, in Kent, led to a geographically based collection of over 2000 samples from volunteers, but produced no match in the UK National DNA Database (NDNAD). However, all of the next four rapes, occurring in other districts, were connected by DNA to the first one. An additional 1000 volunteers were sampled and the perpetrator was named after the road that he supposedly travelled: 'the M-25 rapist'. But only a drawing of the rapist, based on the sixth victim's description, finally led to his identification. The UK Forensic Science Service summarised its expenses on this case: more than 100 scientists and technicians were involved, and more than £2 million was spent. Because the expense of evaluating large quantities of DNA samples is so high, issues have been raised related to optimising budget distribution between DNA dragnets and other forensic and investigative means (Walker and Harrington 2005). The *Anthoni Imiela* case is one of many examples where geography was the main factor in determining the scope of individuals to be sampled in a dragnet. Rothstein and Talbott (2006: 156) also point out that 'sample population frequently consists of members of a single – often minority – racial or ethnic group'.

The following propositions are drawn from the above-described cases. A DNA dragnet should never be a 'stand-alone' operation, as a sole means to detect suspects. Only in exceptional cases – when the targeted population is well defined and relatively small – should police consider using DNA dragnets as their primary operational tool. In general, police should make sure that dragnets are used in conjunction with other investigative means and relevant information sources, based not only on geographical or ethnic considerations. In most cases where these considerations were the only ones determining the scope of volunteers, the results were poor. As mentioned above, it is advisable that a behavioural scientist or a profiler is involved in constructing lists of volunteers and prioritising them. Generally, it is more advisable to carry out a DNA dragnet when all other means have been exhausted. In addition, the inflationary use of DNA dragnets might unnecessarily damage the quality of police work. Although practical guidelines for effective intelligence-led DNA screening were also published by Interpol (Schuller *et al.* 2001), dragnets will never replace skillful investigative police work. Moreover, during a prolonged DNA dragnet, all considerations leading to the decision to carry it out should be reassessed periodically through careful monitoring. Finally, as well as jeopardising community trust, frequent dragnets can compromise community willingness to cooperate.

THE COLLECTION AND USE OF SAMPLES FROM VOLUNTEERS

In the context of an investigation, a 'volunteer' can be any person asked to submit a DNA sample although not directly suspected as a perpetrator: a victim of the assault, a family member giving a DNA sample for identification purposes or a witness present at the scene. The term 'volunteers' also signifies a specific population defined by the police for the purpose of mass DNA screening. A closer look at the concept of 'volunteer', as well as their rights and duties is the topic of this section.

In most countries, DNA samples from volunteers are obtained following prior 'informed consent' rather than on a court order. In the UK, two different consent forms exist. The first one allows police to use a volunteer's DNA profile only in connection with the investigated case (Williams *et al.* 2004). The second enables police to match the volunteer's profile against the database in so-called future speculative searches. In the UK, this second kind of consent, once given, is irrevocable, thus enabling the police to upload the profile into the NDNAD and also store the sample (see below). Many countries allow the use of voluntarily given samples only in connection with the case under investigation, and others do not allow the voluntary collection at all unless DNA evidence is found on the crime scene (Victorian Parliament 2004; Dierickx 2008).

In the UK, two conditions must be fulfilled in order that a sample taken from a volunteer can be considered as having been provided with proper 'informed consent'. First, samples must be collected without the use of physical or verbal threats, and a volunteer must not be given the impression that refusal to provide a sample would immediately turn him or her into a suspect (Walsh 2005). Second, these samples must only be collected after written consent has been given, along with explanations of all options and possibilities as to how the profile and sample would or could be used. This includes routine speculative searches against the whole database, and information with regard to revoking consent

The first condition, however, can be difficult to implement when a presumably innocent person is called to a police station to give a sample, which can appear as a coercive or threatening situation. Moreover, refusal to provide a sample could be seen as an intention to hide information rather than as an insistence on one's privacy rights. Failure of police to clarify, both to themselves and to the 'volunteer', that refusal will not automatically turn the volunteer into a suspect might damage the individual's presumed innocence, transferring the burden of proof from the police onto the citizen (Rothstein and Tallbott 2006). Nevertheless, there are cases where refusal might have a critical impact on the investigation, for example when there is only a small number of possible volunteers, or when a volunteer meets specific criteria relevant to the investigated case and shared only by a specified group of people.

Two different principles are related to the term 'informed consent': first, the 'limited scope of consent', and second, the 'unlimited' consent (or 'initial voluntariness' as termed by Kaye and Smith (2004)). Some argue that comprehensive explanations regarding possible uses of a DNA sample are not necessary in either case, as this contradict the nature of investigative work (Crouse and Kaye 2004). The Nuffield Council on Bioethics report on *The Forensic Use of Bioinformation* published in 2007 stated that more than 40% of volunteers sign for unlimited consent, but it is unclear whether all are aware of the implications. This illustrates the problematic nature of the 'informed' consent procedure (Parry 2008).

In the USA, DNA sampling of volunteers has been challenged by civil libertarians on the basis of the Fourth Amendment of the US Constitution,[1] which aims to shield the individual against undue governmental intrusions into privacy. Nevertheless, the Fourth Amendment does not, in principle, prohibit sampling of DNA from volunteers as it allows compromises of individual privacy rights on the basis of 'special needs' of the government or 'society protection needs' (Etzioni 2004; Simoncelli 2006; see Chapter 12). So far, most courts have turned down appeals against volunteers' sampling on this basis. The fact that biological samples, given intentionally and

---

[1] An amendment to the Bill of Rights of the American Constitution guarding against unreasonable searches and seizures.

voluntarily by citizens (such as in the context of medical research) or left unintentionally (such as biological traces at crime scenes and elsewhere), could be used for criminal investigation purposes can erode the constitutional grounds on which to oppose the use of voluntarily given samples for law enforcement purposes, at the expense of some infringement of individual's rights (Harlan 2004). Supporters of this view compare DNA dragnets with lawful police operations, for example searching for drunk drivers on the highway. Furthermore, there is the concern that many DNA dragnets tend to be ethnically biased (Rothstein and Tallbott 2006; also Chapter 4). Therefore, population-wide DNA databases might be considered as a legitimate solution for these issues.

POPULATION-WIDE DNA DATABASES

Proponents of population-wide (universal) forensic DNA databases argue they will end ethnic biases in existing databases (e.g. the NDNAD contains some 40% of the black male population of the UK and less than 10% of its white male population) (Privacy International 2007). Such databases are also seen to save police efforts and money in the long run, as well as minimise harassment of the population, as they render case-specific dragnets redundant. Finally, public opposition to selective DNA testing on grounds of privacy intrusion would be diminished, as the whole population would be equally involved (Bikker 2007).

With regard to the latter argument, it is further argued that if these databases only include DNA profiles but not the samples themselves (see Chapter 2), perceived infringements of the public's privacy rights would be further reduced as the information in the DNA profile is useful only for identification purposes as opposed to whatever other purposes are envisaged (see below) (Smith 2006). Several countries seem to be establishing quasi-population-wide databases, with profiles of particular population groups being continuously added to existing databases, for example the UK's NDNAD. Proponents include the UK criminal justice system (Whittall 2007), and the former UK Prime Minister Tony Blair (Jones 2006). Another supporting argument is that population-wide DNA databases might well serve the need to fight terrorism through faster identification of terrorists and associates before, during, or after terror-related events.

The counter-argument is that the expected benefits for law enforcement from population-wide forensic DNA databases will adversely impair the balance between privacy rights and society's needs for security, for example, and will lead to a 'surveillance society' (Nelkin and Andrews 2003; Chapter 13), consisting of a 'population of suspects' (Wadham 2002). In addition, population-wide databases would not likely repair the impaired ethnic balances in any existing law enforcement agencies' attitudes when these are already pronounced (Nuffield Council on Bioethics 2007).

Questions are also being raised about the cost-effectiveness of a population-wide DNA database. Although costs for determining a single DNA profile are decreasing dramatically, a very large budget would be needed to create and maintain a comprehensive DNA database; this could add to existing backlogs in most forensic DNA laboratories, which are already suffering from budget shortages and lack of skilled personnel, instrumentation, laboratory space and accreditation programmes (Norton 2005; Simoncelli 2006). Moreover, the Nuffield Council on Bioethics (2007) found that the 'hit rate' of the UK NDNAD had not increased over a number of years, although the number of subject profiles was steadily growing.[2] This might be the case because the efficiency of DNA databases depends not only on the number of subject profiles but also on the number and quality of profiles from crime scenes (Prainsack 2008; Van Camp and Dierickx 2008). Therefore, in light of budgetary restraints, it might be more beneficial to increase forensic DNA coverage of crime scenes instead of sampling additional populations (and thus further increase the number of subject profiles in the database). In addition, the following question seems fundamental: Should infringements of one's right to privacy be considered only in respect to a well-defined – and not general – need to fight crime and maintain public safety?

THE FATE OF VOLUNTARILY COLLECTED DNA
SAMPLES AND PROFILES

Voluntary DNA sampling is frequently used for elimination purposes, enabling police to considerably narrow the range of potential suspects by excluding those whose DNA was not found at the crime

---

[2] The term hit rate is used not only for matching individuals to scenes of crime but also for scene-to-scene matching, thus generating intelligence about the identification of a possible perpetrator. It should be also noted that not every hit leads to the submission of charges and subsequent conviction.

scene. Many judicial systems lack strict or clear guidelines on mini-
mising infringements on volunteers' privacy rights, not only in the
sampling process and the particular criminal investigation for which
the samples are taken but also with respect to future possible uses of
these samples and profiles.

According to civil libertarians, the retention and further utilisation of DNA samples and profiles give rise to more concern than
does the act of voluntary collection. That concern is very pronounced in the UK, where legal provisions allow indefinite retention
of DNA samples and profiles submitted by volunteers or taken from
suspects acquitted or even not charged. That legislation was enacted
in 2001 after the House of Lords overturned a judge's decision to
exclude DNA evidence based on a sample unlawfully held by police
(Williams *et al.* 2004; Walsh 2005). The House of Lords also found
that this legislation did not conflict with the *European Convention on
Human Rights* (Council of Europe 1950; Privacy International 2007).
These decisions drew widespread criticism and were not adopted in
Scotland, where DNA profiles of volunteers as well as other innocent
individuals are not included in Scotland's database (Johnson and
Williams 2004).

Later, the European Court of Human Rights would conclude (in
the case of *S and Marper* v. *the United Kingdom* 2008) that the practice of
retaining DNA and fingerprints of anyone arrested but not charged
or convicted in England and Wales was a violation of the 'right to
respect for private life' under Article 8 of the *European Convention on
Human Rights* (Council of Europe 1950; see also Chapter 2). In
response, in May 2009, the Home Office issued the paper *Keeping the
Right People on the DNA Database: Science and Public Protection*, which
specifically addressed this question of keeping and using volunteers'
samples and profiles. Its recommendations seem far reaching:

> [E]xisting volunteer samples to be removed from the NDNAD; Future
> profiles and samples to be destroyed when no longer required for
> investigative purposes; Future volunteer samples and profiles to be
> subject to distinct processes from speculative searching on the NDNAD.

Turning again to the USA, in several states there are no strict policies
regarding the retention of volunteers' DNA samples and profiles (Cho
and Sankar 2004; Rothstein and Tallbott 2006). Only eight US states
have explicit regulations; for example, California allows the retention
of volunteers' profiles and their possible use for speculative searches
but only for a period of two years (Simoncelli and Steinhardt 2006).

Only two US states (Wisconsin and Vermont) strictly forbid volunteer samples in DNA databases (Gaensslen 2006). Other states allow retention of DNA samples and/or profiles as long as required for investigation or prosecution. Several states differentiate between samples and profiles, though others leave this issue vaguely addressed. Crouse and Kaye (2004) surveyed forensic laboratories at local and state levels in 2000 and found that most feed profiles into local or state DNA index systems but do not have clear policies distinguishing samples and profiles of volunteers from those of suspects. Notably, some laboratories involved in DNA dragnets classify non-matching profiles of volunteers as existing for 'elimination purposes' and will not include them in the database, while others classify these profiles as 'suspects' profiles', and include them although this is not allowed by law. Such variation among the different jurisdictions is creating a great deal of unease in the US legal community. At the same time, the US National DNA Database (Combined DNA Index System or CODIS) strictly regulates the nature of profiles entering it and, in accordance with the law, only convicts are included.

Turning to other countries, a great variety in legal and practical configurations is also evident in this issue. For example, Canadian national law is not specific about the retention of profiles and samples taken from volunteers, although these profiles are not, in fact, included in the national database (Phillips 1998). The policy of the Royal Canadian Mounted Police is to destroy the samples, but not the profiles, when the case has ended. In one case, the Canadian Supreme Court exonerated a person convicted of sexual assault as he had not been informed that the DNA sample he had once consented to give might be used in another investigation (Bieber 2004). In Australia, profiles and samples taken from volunteers are discarded after 12 months but can be used during this period according to the degree of consent given (Victorian Parliament 2002, 2004; see also Chapter 13). In many European countries, all DNA samples (not only those taken from volunteers) are destroyed after processing. Volunteers' profiles not matching any originating from crime scenes are either deleted or saved separately for different periods of time depending on the consent given, police requests and court considerations.

Many judicial systems differentiate between DNA samples and profiles derived from them because individual genetic information not necessary for identification purposes is inherent in the samples (Van Camp and Dierickx 2007). Concerns have been raised that samples might be used in different kinds of research programme

(Prainsack and Gurwitz 2007). Two channels of this expanded use, which has been called function creep, may be considered: the utilisation of information stored in a given database for further development of means serving its initial general purpose, and its utilisation for completely different purposes (Asplen 2006). Indeed, 24 US states allow the use of DNA profiles and samples for 'humanitarian purposes' or federally funded research (Steinhardt 2004; Simoncelli 2006); and 48 states also allow access to their DNA databases (including samples) for purposes related to law enforcement other than mere identification. This implies that samples might also finally be used for research in genetic criminology, which seeks to develop various genetic means for identification of criminal potential in individuals. In the UK, by the end of 2006, 33 research requests had been submitted to the NDNAD; however, Whittall (2007) found that no adequate information regarding the nature of the associated research projects could be obtained.

Fears of function creep are also based on the opposite scenario, namely the use of information and samples from non-forensic DNA databases for criminal investigation purposes. For example, over three million samples are stored in the US Department of Defense DNA Registry, established in 1992 with the explicit purpose of identifying soldiers killed in action. The Department of Defense regulations allow law enforcement agencies, subject to a court order, to use DNA profiles and samples stored in the DNA Registry for criminal investigation purposes (Bieber 2004; Williams and Johnson 2006). Such use suggests a violation of the consent given by the soldiers or of their understanding of their consent. However, attempts to withdraw samples from the database have so far failed (Nelkin and Andrews 2003).

Similar concerns apply to medical DNA databases (so-called biobanks) where samples are provided by volunteers. In Scotland, a judge authorised the use of information stored in a database of HIV-positive volunteers to seek a conviction (McCartney 2004). In the UK, data from biobanks can be used by police under the Police and Criminal Evidence Act 1984 in 'exceptional circumstances' and upon authorisation by a circuit judge. In Sweden, information stored in the Phenylketonuria Register (biobank containing blood samples of most newborns since 1975, which supports research on severe metabolic disorders) was used in the 2003 investigation of the assassination of Foreign Minister Anna Lindh. Subsequently, hundreds of donors withdrew their samples from the database. Later, in the context of a dragnet, a court refused a request by the Swedish police to retrieve information on people who had withdrawn their samples from the Phenylketonuria Register (Ansell and Rasmusson 2008).

In reflecting upon these examples, it seems that concerns on the part of volunteers who submit DNA samples about possible uses of their DNA are well founded. Medical biobanks obtain samples only upon informed consent of their donors, reserving to the donors the right to withdraw samples and information derived from them in a wide range of circumstances. This is not always the case with voluntarily provided forensic DNA samples (Levitt 2007). Therefore, the Nuffield Council on Bioethics (2007) suggests that issues including research policies, exchange of information between databases and conditions of withdrawing consent should be clearly regulated.

The main argument used in favor of DNA *sample* retention is the need to update databases because of anticipated technological developments in forensic DNA profiling (Brettell *et al.* 2007). This argument is not easily dismissed, given that the techniques (and kits) used in the generation of DNA profiles from samples has changed quite considerably since the mid 1980s. For example, single nucleotide polymorphism technology was first used en masse for degraded DNA analysis after the 9/11 terror attack (Gaensslen 2006). Moreover, forensic Y chromosome analysis has been gaining more importance recently (Bieber 2004); and the German Federal Criminal Police (*Bundeskriminalamt*) has used, for the first time, nuclear DNA in conjunction with mitochondrial DNA in a DNA dragnet (Szibor *et al.* 2006). In 2005, an Australian group suggested a completely new approach, the genomic matching technique, that could save up to 85% of the costs in large dragnets (involving thousands of people). This technique is based on the donor–recipient matching screening methods used in bone marrow transplantations and could potentially be used for elimination purposes, dramatically lowering the number of samples that now need to be processed in the relatively expensive commonly used short tandem repeat technology (Laird *et al.* 2005). All these are examples where new technologies might lead to possibilities for further processing of existing samples stored in databases. The question arises, though, whether it will be financially feasible to upgrade a whole DNA database comprising millions of profiles whenever technological breakthroughs occur (see Chapter 6). A possible answer is that only specific portions of the databases (e.g. samples of those convicted of severe crimes or recidivists) will be re-analysed with a new technology. Volunteer populations would undoubtedly be last in line to be re-analysed, as their contribution to solving crimes is considered to be much lower than other populations in the database.

Concerns have also been raised about volunteers' DNA *profiles* retained by authorities. This seems a less difficult issue since the (numerical) information contained in a DNA profile is only useful for identification, similar to fingerprints. However, while polymorphic DNA contains no disease or trait-related information to our knowledge today, it might be found to do so in the future. DNA profiles voluntarily submitted by an individual might also be used for familial searching (genetic proximity testing) (Williams and Johnson 2006; see Chapter 2). Therefore, by volunteering a sample, an individual also submits to the police valuable information about their direct relatives, which brings up privacy concerns about the fate of volunteer DNA profiles.

Not unexpectedly, volunteers may encounter obstacles placed before them by law enforcement agencies when trying to withdraw their DNA samples or delete their profiles after exclusion of a match with crime scene evidence. In the UK, DNA records can only be removed by a committee appointed by the Association of Chief Police Officers, which is in charge of the procedures for removal of three types of information from police records: DNA, fingerprints and Police National Computer (criminal) records (Rodrigues 2007). Since one cannot request deletion of DNA records only but must request deletion of all three types of information, convincing the committee to approve the removal of *all* records is difficult, leading most applicants to withdraw their requests upon denial of their first motion. Furthermore, a senior police officer, who arguably can never be completely 'objective', decides upon the request in the first instance. Only recently has the House of Lords voted to amend the law to help innocent people to have their DNA samples removed from the NDNAD (Out-Law News 2008).

In the US, many such requests access the courts. Authorities argue that once a sample has been lawfully collected, they have the right to hold and further use it unless explicitly restricted by a written limited consent (Kaye and Smith 2004). Since the Fourth Amendment (protecting citizens from unreasonable searches and seizures) occasionally failed to stand for a volunteer's right to have his or her lawfully collected sample destroyed, Harlan (2004) suggested the use of argumentation based on the Fifth Amendment, or the 'right of possession'. According to this argument, the DNA sample is considered the volunteer's property and should be returned once the purpose for which it was collected – namely, the generation of a DNA profile – is fulfilled. At this point, the individual's interest in repossessing the sample is greater than an authority's interest in keeping it, as the latter needed the sample only as raw material for profile generation.

This section concludes by looking at some statistical data provided by the Nuffield Council of Bioethics (2007). In March 2007, the NDNAD included about four million DNA profiles. This number included over 200 000 profiles of innocent people: suspects not charged, acquitted defendants or volunteers. This 'innocent' population generated 14 000 'cold hits' during the years of their inclusion in the database. The total annual average of 'cold hits' in the NDNAD is approximately 50 000. Obviously, the contribution of the 'innocent population' to the success of the NDNAD is relatively small. Since volunteers make up only 10% (approximately 20 000 people) of this population, their contribution to 'cold hits' is very small and probably much lower than their percentage in the 'innocent population'. Since only approximately 50% of the 'cold hits' eventually lead to the apprehension of suspects, and a much smaller percentage to convictions, the issue at hand is about balancing the relatively small gain in crime solving against unknown damage to public and individual perceptions of the state with regard to infringement of individual rights. It is questionable whether suspects, arrestees and individuals charged but not convicted should be included in the DNA database. Although innocent by law, however, they are not necessarily viewed as such by the police and by parts of the society (Williams *et al.* 2004). However, the case of volunteers is entirely different. They should not need to struggle – legally or otherwise – to maintain their status as innocent citizens and, therefore, we can advance that their profiles should never be included in police databases (see also Liberty 2007). We now turn to describing Israeli law concerning DNA databases and provisions with respect to volunteer sampling.

THE ISRAELI FORENSIC DNA DATABASE AND VOLUNTEER SAMPLES: LEGAL FOUNDATIONS

The Israeli forensic DNA database became operational early in 2006, as the outcome of a police initiative started in 1997. Its legal framework is based on the Criminal Procedures (enforcement powers – bodily search and taking identification measures) Act 5756-1996, amended in June 2005 by the Identification Measures Act Amendment (IMA) (State of Israel, 2005). Based on the British model, the Israeli database is a very broad DNA database with wide inclusion criteria and almost no deletion criteria. It includes profiles of suspects and those charged and convicted. The DNA samples and profiles of suspects are deleted seven years after they had been taken if the criminal procedure had ended with no conviction or if no charges were filed. If a new legal cause justifying DNA collection emerges within the seven year period, then profiles and samples are not deleted and a seven-year term will start anew from the day the new cause emerged. Profiles of convicted offenders will be deleted 20 years after their death. Police practice does not distinguish between samples and profiles in this respect.

Israeli law, however, does distinguish between the DNA database and the databases for fingerprints and photographs. Inclusion criteria in the latter two are drawn more broadly than for the DNA database. Only DNA profiles of suspects and those convicted of severe and recidivistic crimes, which have a relatively high probability of producing DNA evidence at the crime scene, can be included. The list of 'qualifying' crimes include sex offences, terror offences, assaults, offences against dangerous drug ordinance, offences against minors, trafficking and most property crimes. The list can be expanded by a decree of the Minister of Internal Security. These provisions create a database having a relatively large growth potential in both populations and types of crime.

The IMA also specifically defines police powers with regard to taking samples from volunteers. Chapter Six provides regulations pertaining to '[b]odily search[es] with consent of a person who is not a suspect'. This regulates the procedure of taking samples from victims and witnesses, and from persons who are not suspects but where there are reasons to believe that they might become suspects. The first article of Chapter Six of the Act (our translation) states the following:

14.   A bodily search of a person who is not a suspect
   (a)   For the purpose of investigating an offense, a police officer is authorized to search the body of a person who is not a suspect . . . if one of the following criteria is met:
      (1)   The person is a victim, a witness, or anyone who is not a suspect and there are reasonable grounds to believe that on his or her body there is necessary evidence for the investigation;

(2)   The search is needed in order to check a possible connection between that person and the investigated offense;

(b)   A police officer will not conduct a search under this chapter, unless s/he has explained to the searched person, in an understandable manner, the purpose of the search, and the possible use of its outcomes. S/he will also mention the individual's right not to agree to the search. The person should give written consent to the conduct of the search.

Refusal to allow the search is not considered a criminal offence and does not justify the use of force in order to obtain a DNA sample. The use of a sample given by a person who is not a suspect is limited under the following provisions:

14A.   Limitations on the use of the outcomes of the search

(a)   No comparison between the profile identifying the searched person and the profiles in the database will be carried out except for the purpose of the investigated case. The profile may be used for the investigation of another case only after written consent was given, on a form specified by the minister. The consent can be given regarding any investigated case separately; classes of offenses; or a single comparison to all cases under investigation.

(b)   The profile identifying the searched person will not be included in the database, unless s/he had become a suspect, an accused or a convict.

This legal provision restricts the use of a voluntarily given sample according to the consent signed by its donor. Nevertheless, on occasion, reality creates a situation that no legislator foresaw. The Farhi case (*Farhi* v. *State of Israel* 2007) demonstrates the gap between theoretical and practical aspects of the Israeli law concerning volunteer DNA samples.

The Farhi case

During the investigation of the murder of Anat Fliner (see above), police collected samples from hundreds of volunteers. One of them was Eitan Farhi, who had consented to provide a DNA sample to be used *only* in connection with the Fliner murder investigation. At the same time, police were investigating the case of a serial rapist who had left sperm traces at three different crime scenes. The Chief

Superintendent in charge of the DNA database laboratory in the Division of Identification and Forensic Science (in police headquarters in Jerusalem) recalled she had encountered a DNA profile having similar relatively rare characteristics to the serial rape cases in an unrelated case. A quick glance at the Fliner case volunteers file disclosed that the profile of the alleged serial rapist was identical to that of Eitan Farhi. This revelation confronted the police with profound difficulties. Should they ignore this discovery and let a suspect of serial rapes escape prosecution (and possibly continue to pose a public threat), or should they go ahead and arrest Farhi on the basis of the DNA match, despite the fact that according to the law, it should never have been found in the first place?

A major obstacle to using Farhi's profile was Article 14A of the IMA. Nevertheless, police decided they could not ignore such vital information. Consultation between high-ranking police officers and the Attorney General's office led to the submission of an application to the Magistrate's Court, requesting a court order for the apprehension of Farhi. The application was approved and an arrest order was issued based on the DNA match between Farhi's subject profile and the crime scene traces. The judge's guiding principle was the 'public interest' (*Farhi* v. *State of Israel* 2007); in his judgment regarding the arrest decision, the Hon. J. Heyman stated: '[t]he police and the court ... cannot ignore the existence of any evidential foundation connecting a person to the execution of such severe offenses beyond a reasonable doubt. I believe that our judicial system principles, which are manifested in the Yissacharov case, do not prevent the arrest of a person in the above mentioned circumstances'.

Farhi was arrested but refused to provide another sample to confirm the DNA match. Unwilling to give up, a police investigator obtained a cigarette butt from Farhi's cell, which generated another DNA profile matching the serial rapist profile.

Farhi argued that the use of the DNA evidence in his case was unlawful, since the main basis for his indictment in the rape cases had been a DNA profile he had provided *only* for the Fliner murder investigation. He argued that Article 14A of the IMA sets a clear exclusionary rule. The District Court dismissed Farhi's complaint and denied a motion to exclude the evidence. The court ruled that the DNA evidence was admissible in accordance with the Yissacharov case (*Yissacharov* v. *Chief Military Prosecutor* 1998). This had dealt with the fundamental issue of the adoption of a doctrine declaring that illegally obtained evidence should be inadmissible in the Israeli legal system (also known as the 'fruits of the poisonous tree' doctrine[3]). Accordingly, a court can exclude illegally obtained evidence if it finds that admitting it in a trial will harm the defendant's rights in a substantial way for an improper purpose, or will violate the fairness of the proceedings to a great extent. Only in such circumstances will permission to admit the evidence in a trial be seen as violation of the constitutional right to dignity, privacy and liberty. To prevent this, the court must declare the evidence inadmissible.

The possibility to exclude evidence in restricted circumstances expresses the proportionality of the right to a fair trial and the profound need to balance this against competing values, rights and interests, such as discovering the truth, fighting crime and protecting public safety and the rights of potential and actual victims of crime. According to the Yissacharov case, in dealing with the question of the admissibility of illegally obtained evidence, courts should take into account a variety of considerations in accordance with the circumstances of the case before them:

The first consideration compels courts to examine whether 'law enforcement authorities made use of improper investigation methods intentionally and deliberately or in good faith' (*Yissacharov v. Chief Military Prosecutor* 1998: 104). In cases where investigation authorities have intentionally violated the provisions of the law, or consciously violated a protected right of a person under investigation by using improper investigation means in collecting evidence, this might constitute a serious violation of due process if the evidence is admitted in the trial. Second, an 'urgent need to protect public safety' (*Yissacharov v. Chief Military Prosecutor* 1998: 105) might be seen as a mitigating consideration, reducing the blameworthiness of the illegality. A third consideration is 'the degree to which the illegal or unfair investigation method affected the evidence that was obtained' (*Yissacharov v. Chief Military Prosecutor* 1998: 106). Courts might consider to what extent the illegality involved in obtaining the evidence is likely to affect the credibility and probative value of the latter. Whenever the credibility of the evidence is questioned, the balance between the value of discovering the truth and the value of protecting the fairness and integrity of due process might be impaired in such a way that may cause its inadmissibility. Courts might also consider whether the nature of the evidence is

---

[3] 'Fruit of the poisonous tree' is a legal metaphor in the USA used to describe evidence gathered with the aid of information obtained illegally.

independent and distinct from the breach of law involved in obtaining it. In *Farhi* v. *State of Israel* (2007), it was held that the improper investigation methods did not affect the content of the DNA evidence, and the court then ruled that evidence could be admitted in trial. However, it is probable that the case law exclusionary rule will not be applicable whenever 'scientifically based' evidence is used, mainly because of its high credibility. The verdict specified clearly that the 'fruit of the poisonous tree' doctrine prevailing in the USA should not be adopted as such in the Israeli judiciary system.

In the Farhi case, the District Court realised that the exclusion of the DNA evidence might excessively harm the public interests of fighting crime and protecting public safety, as well as the interests of the victims of the crime. In these circumstances, exclusion of the evidence might cause a person accused of committing serious offences not to be held accountable for them. This outcome may in itself undermine the administration of justice and public confidence in the legal system. It is, therefore, likely that the more serious the offense is the less willing the court will be to exclude unlawfully obtained evidence. Farhi was convicted in all three rape cases.

Although Farhi's DNA profile was not used for speculative searches against unsolved cases in the database, provisions were defined by the police to avoid such rare unintentional discoveries in the future. As shown above, there are justifications for limiting the possible use of a volunteer's profile, but evidently, there are circumstances compelling its use in specific contexts such as this one. We believe that only the courts should decide whether in any given case public interest overrides the limitations which the originator of the sample placed on its use in the context of the consent procedure.

CONCLUSIONS

The analysis outlined in this chapter has led us to suggest a number of recommendations about intelligence-led DNA mass screenings or dragnets, which are considered indispensable by law enforcement authorities and could possibly lead when properly used to the resolution of severe crimes when all other means have failed. Our first recommendation is that the police should use dragnets only in conjunction with other intelligence and investigation tools, avoid discrimination in their execution and make sure that all functions are operating coherently and are closely monitored.

Second, strict and well-defined rules and policies should be put in place to minimise infringements of volunteers' privacy rights, to ensure that consent is given without coercion and to ensure that samples are used only for purposes justified by law. Like other voluntarily based databases, consent must be revocable and any additional uses of samples or profiles must be accompanied with a renewed consent unless the volunteer is arrested, charged or convicted.

Third, guidelines for use and retention of DNA samples and profiles given voluntarily should be well defined, recognised by the public and all stakeholders (police, laboratories, database administrators, the prosecution and courts) and be carefully followed and regulated. Issues such as familial searches should be clarified by legislation. Any disputes or problems regarding voluntarily collected samples and profiles should only be decided by a court having qualifications to consider all relevant aspects.

We feel that these recommendations would be acceptable in most countries to the vast majority of their population willing to cooperate with the police, particularly in investigations of serious crimes. Enhanced public trust in the governance and use of DNA databases would result, as would cooperation with operational procedures for DNA collection, including dragnets where no other possibility exists. In the case of dragnets, individuals would be more likely to agree to give up their privacy rights to some extent for the sake of the common good if the provisions above, coupled to protection against misuse, were clearly understood. Building the right legal environment and public trust in this regard will thus assure that privacy rights and the welfare of society will not contradict but instead complement each other.

REFERENCES

Ansell, R. and Rasmusson, B. (2008). A Swedish perspective. *BioSocieties*, 3, 88–92.
Asplen, C. (2006). *The Non-forensic Use of Biological Samples taken for Forensic Purposes: An International Perspective.* Boston, MA: American Society of Law, Medicine and Ethics www.aslme.org/dna_04/spec_reports/asplen_non_forensic.pdf (accessed 6 June 2009).
Bieber, F. (2004). Science and technology of forensic DNA profiling: current use and future directions. In *DNA and the Criminal Justice System: The Technology of Justice*, ed. D. Lazer. Cambridge, MA: MIT Press, pp. 23–62.
Bikker, J. (2007). Response submitted to the consultation held by the Nuffield Council on Bioethics. London: Nuffield Council on Bioethics www.nuffieldbioethics.org/fileLibrary/pdf/Jan_Bikker (accessed 6 June 2009).

Brettell, T., Butler, J. and Almirall, J. (2007). Forensic science. *Analytical Chemistry*, 79, 4365–4384.

Burton, C. (1999). The UK NDNAD intelligence led DNA screens: a guide for senior investigating officers. Presented at the *1st Interpol DNA Users Conference*, Lyon, 24–26 November www.interpol.int/Public/Forensic/dna/conference/DNADbBurton.ppt (accessed 6 June 2009).

Cho, M. and Sankar, P. (2004). Forensic genetics and ethical, legal and social implications beyond the clinic. *Nature Genetics*, 36, s8–s12.

Council of Europe (1950). *Convention for the Protection of Human Rights and Fundamental Freedoms*. Strasbourg: Council of Europe http://conventions.coe.int/Treaty/Commun/QueVoulezVous.asp?NT=005&CL=ENG (accessed February 2010).

Crouse, C. and Kaye, D. (2000). *The Retention and Subsequent Use of Suspect, Elimination and Victim DNA Samples or Records*. [National Commission on the Future of DNA Evidence Report.] Washington, DC: National Institute of Justice.

Dierickx, K. (2008). A Belgian perspective. *BioSocieties*, 3, 97–99.

Etzioni, A. (2004). DNA tests and databases in criminal justice: individual rights and the common good. In *DNA and the Criminal Justice System: The Technology of Justice*, ed. D. Lazer. Cambridge, MA: MIT Press, pp. 197–223.

Forensic Science Service (2004). *Antoni Imiela: M25 Rapist Trapped by Crucial Forensic Evidence*. Birmingham: Forsensic Science Service http://213.52.171.242/forensic_t/inside/news/list_casefiles.php?case=23 (accessed 27 February 2010).

Gaensslen, R. (2006). Should biological evidence or DNA be retained by forensic science laboratories after profiling? No, except under narrow legislatively stipulated conditions. *Journal of Law, Medicine and Ethics*, 34, 375–379.

Gill, P., Jeffries, A. and Werrett, D. (1985). Forensic applications of DNA 'fingerprints'. *Nature*, 316, 76–79.

Halbfinger, D. (2003). Police dragnets for DNA tests draw criticism. *New York Times*, 4 January www.nytimes.com/2003/01/04/us/police-dragnets-for-dna-tests-draw-criticism.html (accessed 6 June 2009).

Harlan, L. (2004). When privacy fails: invoking a property paradigm to mandate the destruction of DNA samples. *Duke Law Journal*, 54, 179–219.

Home Office (2009). *Keeping the Right People on the DNA Database: Science and Public Protection*. London: The Stationery Office www.homeoffice.gov.uk/documents/cons-2009-dna-database/dna-consultation? (accessed 6 June 2009).

Jones, G. (2006). DNA database 'should include all'. *Telegraph*, 24 October www.telegraph.co.uk/news/uknews/1532210/DNA-database-should-include-all.html (accessed 6 June 2009).

Johnson, P. and Williams, R. (2004). DNA and crime investigation: Scotland and the 'UK national DNA database'. *Scottish Journal of Criminal Justice Studies*, 10: 71–84.

Kaye, D. and Smith, M. (2004). DNA databases for law enforcement: the coverage question and the case for a population-wide database. In *DNA and the Criminal Justice System: The Technology of Justice*, ed. D. Lazer. Cambridge, MA: MIT Press, pp. 247–284.

Laird, R., Dawkins, R. and Gaudieri, S. (2005). Use of the genomic matching technique to complement multiplex STR profiling reduces DNA profiling costs in high volume crimes and intelligence led screens, *Forensic Science International*, 151: 249–257.

Levitt, M. (2007). Forensic databases: benefits and ethical and social costs. *British Medical Bulletin*, 83, 235–248.

Liberty (2007). *Liberty's Response to the Nuffield Council on Bioethics Consultation*. London: Liberty www.liberty-human-rights.org.uk/pdfs/policy07/bioinformation-ethical-issues.pdf (accessed 6 June 2009).

McCartney, C. (2004). Forensic DNA sampling and the England and Wales national DNA database: a sceptical approach. *Critical Criminology*, 12, 157–178.

Mepham, B. (2006). Comments on the national DNA database (NDNAD). www. nuffieldbioethics.org/fileLibrary/pdf/Professor_Ben_Mepham.pdf (accessed 6 June 2009).

Nelkin, D. and Andrews, L. (2003). Surveillance creep in the genetic age. In *Surveillance as Social Sorting: Privacy, Risk and Digital Discrimination*, ed. D. Lyon, London: Routledge, Taylor & Francis, pp. 94–110.

Norton, A. (2005). DNA databases: the new dragnet. *The Scientist*, 19, 50–56.

Nuffield Council on Bioethics (2007). *The Forensic Use of Bioinformation: Ethical Issues*. London: Nuffield Council on Bioethics www.nuffieldbioethics.org/fileLibrary/pdf/The_forensic_use_of_bioinformation_-_ethical_issues.pdf (accessed 6 June 2009).

Out-Law News (2008). Lords demand amendment to help the innocent get DNA off database. www.out-law.com/page-9564 (accessed 6 June 2009).

Parry, B. (2008). The forensic use of bioinformation: a review of responses to the Nuffield report, *BioSocieties*, 3: 217–222.

Phillips, B. (1998). *Bill C-3, the DNA Identification Act*. [Presentation to the Standing Committee on Justice and Human Rights by the Privacy Commissioner of Canada.] Ottawa: Privacy Commissioner of Canada http://www.priv.gc.ca/speech/archive/02_05_a_980212_e.cfm (accessed 28 February 2010).

Prainsack, B. (2008). An Austrian perspective. *BioSocieties*, 3, 92–97.

Prainsack, B. and Gurwitz, D. (2007). Private fears in public places? ethical and regulatory concerns regarding human genomic databases. *Personalized Medicine*, 4, 447–452.

Privacy International (2007). *PHR2006 – Privacy Topics – Genetic Privacy*. London: Privacy International www.privacyinternational.org/article.shtml?cmd%5B347%5D=x-347-559080 (accessed 6 June 2009).

Rodrigues, R. (2007). Big bio-brother is here: wanting, taking and keeping your DNA. In *Proceedings of the British & Irish Law, Education and Technology Association Annual Conference*, 16–17 April, Warwick, UK.

Rothstein, M. and Tallbott, M. (2006). The expanding use of DNA in law enforcement: What role for privacy? *Journal of Law, Medicine and Ethics*, 34, 153–164.

Sanders, J. (2000). *Forensic Casebook of Crime*. London: True Crime Library/Forum Press.

Schuller, W., Fereday, L. and Scheithauer, R. (eds.) (2001). *Interpol Handbook on DNA Data Exchange and Practice*. www.interpol.int/Public/Forensic/dna/HandbookPublic.pdf (accessed 6 June 2009).

Simoncelli, T. (2006). Dangerous excursions: the case against expanding forensic DNA databases to innocent persons. *Journal of Law, Medicine and Ethics*, 34, 390–397.

Simoncelli, T. and Steinhardt, B. (2006). California's proposition 69: a dangerous precedent for criminal DNA databases. *Journal of Law, Medicine and Ethics*, 34, 199–213.

Smith, M. (2006). Let's make the DNA identification database as inclusive as possible. *Journal of Law, Medicine and Ethics*, 34, 385–389.

State of Israel (2005). *Identification Measures Act Amendment*, 19 June 2005 [Book of Laws, in Hebrew].

Steinhardt, B. (2004). Privacy and forensic DNA data banks. In *DNA and the Criminal Justice system: The Technology of Justice*, ed. D. Lazer. Cambridge, MA: MIT Press, pp. 173–195.

Szibor, R., Plate, I., Schmitter, H. *et al.* (2006). Forensic mass screening using mtDNA. *International Journal of Legal Medicine*, 120, 372–376.

Van Camp, N. and Dierickx, K. (2008). National forensic DNA databases: current practices in the EU. *European Ethical–Legal Papers*, No. 9, Leuven: Centre for Biomedical Ethics and Law.

Van Camp, N. and Dierickx, K. (2007). The expansion of forensic DNA databases and police sampling powers in the post 9/11-era: ethical considerations on genetic privacy. *Ethical Perspectives*, 14, 237–268.

Victorian Parliament Law Reform Committee (2002). *Report by the Office of the Victorian Privacy Commissioner: Inquiry into Forensic Sampling and DNA Databases*, Melbourne: Victorian Parliament Law Reform Committee.

Victorian Parliament Law Reform Committee (2004). *Forensic Sampling and DNA Databases in Criminal Investigation*. Melbourne: Victorian Parliament Law Reform Committee.

Wadham, J. (2002). *Databasing the DNA of Innocent People- Why it Offers Problems not Solutions* [Press release, 13 September]. London: Liberty.

Walker, S. and Harrington, M. (2005). *Police DNA 'sweeps': A proposed model policy on police request for DNA samples*. Omaha, NE: University of Nebraska, Police Professionalism Initiative www.unomaha.edu/criminaljustice/PDF/dnamodelpolicyfinal.pdf (accessed 6 June 2009).

Walsh, S. (2005). Legal perceptions of forensic DNA profiling part I: a review of the legal literature. *Forensic Science International*, 155, 51–60.

Whittall, H. (2007). DNA profiling: invaluable police tool or infringement of civil liberties? *Bioethics Forum*, 15 October www.thehastingscenter.org/Bioethicsforum/Post.aspx?id=648 (accessed 6 June 2009).

Williams, R. and Johnson, P. (2006). Inclusiveness, effectiveness and intrusiveness: issues in the developing uses of DNA profiling in support of criminal investigations. *Journal of Law, Medicine and Ethics*, 34, 234–247.

Williams, R., Johnson, P. and Martin, P. (2004). *Genetic Information and Crime Investigation: Social, Ethical and Public Policy Aspects of the Establishment, Expansion and Police Use of the National DNA Database*. London: Welcome Trust.

CASES

Farhi v. *State of Israel* (2007). Serious Crime File 1084/06, Tel-Aviv District Court, April 26 (Hon. J. Ophir-Tom).

S and Marper v. *the United Kingdom* (2008). A summary of the judgment is available from http://cmiskp.echr.coe.int/////tkp197/viewhbkm.asp?action=open&table=F69A27FD8FB86142BF01C1166DEA398649&key=74847&sessionId=skin=hudoc-en&attachment=true&16785556 (accessed 6 June 2009).

State of Israel v. *John Doe* (2007). Serious Crime File 402/07, Haifa Juvenile District Court, November 15, (Hon. J. Berliner).

State of Israel v. *John Doe* (2008). Serious Crime File 206/08, Tel-Aviv District Court (still pending).

Yissacharov v. *Chief Military Prosecutor* (1998). CrimA 5121/98. http://elyon1.court.gov.il/files_eng/98/210/051/n21/98051210.n21.pdf (accessed 6 June 2009).

4

Base assumptions? Racial aspects of US DNA forensics

INTRODUCTION

After two decades of acceptance in US courtrooms, forensic DNA analysis remains plagued with flaws even as its use burgeons. Instead of fomenting a dialogue with the public, US lawmakers have invoked the spectre of violent crime to promulgate the passage of legislation that permits the coercion of DNA samples from ever-expanding segments of society. This leaves US citizens, who prize both privacy and security, to confront momentous policy decisions without the benefit of comprehensive public education or debate.

This chapter largely focuses on events that encapsulate many of these issues, albeit in microcosm: the US conduct of DNA sweeps. Also called DNA dragnets or DNA mass screenings, this method is a species of 'cold hit' in which law enforcement essays to match DNA left by an unknown miscreant with the person who left it by obtaining samples from members of the community thought to contain the criminal. The discussion will explore how, via the use of DNA sweeps, local police exploit laws in order to expand the scope of DNA profiling, collection and storage to allow the apprehension of unknown miscreants on the strength of non-specific physical descriptors. However, the ethnically heterogeneous nature of US society and the overwhelming racial disparities in arrest and incarceration present challenges that have been largely ignored.

How do ethnic issues, and in particular the tangled calculus of race, inform the debate on DNA use and governance for the purposes of law enforcement? Are such questions particularly relevant for the US context or do other nations share these challenges? By discussing such concerns in the context of racialised DNA sweeps, I hope to contribute to the important discussion about how genetic databases should be

Genetic Suspects: Global Governance of Forensic DNA Profiling and Databasing, ed. Richard Hindmarsh and Barbara Prainsack. Published by Cambridge University Press. Copyright © Cambridge University Press 2010.

designed and governed to maximise citizens' security while protecting privacy, autonomy and social justice.

DNA NARRATIVES

The apparent omnipotence of DNA technologies to mediate justice has captured the popular imagination, with evidence of this abounding on television and film screens, in mystery novels and newspaper accounts. Such dramas broadcast the conviction that crime can no longer hide from the unerring Argus eyes of DNA detectives. Nightly, we absorb an armchair DNA education, accurate or otherwise, as we are treated to a staggering variety of plot twists that are unequivocally unravelled by the helical molecule of truth.

Of course, forensic medicine programmes such as *CSI, Crossing Jordan* and *Bones* are fiction, with a focus upon drama rather than on facts, imperfectly objective and brief with nuance and complexity. Such entertainment is pervaded with assumptions that the collection, rapid processing and interpretation of DNA evidence are ubiquitous and infallible. Such shows tend to ignore the untidy, uncomfortable truths of bureaucratic practice, of delay and deceit and of unduly violated privacy. They routinely overlook or even reinforce class and racial biases, and they turn a blind eye to the reality that some of the 'white hats' are cursed with feet of clay, capable of flouting laws or testifying falsely.

Instead, a 'good guys versus bad guys' mentality rules televised dramas and films. Police officers and prosecutorial teams are portrayed in environments that telegraph their sterling characters – brightly lit, orderly laboratories, manicured lawns and beautiful homes in clean, quiet, peaceful neighbourhoods populated by caring, virtuous, law-abiding people, most of them white. The acclaimed Home Box Office television series *The Wire* does present a unique exception to this rule as it deftly evades the pervasive racial assumptions and stereotypes, but it is far from a normative portrayal and the series makes fewer references to genetic technologies.

Most television programmes utilise environmental differences to alert viewers that they have entered an area where 'bad guys' live and where suspects abound. The palette turns dark or lurid, music becomes cacophonous and hostile people of colour with menacing scowls replace the 'good guys'. Denizens of these dirty, dangerous streets populate an unrelieved landscape where a bestiary of 'suspects' ply illegal trades, unlike real neighbourhoods of the poor and black, which are mixtures of hard-working strivers, the law-abiding poor, criminals and others. One expects nuance, precision and even facts to be sacrificed to television drama, but drama is not the exclusive province of the screen. I submit that the most momentous influence of forensic DNA mythology has transpired in a different entertainment arena – the news media.

Readers in the USA are likely to learn most of what they know about forensic DNA strategies from their newspapers and magazines, but this information reflects the errors, unsupported assumptions and insufficiently examined claims of writers, law enforcement officers and of some scientific experts as well (Scheck and Neufeld 2007). Stories are spun tightly without confusing the plot line with messy hanging threads, possibly unreliable eyewitnesses, hyperbolic experts and fractious data or less-than-sterling mores and motives on the part of the good guys. Daily newspaper accounts, like television series, subscribe to the geography of evil, and when a suspect is encountered in a neighbourhood that readers have been taught to 'recognise' as a crime hotbed, be it Harlem, East Palo Alto or Compton, this telegraphs latent criminality.

As in *CSI*, a newspaper's 'DNA opera' tends to deliver an orthodox climax of unambiguous justice, limning a world where DNA never offers up ambiguity, never becomes degraded or lost and is never subverted or misrepresented by good guys gone bad. In the USA, however, these things happen with appalling frequency (e.g. Anon. 1993; Scheck and Neufeld 2007). Moreover, newspapers, magazines and news programmes, unlike television dramas, are extensively relied upon as credible information sources. Therefore, their failure to address certain ethical, legal and social consequences of forensic DNA practices has greater real-world repercussions. Lay people who derive their understanding of the issues from news accounts use their votes to usher in policies that have accelerated forensic DNA collection and have permitted the racialised DNA sweeps upon which this chapter will focus (Cole 2007). In short, like *CSI*, the news media tend to portray DNA analysis as an unalloyed tool of justice.

Sometimes DNA is exactly this – as in the case of DNA exonerations.

DNA EXONERATION: AN AMERICAN JANUS

On January 14, 2008, US newspapers announced that Ronald Gene Taylor, who was serving a 60-year sentence in a Texas prison, had become the 220th American to be exonerated of his crimes and freed from prison by DNA testing (Tolson and Khanna 2007). Since 1989, DNA testing prior to conviction has proven that tens of thousands of prime suspects were wrongly accused, wrongly identified and wrongly pursued. But those who, like Taylor, are wrongly convicted and sent to jail serve an average of 12 years before being released: Taylor served 14 years (Tolson and Khanna 2007).

Like most of the imprisoned who have found liberation in DNA testing, Taylor was convicted of a violent sexual assault, and like most of those liberated, he is black. Each year since 2000, between 50 and 70% of the incarcerated men freed by DNA technology have been black or Hispanic. Most of the convictions disproved by DNA evidence involve African American men wrongfully convicted of assaulting white women (P. Neufeld, personal communication). The pertinent ethnic crime statistics are discussed below, but, first, I will consider the fact that the unambiguously celebratory news media coverage suggests that this forensic use of DNA is an unalloyed blessing for black men.

If forensic DNA identification was such a blessing, it would be an anomaly, because historically, genetic technology has had a checkered past among black Americans (Bowman 1977; Guthrie 1998). Every key advantage in disease protection, identification or in detection imparted by genetic technology seems to have spawned a doppelgänger that bears racially mediated error, punitive effects and/or stigmatization. As a result, fears abound that currently tested or employed identification techniques, applied in a highly racialised context, may share these racial-bias errors. Such errors also threaten to perpetuate the punitive effects and the stigmata (Washington 2007: 299–324).

These historical attributes of genetic innovation are quite important when discussing racial applications, for three reasons. The first is that iatrophobia (fear of medical applications and treatment) is a response of many African Americans to genetic technologies; this response has its origins in the systematic harms that have emanated from clumsy or biased application of genetics to medicine (Bowman 1977; Bowman and Murray 1990; Washington 2007: 299–324). The second reason that this history is pertinent is that scrutiny of past US genetics research and practice reveals a tendency toward scientific errors or unsupported assumptions that enshrine assumptions of black difference, inferiority and criminality (Kahn 2004; Washington 2007: 21, 299–324).

The final reason is that DNA profiling, like other earlier genetic technologies, also risks the reification of racial assumptions should it not be analysed with scrupulous logic without assumptions that spring from ethnic bias (Bowman and Murray 1990; Kahn 2004). For example, profoundly flawed intelligence testing has long been used to promote the heritable intellectual 'genetic inferiority' of African Americans, supporting other cherished social agendas, such as racially selective sterilization, on a specious logical basis. For many, this further impugns the credibility of research labeled 'genetic' in toto (Gould 1992; Guthrie 1998).

To appreciate this, it is important to understand that, quite obviously, genetic testing for disease risks or susceptibilities such as sickle cell disease or phenylketonuria, or even, farther afield, for intelligence-quotient testing, can utilise very different techniques from the DNA profiling employed for purposes of identification or exclusion. However, all these assessments employ the analysis of genetic information in a context that carries a high risk of stigmatization, whether in diseases that code for racial status (such as sickle cell disease) or for 'identification' tests that use not only legal but also medical paradigms in order to narrowly focus upon members of a single race, even purporting to identify the race of an unknown suspect. Each of these assessments has also been conducted within a politicised context, and their results have served to bolster questionable social policies hostile to African Americans (Bowman 1977; Washington 2007).

Furthermore, in the contexts under discussion here, the vaunted differences among different types of genetic technology have far less impact on the general public, whose votes drive policy, than do the overarching labels 'genetic' or 'DNA'. From the viewpoint of the lay patient–consumer–voter, these two labels powerfully convey either infallibility ('DNA doesn't lie') or untrustworthiness ('Faulty genetic research has erred in labeling blacks as "unintelligent", "ridden with sickle cell disease" and "violent": Why trust it now?') depending upon that person's sociological experience (Duster 2006). For this long, consistent history of misinterpretation, misdiagnosis and stigmatisation in African Americans also bolsters profound distrust of genetic technologies by the affected population, which tends not to make distinctions between medical testing and medically mediated identification when considering whether to embrace novel genetic technologies such as DNA profiling (Bowman 1977; Washington 2007). Consider, for example, the overwhelmingly negative reaction *ab origine* among African Americans to DNA sweeps in municipal sites such as Charlottesville (Glod 2004) and Ann Arbor (Grand 2002).

A long scientific tradition in the USA links blacks and a hereditarian view of criminality. This includes the nineteenth century work of the American School of Ethnology (Johnson and Mead 1934; Washington 2007: 246–251). By the early twentieth century, the forensic psychology of Cesare Lombroso perhaps did most to provide hereditarian biological underpinnings to the ascendant medical view of blacks as 'born criminals'. Lombroso anointed southern Africa's Dinka tribe as the iconic exemplars of his 'criminal man' and he wrote: 'There exists a group of criminals, born for evil, against whom all social cures break as though against a rock' (Lombroso 1911).

Drawing upon the influential work of Lombroso and others, US medicine has long stigmatised blacks as harbouring marked criminal tendencies. Today, this stigmatisation continues unabated in some quarters. Much US research has given short shrift to environmental factors such as readily available guns and drugs, racial and financial inequities and a culture that glorifies violence in favor of a Quixotic search for putative genetic predictors of violent behaviour – the quest for an elusive 'mean gene' (Balaban et al. 1996). This search has been focused upon African American populations (Katz 1972; Washington 2007: 271–293). In 1969, the National Institutes of Mental Health's Center for Crime and Delinquency awarded a three-year US$300 000 grant to Digamber Borgaonkar. Under the aegis of Johns Hopkins University, Borgaonkar scrutinised the genomes of approximately 15 000 Baltimore boys, with about 85% of them black, for the XYY chromosomal anomaly that was then associated with criminality (Katz 1972; Washington 2004). About 30 years later, in the late 1990s, New York City researchers gave fenfluramine to black boys (white boys were specifically excluded by the research protocol) in a parallel attempt to indirectly identify markers for genetically medicated violent behaviour (Cherek 1999; Washington 2007: 271–278).

Such studies share the foci of the Violence Initiative, a government-funded matrix of studies that ostensibly proposed to study violence in 'inner cities' – a phrase that narrowly denominates black communities. The Violence Initiative and similar projects were planned to avoid dramatic environmental, social and financial stressors in order to investigate a possible genetic link between violence and black children, especially boys. The initiative attained national visibility in 1992 when Director Frederick Goodwin of the National Institute of Mental Health's Alcohol, Drug Abuse, and Mental Health Administration appeared before the National Health Advisory Council to champion it. He did so by comparing young black boys to 'hyper-sexed', violent rhesus monkeys in the jungle, which outraged many of his auditors (Hilts 1992; Leary 1992; Marks 1995: 231–234). Goodwin's remarks championed genetically mediated medical testing for violent propensities and dwelt upon the importance of not *treating* but of genetically *identifying* future violent criminals.

Critics have challenged both the factual basis and constitutional validity of the now-defunct Violence Initiative, but although many similar studies do not exactly mimic its funding and organisation paradigm, they share its stigmatising features and its silence on non-genetic risk factors (Sellers-Diamond 1994). Such issues of forensic genetic determinism with children are not unique to the USA. In the UK, Scotland Yard forensics chief Gary Pugh evoked similar concern when he suggested in March 2008 that DNA testing should be employed to identify those children who will become violent criminals. Such modest proposals provide examples of the persistent associations drawn between genetic identification and diagnosis in medico-forensic theory and practice (Page 2008).

Is DNA exoneration the purely benign exception to the cavalcade of Janus-faced genetic technologies? Not in the view of some legal scholars. 'These [exonerated inmates] are mostly African American men convicted of raping white women', says Peter Neufeld, a professor at the Cardozo School of Law in New York: 'Only 10 percent of reported sex assaults are allegations of white women attacked by black men. Yet 54% of all unjust conviction cases involve African American men wrongfully convicted of assaulting white women. This is a crime that seems associated with many wrong convictions.' The emphasis, Neufeld says, should be on the many men, disproportionately black and Hispanic, who will never be freed by DNA. 'The real significance is not that DNA got them out, but that DNA provides a window into the criminal justice system to see what went wrong with the system to let so many innocent people be convicted' (P. Neufeld, quotations from a telephone interview in 2001) (Washington 2001).

COLOR-CODED JUSTICE

What has gone wrong? The USA, which imprisons a larger percentage of citizens than any other nation, has seen the proportion of its black and Hispanic prisoners balloon over the past century (Sampson and Lauritsen 1997). Blacks currently constitute only 12.9% of the nation's population but more than 40% of those behind bars: together, blacks and Hispanics make up 60% of prisoners. Therefore, any discussion of US incarceration must address race.

The burgeoning imprisonment rates of dark-skinned minorities are driven not by rapes, murders or other violent crimes, but by a racially inequitable response to drug abuse (Human Rights Watch 2008). Prison rolls have grown threefold since the late 1970s in a manner that targets blacks because the harshest penalties for drug use are not colourblind. For example, the smokeable 'crack' form of cocaine is used by black addicts at twice the rate of whites, and much harsher penalties for crack cocaine are mandatory, forcing judges to impose incarceration even for the possession of small amounts of these drugs. Penalties for the powdered cocaine preferred by whites include drug treatment, probation or even suspended sentences (Beiser 2001; Amnesty International 2004: 39). Black women, who constitute the fastest-growing group in prisons, abuse drugs at the same rate as whites (Chasnoff et al. 1990) but are 10 times more likely to be incarcerated for 'drug use while pregnant' (Smith and Dailard 2003: 97–108).

Also, although 80% of US cocaine users are white (Harris 1999b: 3; Washington 2007: 300–307), law-enforcement tactics focus on the inner city (Levine 2008) and culminate in more frequent, longer sentences for blacks and Hispanics. This inequity fosters a perception that blacks make up the majority of drugs users (Chasnoff et al. 1990; Roberts 1997).

SOURCES OF ERROR

Judicial error also drives the incarceration rate of black Americans, particularly laboratory error, eyewitness identifications, false confessions and jailhouse informants (whose testimony is likely to be false). Yet the celebratory press coverage fails to ask why most of the exonerated are black, or to ask about the many other innocent men who will never be freed because of DNA samples that have been lost, degraded or whose very existence technicians and experts deny. An independent review of the Houston Police Department Crime Laboratory found 275 cases in which biological material was detected but never accurately tested.

A subsequent audit uncovered deficiencies within a section of Houston's DNA Laboratory that resulted in its closure in 2002; another independent review found hundreds of other affected cases (Khanna and McVicker 2007). Nationwide, laboratory error and junk science contribute to 65% of cases being reversed by DNA evidence. Factual and numerical errors also abound, including the erroneous matching of DNA samples or inflating the odds against a DNA match with someone other than the criminal (Ungvarsky 2007).

Eyewitness identification provides the most common source of racialised error: 48% are transracial, yet studies suggest that persons are less able to recognise faces not of their own race (Rutledge 2001; Brigham *et al.* 2007). Fully 77% of DNA-reversed convictions are attributable to mistaken eyewitness identification. False confessions, often delivered under duress, drive 25% of reversed convictions; of these, 35% are procured from the mentally disabled or from children under 18 years. Finally, intentional fraud is not unknown in the nation's state and municipal forensic laboratories (Innocence Project 2009).

Therefore, a quality-control crisis pervades America's forensic DNA laboratories, resulting in justice that is delayed, subverted or pressed into service to exacerbate racial bias.

DNA SWEEPS AND RACE

The DNA technology utilised to provide freedom for the fortunate innocent has a hideous obverse for African Americans. The same genetic technologies used for exculpation can compound the trend toward racialised incarceration, because DNA technologies, in themselves neutral, target blacks when applied through racial filters in forensic settings. One such filter is the racialised DNA sweep. This DNA sweep or dragnet (intelligence-led mass screening) is an especially fraught species of 'cold hit' in which law enforcement attempts to match DNA left by an unknown assailant with the person – typically but not always a man – who left it (Matejik 2008). If police find no match in available databases they can resort to fanning out through a community that is thought to contain the criminal, confronting large numbers of men on the street, in their homes or on their jobs. Police 'persuade' each man in the targeted community to undergo a buccal swab – a scraping from the DNA-rich interior of the cheek – to be tested against the crime-scene sample. The Fourth Amendment of the US Constitution, which protects against unreasonable search and seizure, makes forcing persons who have not been arrested or convicted to surrender their DNA illegal in the absence of compelling evidence against *the individual* – not a group. Therefore, such sweeps hinge upon police ability to persuade, not to compel – at least in theory (Matejik 2008).

Citing the per capita cost of obtaining and testing samples as prohibitive, police often narrow the search by race (see Chapter 3). In the frequent absence of reliable eyewitness accounts that could provide detailed phenotypic information, the police work from a racial identification that is highly speculative. Of the 18 major municipal US DNA sweeps undertaken and studied, the very first one in the USA was designed to test black men only (Walker 2004). In 1990, police in San Diego tested more than 800 African American men in an attempt to identify the serial intruder who stabbed six people to death in their respective homes (Chapin 2005). Sharp *intentional* racial disparities are applied as police target a municipality's 'Hispanic community' or 'black community' as they search for a suspect. Police also erroneously use race as a proxy for ancestry, by which I mean they approach persons whose racial or ethnic identity is often undefined while seeking a suspect whose race is often unknown. Meanwhile all they know about this suspect is what their DNA sample can reveal, which is information about ancestry but not racial identity (Duster 2006). To make matters worse, instead of referring to a genetic distinction such as the presence or absence of specific alleles that are particularly common or uncommon within an ethnic group, police often rely upon eyewitnesses who divine race from features or even from images of inadequate resolution (M'charek 2008). Moreover, said features are frequently ambiguous and such eyewitness identifications rest upon pronouncements of race or upon racial criteria that are devoid of definition, and are, as we will see, frequently wrong.

Racial categories themselves are confusing, fluid and overlapping, and such interpenetrance sometimes renders racial labels meaningless or misleading. This is the case particularly within the ethnically heterogeneous borders of the USA. Phenotypic characteristics such as hair texture, lip and nose width and breadth and skin shade are far from definitive of race, which often is, as noted above, itself ill-defined. Scholars have observed that even such identifiers of nationality as Moroccan or Turkish, are often mistaken for ethnicity and race (M'charek 2008). In the USA, police officers and law enforcement data often make racial-identification distinctions between blacks and Hispanics, although the racial conventions of the US dictate that Hispanics constitute not a race but an ethnic group. Therefore, a dark-skinned Hispanic man is also black, but the categories typically are treated as mutually exclusive.

Television programmes and daily newspapers reinforce the perception of geography as destiny, and this cultural context informs a DNA sweep. So, as police descend upon the black areas so familiar from television and newspaper accounts as 'crime saturated', the resultant ethical breaches are rarely inveighed against by the news media even as the character of every black man in the neighbourhood is tacitly impugned, creating a collective presumption of guilt (Cole 2007).

DNA sweeps often arise from crime scenarios that preclude a useful eyewitness identification – darkness, a blindfolded victim, an attack from behind or a masked assailant. Yet, in such conditions, witnesses have still averred that the crime was committed by a black or dark-skinned man, and police investigators have accepted this 'description' (Walker 2004), despite a large number of high-profile cases in which a black male assailant was found to have been an invention, often made up by the actual criminal (Terry 1994; Bell 1996; John-Hall 2009). In fact, interpretation of genotypic (DNA) evidence sometimes *creates* the 'black' phenotype. In December 2002, the now-defunct for-profit firm DNAPrint Genomics contacted Louisiana homicide investigators to inform them, and the news media, that their search for a serial killer, predicated upon a detailed psychological profile producted by the Federal Bureau of Investigation (FBI), was misguided (Silver 2004). They should be looking for a *black* serial killer. The police accepted the company's offer to collaborate and gave DNAPrint a sample. DNAPrint's geographic analysis indicated an ancestry that was 85% sub-Saharan African and 15% Native American, and the company even portrayed for police the killer's putative skin shade. Here it should be noted that skin-colour assessments based upon African ancestry actually are illogical because they map erratically, even poorly onto genotype. Many people who appear phenotypically 'white' share a genetic complement that is largely African or otherwise 'non-white', and vice versa.

Nonetheless, a black man was duly convicted of the crimes (Lowe *et al.* 2001; Sachs 2004; Cho and Sankar 2004). But were DNAPrint's assessments really predictive? DNAPrint's DNAWitness programme uses genetic mutations called single nucleotide polymorphisms (SNPs), which occur more frequently in certain ancestral groups than others owing to a group's geographic separation, intermarriage or other genetic pressures. DNAPrint insisted that SNPs are 'highly informative of ancestry', but other scientists are loath to make phenotypic predictions and doubt whether a DNA screen can tell you anything more than whence one's ancestors probably hailed. Even the latter assessments must be informed by historical information to discern, for example, whether an SNP is suggestive of East Indian or of Native American ancestry. The point is that, accurate or not, DNAPrint's claims gave a scientific imprimatur to racial biases entrenched within the US justice system (Henig 2004).

In Charlottesville, Virginia police searching for a serial rapist who attacked six women between 1997 and 2003 targeted 690 black men in the Charlottesville area and asked those black men whose samples were not already in the database to provide genetic samples. They often prefaced their request with a claim that the man had been looking or acting 'suspiciously', thus providing a putative, surely convenient, basis for individual rather than purely racial suspicion. A *Washington Post* story related how Charlottesville police confronted Jeffery Johnson at the restaurant where he worked as a cook. In front of his supervisor and customers, police informed him that he was a suspect but that he could easily clear himself by submitting a DNA sample on the spot. He complied but understandably was enraged (Glod 2004). The DNA sweep inflamed racial tensions throughout the city as many other black men complained that their civil liberties were curtailed as the sweep stigmatized and robbed them of basic human rights (Finer 2005) and their dignity.

By targeting black men, who constitute a mere eighth of the nation's male population, the police compile databases that largely exclude white men, the majority group. These sweeps miss most criminals. As Rebecca Sasser Peterson points out in the *American Criminal Law Review*: 'Optimal effectiveness, however, would require a universal DNA database that contains DNA fingerprint of *every* citizen, otherwise potential matches would be missed' (Peterson 2000). Consequently, racial profiling contributes to the ineffectiveness of DNA sweeps.

DNA sweeps share troubling features that mingle racial bias, elements of coercion, incomplete disclosure and a disregard for the privacy of the selected subjects (Duster 2006). In 1994, police descended upon black communities, businesses and homes in Ann Arbor, a Michigan college town with a small black population, their aim being to acquire 'consensual' DNA samples from black men only. They collected 160, even though the Fourth Amendment is supposed to protect an individual against 'unreasonable searches and seizures'. However, imprecision of language has led to frequent legal skirmishes. The Supreme Court has clarified: 'A search or seizure is ordinarily unreasonable in the absence of individualized suspicion of wrongdoing' (Esmaili 2007; Anton 2008). This better defines the parameters within which sweeps may be conducted because by definition the sweep conveys no individualised suspicion; potential suspects must submit voluntarily.

Police have confronted male suspects on the streets, in their homes, in restaurants and bars and in their workplaces to procure DNA samples, and many black residents complained that they had been coerced by police officers who ignored their alibis and threatened to prosecute them if they refused to submit. Significantly, the Ann Arbor killer refused to provide police with a DNA sample and was later identified only after he was arrested for an unrelated crime, after which he then could be forced to give a sample (Grand 2002). All the Ann Arbor men who gave samples proved innocent, but police still stored their DNA data in local databases to be tapped when next seeking a perpetrator. Although the 2004 Justice for All Act continues the proscription against depositing 'voluntary' DNA data from state databases into the federal Combined DNA Index System (CODIS),[1] this law is sometimes flouted, as Louisiana did when it inserted data from 120 men garnered during a DNA sweep. San Diego police similarly pressured 800 black men in order to catch a serial killer described only as dark-skinned (Esmaili 2007).

Not one of the black men in Ann Arbor who were induced to surrender DNA during the sweep was guilty, and this fits the national pattern. It is the innocent who are cajoled, intimidated or coerced into yielding their DNA. Of the more than 7000 DNA samples obtained by US sweeps between 1995 and 2002, only one identified a suspect and that one came from an atypical, relatively tiny sweep of only 25 people in a nursing home (Esmaili 2007). This makes the DNA sweep an ineffective but very expensive forensic technique (see also Chapter 3).

Error and fraud are pervasive throughout the USA; for example, a probe that began in 2003 found 180 cases by 2007 at the Houston Police Department Crime Laboratory that were marked by 'major issues' in both criminal and administrative violations, including improper record keeping and false and scientifically unsound reports; this resulted in forced resignations and the suspension of the laboratory's operations (Khanna and McVicker 2007). Another example was in 1988, when a Los Angeles County Sheriff's Department expert from the California State Laboratory at Riverside fraudulently characterised DNA evidence, which resulted in the erroneous conviction of Herman Atkins of Riverside County for a rape and robbery he did not commit (Neufeld and Scheck 2007). According to the West Virginia Court of Appeals, Fred Zain, the former director of the West Virginia State Crime Laboratory had testified for the prosecution in 12 states, but he

[1] Federal DNA Index System. Available from: www.fbi.gov/hq/lab/codis/national.htm.

fabricated results and offered false testimony in hundreds of cases (*Philip A. Ward* v. *George Trent* 1999). Zain was on trial for fraud when he died in 2002 (Ross and Castelle 1993; US District Court for the Southern District of West Virginia 1999; Scheck and Neufeld 2001).

Federal laws have focused upon expanding the FBI's national CODIS database, which began in 1990 as a pilot scheme between 14 laboratories and was initiated with DNA samples from 8000 unsolved crimes (National DNA Index System). Initially, only one particularly repugnant breed of criminal – convicted child molesters – was compelled to produce DNA samples for the database. But within a decade, CODIS expanded to require samples from certain categories of convicted felons, and in 2002 the US Attorney General ordered the FBI to generate a plan to expand CODIS from 1.5 million to 50 million profiles (Simoncelli and Steinhardt 2006). Gradually, in an instance of function creep, which describes the tendency to expand the use of sensitive, narrowly applied technology to progressively broader uses, legislators successfully marketed the widening compulsion of DNA samples as a measure to protect women and eventually even larger groups of the population. At the same time, these legislators and the news media tend to maintain silence regarding the ethical cost of such laws (see Chapter 12).

Lawmakers have also ignored the communitarian dangers of publicly approaching all of an area's black men as potential criminals (Rushlow 2007). They ignore the possibility that DNA databases will be racially skewed by police stop-and-search policies that target blacks and Hispanics, resulting in a heavily black-based database that constitutes a collective presumption of guilt. Proponents also ignore the freighted cultural context: The rationale of protecting women taps into a prominent racial trope that recalls the history of deploying officially sanctioned violence (Allen *et al.* 2000) against black men who have been accused of sexually assaulting white women.

In early 2008, New York City Mayor Michael Bloomberg proposed that *everyone* arrested for any crime whatsoever in New York State should be compelled to provide a DNA sample (McGeehan 2007). The laws in New York City are important because its forensic policies and techniques tend to become models for the nation. But a single infraction can inflate the database: marihuana possession, for which New York City has arrested 362 000 people since the mid-1990s. Of these, 55% are black and nearly 30% are Hispanic. Fewer than 15% are white, because police target poor minority neighbourhoods while ignoring college students and other whites likely to have marijuana (Levine

2008). In April 2008, the US Department of Justice announced its plans to collect DNA samples from each of the 140 000 people it arrests each year. This escalation is billed as a measure to prevent violent crime, but it raises concerns about the privacy of innocent people as well as that of the non-violent shoplifters, loiterers, marihuana users and jaywalkers who would be coerced into surrendering DNA. The anti-gun campaign run by the New York City police force incorporated pervasive racial profiling of pedestrians between 1998 and 1999, and again in 2007: 51% of all those stopped were black and 33% Hispanic, but few arrests resulted (Baker and Vasquez 2007). This phenomenon is not confined to the USA. In the UK, blacks are five times more likely to be stopped than whites, but only 1% of these stops resulted in arrests (Open Society Justice Initiative 2006). A variety of studies conducted in disparate manners by different researchers has yielded consistent results. Police officers tend to target blacks and Hispanics, not because of their actions but because of their race.

The most frequent site of encounters between police and civilians, traffic stops, is also commonly racialised in the USA. Between January 1995 and September 1996, David A. Harris determined that 70% of the 823 citizens detained for drug searches on a particular highway, I-95, were African American (Harris 1999a). In 2002, a larger study by Harris (2005) verified this as a national trend. In 2005, police stopped approximately 17.8 million US drivers (Glover 2005; MSNBC 2007) Of these, John Lamberth, who directed the Ethnic Profiling in the Moscow Metro study, found blacks to be five times more likely to be stopped than whites despite the fact that large controlled studies have found no racially-based differences in motorist behaviour, and despite the fact that blacks are less likely than whites to even have a car (Open Society Justice Initiative 2006).

In a landmark 1996 case brought by 17 African-American defendants (*State* v. *Soto* 1996), Judge Robert E. Francis, a Superior Court judge in a Gloucester County, New Jersey court, was convinced that the New Jersey State Police were engaging in unlawful racial profiling on the basis of statistical evidence revealing a wildly disproportionate number of traffic stops involving dark-skinned men that resulted in a paucity of demonstrated infractions. Francis ruled that state police troopers were targeting black and Hispanic motorists on the New Jersey Turnpike, stopping them simply because they had dark skin and searching their cars, harassing and threatening them, and in some cases assaulting them and then charging them with everything from traffic infractions to drugs offenses (Hefler 2009). In the wake of *State* v. *Soto*, charges against nearly 300 motorists who had been improperly detained were dropped and the US Department of Justice imposed monitoring of the state's traffic stops (*State* v. *Soto* 1996).

This stricter scrutiny of blacks is rationalised by police who opine that blacks commit most crimes and that police are targeting the right people. 'Unfortunately, on the street the police perception is "The criminals are black"', says John C. Connolly, Chief of Police in Manchester, Missouri: 'Not that blacks are criminals: They *think* the criminals are black. So that is where they put their attention nine times out of ten, so that people are detained and arrested inappropriately' (Connolly, interview with the author 2008). Yet racially targeted stops yield lower hit rates than do stops that utilise no racial profiling. As with DNA sweeps, this inefficiency results from the fact that in ignoring whites, who constitute the majority, police are missing most criminals (Harris 2005: 68).

Non-genetic ethnic profiling of despised ethnic minority groups by law-enforcement authorities does not stop at US borders. Investigations in Bulgaria, Hungary and Spain document that police conduct frequent raids on Romani communities and subject immigrant neighbourhoods to intensive surveillance and searches. Their reports also describe complaints of selective police violence against ethnic minorities and the markedly disproportionate confrontation, harassment and arrest of minorities during police stops in Russia. For example, the Moscow Metro Monitoring Study found that while persons of non-Slavic appearance made up only 4.6% of the riders on Moscow's Metro system, they formed 50.9% of persons stopped by the police at Metro exits as part of their security surveillance (Open Society Justice Initiative 2006). This means that Moscow police are more than 10 times more likely to stop non-Slavs than Slavs, an extreme degree of harassment, particularly when compared with the fivefold greater stop rate of blacks in the USA.

Racial profiling in the development of forensic DNA databases is also not limited to the USA. In fact, the UK's database, the earliest and the largest in the world, was established in 1995 and holds DNA profiles of 37% of the nation's black men, compared with only 13% of its Asian men and a mere 9% of its white majority (Randerson 2006).

'STOP 'N SWAB': A SYNERGY OF BIAS

Police departments in the USA are eager to combine traffic stops and DNA collection, despite the pervasive racial profiling characterising both. When he served as New York City Police Commissioner, Howard Safir vociferously supported DNA testing of suspects immediately upon their arrest, and after he resigned, Safir joined those vendors, assuming the CEO positions at Bode Technology and at Safir Rosetti (Smith 2007). Safir extols the virtues of forensics DNA analysis, including combining traffic stops with DNA sweeps (Safir 2007). The most popular proposed model has sought to employ a one-person ad hoc laboratory staffed by individual police officers and providing Record of Arrest DNA Testing (RADT) (Sosnowski 2006). Nanogen Corporation received a federal grant to develop a 'chip-based genetic detector for rapid identification of individuals' that allows a police officer to stop a motorist, take a buccal swab and then place it in on a credit-card-sized chip.[2] Inserting the chip into a device the size of a CD player creates a DNA profile within a few minutes. The police officer then transmits this information to a central database, which requires minutes to report whether the sample 'matches' any in the targeted database (Sosnowski 2006). Today, its website warns customers that this sort of device is no longer being supported by Nanogen, but other candidate devices have been explored (Anon 2004), including a palm-sized 'DNA fingerprinter' from the Whitehead Institute for Biomedical Research, which has been developed with a $7 million federal grant (Philipkoski 1998).

NON-DISCRIMINATION POLICIES

Unlike the situation in Netherlands, which has adopted prescient protective legislation ahead of the policy curve (M'charek 2008), US legislation often trails the adoption of database policies. For example, the federal Genetic Information Nondiscrimination Act (GINA), enacted in May 2008, bars employers and health insurers from penalizing those persons with flaws, anomalies or atypical disease risks that are revealed by genetic testing. It does not, however, prohibit life insurance or disability insurance companies from considering genetic data in making coverage decisions; neither does it extend protection to forensic applications of DNA testing. As such, GINA represents a significant boon to privacy rights but does so while continuing the long-term trend of enshrining legal protections to the *medical*

---

[2] Nanogen, Inc. Briefcase-sized system for accurate, cost-effective DNA diagnostics (a portable genetic analysis system). Available from: www.nanaogen.com (accessed November 2007).

applications of genetic knowledge while failing to address *forensic* applications with parallel statutes (Matejik 2008).

There are also neglected consequentialist concerns for police officers who carry out racialised sweeps and traffic stops. Ethical analyses often overlook the brutalising effects of unjust coercion, violence or threatened violence on the perpetrators of such behaviours. The social-justice violations resulting from racialised DNA sweeps also nullify the supererogatory virtues associated with police officers. We expect police, as guardians of the law, to exhibit not only strength and authority but also a greater than usual degree of truthfulness, fairness and emotional maturity and to be motivated by a dedication to the protection of the public, not by racial hatred. When police officers harass persons because of their race, this causes members of the public, black and white, to lose faith in and respect for them, and such behaviour ultimately sabotages the police's effectiveness on the streets and their credibility in court (Beauchamp and Childress 2001; Harris 2005: 69).

CONCLUSIONS: POLICY RECOMMENDATIONS

This chapter raises concerns regarding DNA profiling and databasing. In seeking better governance, the primary recommendation for addressing racial inequities that haunt forensic DNA technology is simple: better public education and awareness concerning DNA forensics, which is a necessary prelude to a wider public policy debate. It is the US public that risks life in a genetic dystopia or amidst genetically mediated racial repression. The public must be informed and invited into the conversations and policy dialogue about race, security and genetic science (Neufeld and Scheck 2007).

By association, another recommendation is that voting on future referenda, laws or policy decisions, unlike the passage of prior legislation, must be based upon fuller and more objective presentations of the facts and potential pitfalls of expanding the use of DNA data and samples in forensic settings (Secko *et al.* 2009). Similarly, jurists and jurors should be required to complete courses that will allow them to better evaluate (and especially to detect hyperbole in) DNA testimony (Ungvarsky 2007). In addition, a moratorium should be imposed on the proposed marriage of DNA sweeps and traffic stops, and on federal funds to investigate such a marriage. This combination threatens to create a dangerous synergy of two technologies, the application of both being demonstrably fraught with profound racial bias.

It was a full decade ago that Ron Paul, Republican Congressional Representative for the 14th district of Texas, sought to halt all biometric profiles of US Americans in the form of DNA databases, photographs and retinal scans. But an absolute ban is not the answer. As illustrated by the exoneration of the innocent, by the identification of remains and by logical, unbiased forensic applications, varieties of DNA analysis technology unquestionably offer great promise in the forensic arena.

Instead, the challenge for good governance lies in determining how best to exploit genetic power without abusing it. One place to start is to abandon racialised DNA sweeps as inefficient, expensive, scientifically inaccurate and, most of all, as dramatic violations of social justice.

REFERENCES

Allen, J. (ed.), Lewis, J., Litwack, L. F. and Als, H. (2000). *Without Sanctuary: Lynching Photography in America*. Sante Fe, NM: Twin Palms.

Amnesty International USA (2004). *Threat and Humiliation: Racial Profiling, Domestic Security, and Human Rights in the United States*. Ridgefield Park, NJ: US Domestic Human Rights Program.

Anon. (1993). Court invalidates a decade of blood tests results in criminal cases *New York Times*, 12 November, 20A.

Anon. (2004). Tabletop DNA test lab developed. *Lab Business Week*, 14 November, 75.

Anton, L. (2008). How far to catch killer? *St Petersburg Times* (Florida), 1A.

Baker, A. and Vasquez, E. (2007). Police report far more stops and searches. *New York Times*, 3 February, 1A.

Balaban, E., Alper, J. and Kasamon, Y. (1996). Mean genes and the biology of aggression: a critical review of recent animal and human research. *Journal of Neurogenetics*, 11, 1–43.

Beauchamp, T. and Childress, J. (2001). *Principles of Biomedical Ethics*, 5th edn. Oxford: Oxford University Press.

Beiser, V. (2001). How we got to two million: How did the land of the free become the world's leading jailer? *Mother Jones*, 10 July, 2.

Bell, D. (1996). The Foulston & Siefkin lecture: racial libel as American ritual. *Washburn Law Journal*, 36, 1–7.

Bowman, J. (1977). Genetic screening programs and public policy. *Phylon*, 38: 117–142.

Bowman, J. and Murray, R. (1990). *Genetic Variation and Disorders in People of African Origin*. Baltimore, MD: Johns Hopkins University Press.

Brigham, J., Bennett, L., Meissner, C. et al. (2007). The influence of race on face identification. In *Handbook of Eyewitness Psychology*, Vol. 2, ed. R. Lindsay. Mahwah, NJ: Lawrence Erlbaum, pp. 257–282.

Chapin, A. (2005). Arresting DNA: privacy expectations of free citizens versus post-convicted persons and the unconstitutionality of DNA dragnets. *Minnesota Law Review*, 89, 1842–1874.

Chasnoff, I, Landress, H. and Barrett, M. (1990). The prevalence of illicit-drug or alcohol use during pregnancy and discrepancies in mandatory reporting in Pinellas County, Florida. *New England Journal of Medicine*, 322, 1202–1206.

Cherek, D. and Lane, S. (1999). Effects of DL-fenfluramine on aggressive and impulsive responding in adult males with a history of conduct disorder. *Psychopharmacology*, 146, 473–481.

Cho, M. and Sankar, P. (2004). Forensic genetics and ethical, legal and social implications beyond the clinic. *Nature Genetics 36*, 11(Suppl 36), S8–S12.

Cole, S. A. and Dioso-Villa, R. (2007). *CSI* and its effects: media, juries, and the burden of proof. *New England Law Review*, 41, 435–470

Duster, T. (2006). Explaining differential trust of DNA forensic technology: grounded assessment or inexplicable paranoia? *Journal of Law, Medicine and Ethics*, 34, 293.

Esmaili, S. (2007). Student note: searching for a needle in a haystack – the constitutionality of police DNA dragnets, *Chicago-Kent Law Review*, 82, 495–497.

Finer, J. (2005). Baffled police try DNA sweep: towns men asked to give samples in murder case, *Washington Post*, 12 January, A03.

Glod, M. (2004). DNA dragnet makes Charlottesville uneasy: race profiling suspected in hunt for rapist. *Washington Post*, 14 April, A01.

Glover, K. (2005). Racial profiling and the pretextual traffic atop: a critical look at the US Supreme Court's Whren decision. In *Proceedings of the American Sociological Association Annual Meeting*, Philadelphia, 12 August.

Grand, J. (2002). The blooding of America: privacy and the DNA dragnet. *Cardozo Law Review*, 23: 2277.

Gould, S. (1992). *The Mismeasure of Man*. New York: Norton.

Guthrie, R. (1998). *Even the Rat Was White: A Historical View of Psychology*, 2nd edn. Boston: Allyn and Bacon.

Harris, D. (1999a). *Driving While Black: Racial Profiling on our Nation's Highways.* [*American Civil Liberties Union Special Report.*] New York: American Civil Liberties Union www.aclu.org/racialjustice/racialprofiling/15912pub19990607.html (accessed August 2009).

Harris, D. (1999b). The stories, the statistics, and the law: why driving while black matters. *Minnesota Law Review*, 84, 265–326.

Harris, D. (2005). *Good Cops: The Case for Preventive Policing*. New York: New Press.

Hefler, J. (2009). N.J. bill would put state in charge of monitoring troopers. *Philadelphia Inquirer*, July 2, B1.

Henig, R. (2004). The genome in black and white (and gray). *New York Times*, 10 October, 47.

Hilts, P. (1992). Federal official apologizes for remarks on inner cities. *New York Times*, 22 February, 6.

Human Rights Watch (2008). Targeting blacks: drug law enforcement and race in the United States. Washington, DC: United Nations Foundation http://www.hrw.org/en/node/62236/section/1 (accessed August 2009).

Innocence Project (2009). *Website.* www.innocenceproject.org/ (accessed 8 January 2009)

John-Hall, A. (2009). Time to stop demonizing 'black men'. *Philadelphia Inquirer*, 29 May, A13.

Johnson, C. and Horace, M., (1934). The investigation of racial differences prior to 1910. *Journal of Negro Education*, 3, 333.

Kahn, J. (2004). How a drug becomes ethnic: law, commerce, and the production of racial categories in medicine. *Yale Journal of Health Policy, Law and Ethics 4*, 1–46.

Katz, J. (1972). *Experimentation with Human Beings: The Authority of the Investigator, Subject, Professions, and State in the Human Experimentation Process*. New York: Russell Sage Foundation.

Khanna, R. and McVicker, S. (2007). Troubling cases surface in report on HPD crime lab. *Houston Chronicle*, 17 June, 1.

Leary, W. (1992). Struggle continues over remarks by mental health official. *New York Times*, 8 March, 34.

Levine, H., Gettman, J., Reinarman, C. *et al.* (2008). Drug arrests and DNA: building Jim Crow's database. *GeneWatch, 21*, 9–11.

Lombroso, C. (1911). *Crime, Its Causes and Remedies.* [Trans. Henry P. Horton.] Boston, MA: Little, Brown, pp. 428, 447–448.

Lowe, A., Urquhart, A., Foreman, L. *et al.* (2001). Inferring ethnic origin by means of an STR profile. *Forensic Science International*, 119, 17–22.

Marks, J. (1995). *Human Biodiversity: Genes, Race, and History.* New York: Aldine Transactions.

Matejik, L. (2008). DNA sampling: privacy and police investigation in a suspect society. *Arkansas Law Review*, 61, 53–58.

McGeehan, P. (2007). Spitzer wants DNA sampling in most crimes. *The New York Times* 14 May, 1.

M'charek, A. (2008). Contrasts and comparisons: three practices of forensic investigation. *Comparative Sociology*, 7, 387–412.

MSNBC (2007). Black, Latino drivers fare worse in traffic stops. *MSNBC.COM*, April 29, http://74.125.113.132/search?q=cache:rhY_WIURuo4J:www.msnbc.msn.com/id/18383182/+MSNBC+driving+while+black+2007&-cd=1 &hl=en&ct=clnk&gl=us&client=firefox-a (accessed 5 March 2010).

Neufeld, P. and Scheck, B. (2007). *CSI* isn't like real-world crime labs. *Los Angeles Times*, 2 February, 18.

Open Society Justice Initiative (2006). Ethnic profiling in the Moscow metro. New York: Open Society Institute http://www.soros.org/initiatives/justice/focus/equality_citizenship/articles_publications/publications/profiling_20060613 (accessed 15 March 2010).

Page, L. (2008). Scotland Yard criminologist: DNA-print troublemaker kids Give me the child, I'll lock-up the man. *London Observer*, 17 March, 1.

Peterson, R. (2000). DNA databases: when fear goes too far. *American Criminal Law Review*, 37, 1219-1230.

Philipkoski, K. (1998). A crime sniffing network. *Wired News*, 11 August www.wired.com/news/technology/0,282,14231,00.html (accessed July 2002).

Randerson, J. (2006). DNA of 37% of black men held by police: Home Office denies racial bias. *Guardian*, 7 January www.guardian.co.uk/world/2006/jan/05/race.ukcrime (accessed August 2009).

Roberts, D. (1997). *Killing the Black Body: Race, Reproduction and the Meaning of Liberty.* New York: Vintage Books.

Ross, A. and Castelle, G. (1993). In the Matter of an Investigation of the West Virginia State Police Crime Laboratory, Serology Division (Report No.21973) In the Supreme Court of Appeals of West Virginia.

Rutledge, J. (2001). They all look alike: the inaccuracy of cross-racial identifications. *American Journal of Criminal Law*, 28, 207–228.

Rushlow, J. (2007). Rapid DNA database expansion and disparate minority impact. GeneWatch 20, 3–11.

Sampson, R. and Lauritsen, J. (1997). Racial and ethnic disparities in crime and criminal justice in the United States. *Crime and Justice*, 21, 311- 326.

Sachs, J. (2004). DNA and a new kind of racial profiling. *Popular Science* 21 April, 16–20.

Secko, D., Preto, N., Niemeyer, S. *et al.* (2009). Informed consent in biobank research: a deliberative approach to the debate. [Paper No. GE3LS,

W. Maurice Young Centre for Applied Ethics.] *Social Science Medicine*, 68, 781–789.

Sellers-Diamond, A. (1994). Disposable children in black faces: the violence initiative as inner-city containment policy. *University of Missouri Kansas Law Review*, 62, 423–428.

Simoncelli, T. and Steinhardt, B. (2006). California's Proposition 69: a dangerous precedent for criminal DNA databases. *Journal of Law, Medicine and Ethics*, 34, 199–213.

Safir, H. (2007). DNA technology as an effective tool in reducing crime. *Forensic Magazine*, October/November www.forensicmag.com/articles.asp?pid=168 (accessed August 2009).

Scheck, B. and Neufeld, P. (2001). Junk science, junk evidence. *New York Times*, 11 May, 35.

Silver, J. (2004). DNA technology advancing to produce more information faster for police. *Pittsburgh Post-Gazette*, 12 December, A3.

Smith, B. and Dailard, C. (2003). Black women's health in custody. In *National Colloquium on Black Womens' Health*. Washington, DC: Black Womens' Health Imperative, pp. 97–108.

Smith, G. (2007). Cash ties that bind: possible violation of city bidding rules for DNA lab linked to Safir. *New York Daily News*, 20 August www.nydailynews. com/news/2007/08/20/2007-08-20_cash_ties_that_bind-2.html (accessed August 2009).

Sosnowski, R. (2006). *A Chip-Based Genetic Detector for Rapid Identification of Individuals.* [Document No. 213911.] Washington, DC: National Institute of Justice.

Terry, D. (1994). A woman's false accusation pains many blacks. *New York Times*, 6 November, 32.

Tolson, M. and Khanna, R. (2007). Mix-up on DNA deals HPD lab another blow: man exonerated 14 years after rape conviction. *Houston Chronicle*, 4 October, 1.

Ungvarsky, E, (2007). What does one in a trillion mean? *GeneWatch*, 20, 1.

Walker, S. (2004). *Police DNA Sweeps Extremely Unproductive: A National Survey of Police DNA Sweeps.* Omaha, NE: University of Nebraska Police Professionalism Initiative, Department of Criminal Justice.

Washington, H. A. (2001). Gene blues. *Essence*, 32, 88.

Washington, H. A. (2004). Born for evil? Stereotyping the karyotype: a case history in the genetics of aggression. In *Twentieth Century Ethics of Human Subjects Research: Historical Perspectives on Values, Practices and Regulations*, eds. V. Roelcke and G. Maio. Stuttgart: Franz Steiner, pp. 319–334.

Washington, H. A. (2007). *Medical Apartheid: The Dark History of Medical Experimentation on Black Americans from Colonial Times to the Present.* New York: Doubleday.

CASES

Philip, A. Ward v. *George Trent* (1999). Appeal from the United States District Court for the Southern District of West Virginia at Huntington 9No. 98-7267). Robert C. Chambers, District Judge. (CA-97-1107) Decided April 23, 1999.

State v. *Soto* (1996) 734 A.2d 350 N.J. Super. Ct. Law Div.

5

Health and wealth, law and order: banking DNA against disease and crime

INTRODUCTION

In the 1990s, genetic databases moved into the mainstream, no longer limited to disease registers of affected families and other collections for specific research projects. The creation of large population-based collections of DNA samples linked to other information was supported by significant economic and political investment. This investment was driven by the promise that these new techno-scientific initiatives would deliver a variety of benefits for society, including economic competitiveness, improved public health, more effective law enforcement and greater security. In the field of criminal justice, forensic science organisations and police forces established DNA databases containing samples from offenders or suspects. In the medical realm, research institutions and governments created databases, more usually called biobanks, for the purposes of undertaking biomedical research. Until recently, forensic databases have not attracted particular attention from the media, public or social researchers, whereas biomedical databases have been seen by ethicists, politicians, policy advisors, lawyers and social scientists to give rise to many issues of public concern such that their formation has been the subject of intense scrutiny.

Some commentators have drawn comparison between the two scientific, technological and social arenas of law enforcement and biomedicine to highlight differences and incongruences between forensic and medical research databases and their governance. They point out, for example, that owing to a focus on concerns such as consent and privacy by those writing about biomedical databases, forensic databases have not received as much attention (Williams 2005; Levitt 2007). At the level of policy advice, however, connections

Genetic Suspects: Global Governance of Forensic DNA Profiling and Databasing, ed. Richard Hindmarsh and Barbara Prainsack. Published by Cambridge University Press. Copyright © Cambridge University Press 2010.

between the two have been made repeatedly. The Human Genetics Commission, for example, created by the UK Government to advise on matters relating to human genetic science and technology, has successfully sought to have police use of DNA included under its remit alongside the use of human genetic technologies in the biomedical arena. The 2007 Nuffield Council on Bioethics investigation into the forensic use of bioinformation made repeated comparisons between the medical and forensic arenas, and the membership of the newly formed Ethics Group for the UK National DNA Database (NDNAD) includes five individuals (out of a total of 12) with backgrounds in medicine.

Given this 'trade' or flow across the boundaries between the medical and the forensic, this chapter sets out not simply to enumerate the similarities and differences between the way police databases and medical research databases have developed but also to reflect on how we conceptualise their different sociotechnical configurations. How do we begin to analyse the simultaneous development of forensic DNA databases and the establishment of biomedical biobanks? One type registers (by compulsion) profiles and samples of the arrested, accused and convicted, for policing purposes, and the other relies on the voluntary participation of individuals and has been subject to extensive ethical review, anxieties about informed consent and regular public consultation. One response to this question is suggested by the notion of 'biolegality', coined by Lynch and McNally (2009: 284) to describe how 'developments in biological knowledge and technique are attuned to requirements and constraints in the criminal justice system, while legal institutions anticipate, enable, and react to those developments'. They posit this concept as a variant on others, such as Paul Rabinow's (1996) notion of 'biosociality' and Nikolas Rose and Carlos Novas' (2005) idea of 'biological citizenship' (see also Chapter 13), but one with specific reference to the forensic as opposed to the biomedical context. However, our purpose here is to explore how these influential analyses, developed by social scientists to address biomedical developments, might also illuminate issues posed by the parallel development of genetic databases for policing purposes. We centre on three sets of contributions to the social debate about biomedicine, biotechnology and biobanks. The first uses the sociology of expectations and promise to consider the dynamics of contemporary technological innovation; the second addresses the 'politics of legitimation' in the governance of biobanks; and the third discusses wider changes in contemporary citizenship practices and how individuals are governed and govern themselves in light of scientific and technical change. These discussions will illuminate some of the potential convergences and divergences between developments in forensic and biomedical DNA databases.

In this endeavour, we focus primarily on the UK context, because this vividly illustrates the differences between databases in the policing and medical contexts. In 1995, the UK became the first country worldwide to establish a national forensic DNA database (or as we prefer to call it, a 'police DNA database' to emphasise its social, political, ethical, legal, cultural scientific and technical dimensions, which the term forensic does not capture adequately). Four years later, the UK also became one of the first countries to propose setting up a large-scale database to examine causes of common complex diseases. Of these two DNA databases, the second will eventually contain biological samples and medical and lifestyle information from half a million adult volunteers, to be used only for medical research, with all research proposals subject to the usual ethical review procedures. The first (forensic) database already contains the DNA profiles from over four million children and adults, together with personal information and retention of the original biological sample. Samples can be taken without consent from anyone over the age of 10 years who has been arrested on suspicion of a recordable offence. This sample range has been expanded since 1997 to include minor offences such as begging and taxi touting. Profiles remain on the forensic database for life, with no right of withdrawal, even if an individual is never charged with an offence or is charged but later acquitted.[1] Samples gathered for the purposes of elimination, including those of crime victims, may also be entered on the database, in this case with consent, and are then treated in the same way as other profiles and are subject to speculative searching of the whole database for matches with crime scene samples. The recent judgment in the *S* and *Marper* appeal to the European Court of Human Rights was that the retention of DNA from S, who was arrested at age

---

[1] Scotland exports profiles, not samples, to the NDNAD. Scottish profiles and samples are destroyed if the suspect is subsequently not charged or is acquitted, except for some serious offences where they are retained for five years. Volunteers may consent to comparison of their profiles with the crime scene profiles only or to retention on the database. Consent can later be revoked in Scotland, but not in England or Wales.

11 but acquitted, and Marper, whose case did not come to court, violated their human rights (European Court of Human Rights 2008, *S and Marper* v. *the United Kingdom* 2008; see also Chapter 2).

DATABASES AND THE POLITICS OF PROMISE

Brown *et al.* (2006) emphasise the way that investment in particular sciences and technologies is made on the basis of certain visions, promises and expectations about the future. They observe that 'imagination, expectation and promise have all recently been the focus of concerted critique ...', making the point that there are 'observable "dynamics" to the way that futures are mobilized in scientific and technological innovation' (Brown *et al.* 2006: 332). The field of pharmacogenetics has been a particular focus of enquiry as an example of a 'promissory science' in which various futures are being created – for example, the vision of personalised medicine – which then shape the direction of scientific enquiry and the investment of funds by the pharmaceutical industry. Hedgecoe and Martin (2003) argue, in their work on pharmacogenetics, that understanding visions and expectations is central to the analysis of new technologies. Brown *et al.* (2006) also examine the stem cell as 'promissory matter', around which coalesce a number of powerful and hopeful anticipations about its future therapeutic uses. Situating these current expectations about stem cells within a history of previous expectations about this science that stretch back more than half a century, they remind us that our futures have a past that needs to be appreciated when looking at contemporary developments.

The articulation and effects of expectations and promissory discourses, though, are not limited to contemporary scientific practices. Public policy is another important arena in which they can be analysed. In a widely cited paper, Nightingale and Martin (2004) outline that a range of actors from policy advisors to industrialists to academics have been advancing a revolutionary model of technical change in medical biotechnology. For instance, 'this 'revolutionary' model has generated widespread expectations that biotechnology has the potential to create an increased number of more-effective drugs and bring about radical changes in healthcare' (Nightingale and Martin 2004: 564). However, these authors argue that the empirical evidence, in the form of data on the number and scope of new drug approvals, does not support the revolutionary and transformative claims made about current biotechnology. Instead, innovation in the biotechnology sector is slow and incremental. Therefore, it is important not to foster unrealistic expectations about the speed of innovation, as such expectations can 'lead to poor investment decisions, misplaced hope, and distorted priorities, and can distract us from acting on the knowledge we already have about the prevention of illness and disease' (Nightingale and Martin 2004: 568).

However, despite the slow speed of innovation in biotechnology, governments see that investment in biobanking could boost the competitiveness of the national pharmaceutical industry, deliver public health benefits and reduce healthcare expenditure. In turn, disease advocacy organisations emphasise the value of biobanks to accelerate the process of medical innovation, aiding what is called translational research, which is to speed up the development of new drugs from the laboratory bench to the hospital bedside. In contrast, some epidemiologists – while mindful of the new understandings of disease aetiology and improved risk prediction that could be gained from investment in biobanks – are weary of the breakthrough model when it comes to the study of common, complex diseases (Tutton 2007). In the case of population biobanks, the data collected are promised to enable researchers to improve their understanding of the interaction of genetics, environment and lifestyle. This interaction may hold the key to understanding common disease but, of course, the value of any conclusions depends on the quality of the non-genetic data collected. Consequently, 'high-tech' or speedy solutions are not so anticipated in this area (Nightingale and Martin 2004: 568).

The analytical framework of imagination, expectations and promise allows us also to see the ways that police DNA databases are also 'promissory objects', with their creation, growth and governance shaped by certain expectations about the future of policing, crime detection and security. In popular culture, forensic data is valorised in American and British television programmes such as *Crime Scene Investigation (CSI)* and *Silent Witness*, and the media celebrate high-profile success stories but at the expense of examining detailed crime detection statistics. As a result, techniques of forensic investigation, especially DNA analysis, have become a significant dimension of the public imagining of contemporary science and policing.

As in the provision of resources for biobank development, much the same rhetoric (or promissory discourse) is evident in the claim of the UK NDNAD in England and Wales to be the first in the 'race' for scientific innovation and accompanying success in this area.

Already, the NDNAD is the world's largest in terms of the proportion of the national population it contains and has the most extensive criteria for inclusion. The UK Government claims that 'this expansion and investment is being closely followed by Europe and America who are keen to emulate the crime-solving successes of the database' (Home Office 2008).

The language of speed and acceleration is also evident in the promissory discourse on the forensic use of DNA. The Nuffield Council on Bioethics report (2007: 50) on the forensic use of bioinformation observes that a DNA match between suspect and crime scene samples can 'dramatically accelerate an investigation and prosecution. Much expense and distress can be spared in reaching a verdict swiftly, especially if fingerprint or DNA evidence can prompt a guilty plea to be entered by the suspect at an early stage'. This language is echoed in government pronouncements, for example, in 1999 when the then Home Secretary Jack Straw claimed: 'breakthroughs in DNA technology mean that offenders can be matched to scenes of crime through microscopic samples of no more than two or three human cells – sometimes years after the event. These advances have the potential to help the police clear up more crimes, free police time and speed communications' (Hartley-Brewer 1999).

Despite the optimistic claims made for the use of DNA in policing, there is no reliable statistical evidence on the efficacy of the NDNAD as currently configured. If DNA profiles are loaded onto the database from a crime scene, detection is more likely. Figures suggest that the overall detection rate for all recorded crime is 26% but this increases to 40% in cases where DNA evidence is available and analysed through the database (Home Office 2005a). GeneWatch UK, a non-governmental organisation that takes a critical perspective on genetic applications and policy making, has used available information from government sources to suggest that in fact only about 0.36% of crimes are solved through DNA evidence and that this figure has remained constant from 2002 to 2006 despite increasing numbers of profiles stored from individuals (GeneWatch UK 2007). With reference to the retention of profiles held on those arrested or charged but not subsequently convicted, UK Government ministers have spoken of the 'enormous number of crimes [solved] because of the data held on people who have had their DNA taken on arrest, whether they were charged or not … including 37 murders, 16 attempted murders and 90 rapes' (Hillier 2008). In a detailed analysis of similar claims made by politicians, with different figures given, GeneWatch

UK (2008) pointed out difficulties in verifying such claims, as they are based on estimates rather than actual tracked crimes, on matches rather than solved crimes and will include matches made with victims and passers-by whose profiles are on the database.

DATABASES AND THE POLITICS OF LEGITIMATION

Linked to the dynamics of expectations that shaped the development of genetic databases is the challenge of establishing political legitimacy. A number of commentators have remarked on the apparent decline in public trust and confidence in the UK as a consequence of various highly publicised incidents in the 1990s involving scientists, politicians and doctors (e.g. the bovine spongiform encephalopathy and varient Creutzfeldt–Jakob disease issue) (Kaye and Martin 2001; Weldon 2004; Petersen 2006). Against this background, medical sociologist Alan Petersen (2006: 272) has suggested that, 'as with others undertaking controversial biomedical research involving the participation of human subjects, the partners of the UK's national medical database, UK Biobank, face the problem of establishing support and legitimacy for their project'.

This question of legitimacy has particular relevance for biobanks because they depend on collecting genetic information from large numbers of individuals enrolled for long periods. Moreover, this information is collected for the purposes of long-term, open-ended research, the precise focus of which cannot be specified in advance. As a consequence of these features of biobanks, questions have been raised about trust, security, confidentiality, consent, discrimination and commercialisation (Tutton and Corrigan 2004; Levitt and Weldon 2005; Salter and Jones 2005; Busby 2006; Petersen 2006).

Given these issues, some argue that population biobanks in the medical arena require novel forms of legitimation (Salter and Jones 2002, 2005). One novel way in which legitimation has been sought is through the use of 'upstream' public consultations. From its conception, the funders of UK Biobank – the Wellcome Trust, the Medical Research Council (MRC, a publicly funded body that supports health research in the UK) and the Department of Health – commissioned market research companies to conduct consultations to explore the views of healthcare professionals, special interest groups and sections of the 'general public' on various aspects of the organisation, governance and ethical issues of the project (Cragg Ross Dawson 2000;

People, Science and Policy 2002). These consultations were 'upstream' in the sense that they were conducted before plans for the database were finalised. However, they were not novel in the sense that the public would have been able to influence the direction or shape of the biobank itself. Indeed, the House of Commons Science and Technology Select Committee's 2003 report on UK Biobank noted that, 'it is our impression that the MRC's consultation for Biobank has been a bolt-on activity to secure widespread support for the project rather than a genuine attempt to build a consensus on the project's aims and methods' (cited in Petersen 2006: 283).

Salter and Jones (2005: 712) argue that when faced with such political uncertainties, policy actors turn to enrolling epistemic communities – 'network[s] of professionals with recognised expertise and competence in a particular domain and an authoritative claim to policy-relevant knowledge' – to negotiate the pathway through these uncertainties. Traditionally, this has seen politicians turning to scientific experts to base their decisions on 'sound', 'impartial' science but, given the fact that incidents and developments such as bovine spongiform encephalopathy, genetically modified food and retention of tissue samples are seen to have undermined public confidence in science (or at least the use of science by politicians), this would lack credibility. Therefore, Salter and Jones (2002) argue that bioethicists have become enrolled into the governance of novel science and technology. For example, following political criticism of its funding and value for money, the UK Biobank partners established an Ethics and Governance Framework and also created a quasi-independent committee called the Ethics and Governance Council, which was seen as 'crucial to securing [public] confidence' (UK Biobank Ethics Consultation Workshop 2002). The Ethics and Governance Council, with a bioethicist as the first chair, was established long before its first participant was recruited. Despite the high profile given to this committee, its powers are somewhat limited. It does not, for instance, have veto power over the use of data or samples for any particular research project; that power rests with the UK Biobank Board of Directors on which the Ethics and Governance Council is not even represented. In summary, the supporters and organisers of biobanks anticipated that they would face particular difficulties with establishing their legitimacy, which led to the extensive use of public consultations in the UK and the enrolment of ethics specialists as an expert community to deal with anticipated public concerns.

The legitimacy of the NDNAD was not initially questioned, presumably because it began by retaining DNA data only from criminals convicted of serious offences and only for the purpose of investigating and solving crime. Given this, the NDNAD did not have to engage at all in a politics of legitimation similar to that of the medical biobanks. Public consultations were not commissioned prior to its establishment to investigate people's attitudes towards providing DNA samples should they be arrested or commit an offence. UK Biobank, as Salter and Jones (2005) point out, is dependent on the active participation of citizens and will fail if the required numbers of participants do not choose to participate. In the case of the NDNAD, the fear of crime and the need for enhanced action from the criminal justice system was already established as a major public issue. Legislative changes needed to increase its size were simply embedded in broader legislation to tackle crime, without public debate. This was reinforced by negative perceptions of young people's anti-social behaviour in the UK compared with the rest of Europe (Margo and Stevens 2008).

But public support may well rest on the (incorrect) assumption that NDNAD is a criminal database, with DNA retained from those convicted of an offence, and not retained for what the public consider minor offences or for those not convicted, mistakenly arrested and never charged or who volunteer samples in screenings (Levitt and Tomasini 2006). This lack of correct public knowledge is supported by research into public attitudes in 2000 carried out on behalf of the Human Genetics Commission. That research also found that a majority of the public was opposed to the taking of DNA samples for shoplifting, for example (Human Genetics Commission 2000: 36). A recent citizens' inquiry conducted on behalf of the Human Genetics Commission, which involved several meetings for participants and question-and-answer sessions with a variety of experts, concluded that an independent statutory authority should control the police database, that those acquitted should be removed from the database, and that there should be a nationwide information campaign to raise public awareness of it (Human Genetics Commission 2008). The European Court of Human Rights came to similar conclusions in their judgment in the case of S and Marper v. the United Kingdom (2008; see also Chapter 2 this volume).

The legitimacy of the NDNAD has been called into question over the continued expansion of the categories of individuals whose samples and profiles are included (Anon. 2003). One of the explicit strategic objectives of the NDNAD, set out when the database went live in 1995, was to 'obtain DNA samples from the active criminal population' (Home Office 2005b). Retaining DNA from those *not* found guilty of a crime was, therefore, not made explicit as an objective but came about through a series of piecemeal changes. For example, in 1994, DNA could be taken from anyone arrested for a recordable offence. The definition of recordable offence was broadened to include 42 extra offences in 1997, another five in 2003 and non-imprisonable offences were added in 2003 including 'begging' and 'touting for car hire services' (Williams and Johnson 2008: 146). In the case of retaining profiles of those arrested but not charged, changes were made in 2001 to legitimate violations of the existing legislation and allow the retention of samples and profiles that should have been destroyed because the person was not in the end convicted of a criminal offence (Williams and Johnson 2008: 84).

What started out as a criminal DNA database has thus grown into something quite different through legal changes contained in broader criminal justice and police legislation (Parliamentary Office of Science and Technology 2006). The gap between the stated objectives of the NDNAD and the criteria for inclusion is increasingly being highlighted in the media, for example in the story of the Home Secretary admitting that the DNA of a one-year-old baby was being held on the NDNAD (Anon. 2009). The laws governing the NDNAD also give no weight to any undesirable consequences for innocent children and adults who object to being on a database that was designed for the active criminal population for their entire lives. Crucially, at present, little evidence exists to demonstrate the value to society of retaining samples from those not charged or convicted of offences (GeneWatch UK 2008).

In the absence of such evidence, however, a key argument often made in support of current policies for the expansion of the NDNAD is that the innocent have nothing to fear from having their profiles loaded on the database. The Nuffield Council on Bioethics says this (simplistic) argument cannot be used by itself to justify the expansion of police powers (Nuffield Council 2007: 34). Instead, 'government always needs to show a strong reason, backed by objective evidence, that there is adequate justification for interfering with the lives of its citizens' (Nuffield Council 2007: 34). One sugges-tion advanced from various quarters now is that, to re-establish its legitimacy, the NDNAD should be expanded to include the entire population. The government, mainly on practical grounds, has at the moment rejected this (see also Chapter 6, with regard to similar rejection in the USA).

Concerns about lack of independent oversight have also been expressed by official government bodies, from the Parliamentary Science and Technology Committee (House of Lords 2001) to the Human Genetics Commission (2002). Responding to the pressure, the UK Government eventually agreed to set up an Ethics Group in 2007 (Levitt 2007). Its remit is to 'advise ministers on ethical issues concerning the NDNAD' (Home Office 2008: 35). In its first annual report, the Ethics Group sought to tackle some of the concerns raised. For example, regarding volunteer samples from those not suspected of a crime but likely to have left their DNA at the crime scene for different reasons, the Ethics Group made the recommendation that these should not normally be loaded on the database but used only in the specific case and then destroyed if found irrelevant. The volunteers should also have a specific consent form, and more information regarding the use and limitations of forensic data analysis should be given to all competent adults whose profiles are taken, whether volunteers or not. A simpler system of appeal against the retention of DNA was also recommended, as was a simpler system to 'reinforce the message that [the police database] is intended only to be used for criminal investigation' (UK National DNA Database 2008: 26). Moreover, the Ethics Group recommended that there should be a formal public announcement that the 'NDNAD will only be used for its currently described purposes (that is, criminal intelligence) and will never transform into a repository for the whole nation's DNA characteristics' (UK National DNA Database 2008: 26).

In summary, the UK Government found itself needing to reassert the legitimacy of the NDNAD similarly to that of medical biobanks by enrolling new epistemic communities outside of forensic science, from philosophy, medicine and research ethics as well as from law and criminology to manage public concerns.

DATABASES AND BIOLOGICAL CITIZENSHIP

In this final part of the chapter, attention is shifted to questions of changing political formations and citizenship practices as a feature of social discussions about developments in biomedicine. The discussion engages primarily with the account given by sociologists Nikolas Rose and Carlos Novas (2005) of contemporary 'biological citizenship' and considers, in light of this, Catherine Waldby's (2002) concept of 'biovalue'.

The concept of 'biological citizenship' has become influential in discussions about the social dimensions of biomedicine. This can be traced in part to Nikolas Rose and Carlos Novas (2005: 440), who, borrowing this expression from Adrian Petryna (2002), describe 'biological citizenship' as including 'all those citizenship projects that have linked their conceptions of citizens to beliefs about the biological existence of human beings'. Specifically, they argue, a 'new kind of citizenship is taking shape in the age of biomedicine, biotechnology and genomics' (Rose and Novas 2005: 439), in which individuals think and speak about themselves in a language that is increasingly biological. In other words, their corporeality is to some degree becoming central to their self-identity. Moreover, individuals are taking responsibility for their own health and seeking to inform themselves of the latest expert advice on lifestyle, pharmaceutical treatments or predictive genetic testing.

Biological citizens also relate to others in biological terms by seeking alliances with other citizens who might share a genetic trait or disease to form biosocial communities such as disease advocacy or support organisations. These groups are no longer passive patients but informed and active consumers, even quasi-professionals (Novas and Rose 2000). To this extent, these groups reflect prevalent notions associated with contemporary personhood in western society, such as choice, autonomy and self-responsibility, which are themselves tied to debates about how citizenship has transformed in the period since the end of World War II, linked to significant social, economic and technological changes. These citizen groups invest in the promise of science to find treatments or cures for conditions from which their members and others suffer, raising not only funds to support ongoing research but also in some cases collecting blood samples, reviewing scientific applications for scientific research grants or participating in clinical trials (Fitzgerald 2008). These groups are involved in what Rose and Novas call a 'political economy of hope', by which biomedicine is not only about the production of truth but also the production of hope in the face of currently incurable conditions (cf. Moreira and Palladino 2005). In summary, biological citizenship is a 'hopeful domain' in which knowledge about biology is not fatalistic or deterministic but something that might be modified through scientific research. Another significant dimension to contemporary biological citizenship is the way in which the vitality of each individual can become a potential source of biovalue.

Catherine Waldby's (2002) notion of *biovalue* refers to how life itself is productive of economic value, linked to the capabilities of biomedicine to visualise and manipulate the body at the molecular level, and to produce patentable products such as new pharmaceuticals. As discussed above, it can be argued that biobank projects in Iceland, Sweden, Estonia, the UK and elsewhere are predicated on the idea of life as productive of economic value, with the promise to utilise the genetic characteristics of their populations as resources for the production of intellectual property.

Forensic DNA databases are also implicated in the production of biovalue. The samples of those included in the database – which, as we must always say in the UK context, are those from both the legally guilty and innocent – are a potential source of new biovalue. For instance, the UK NDNAD, with more than four million samples, is a significant and unique resource that already has been utilised to conduct particular forms of research. As indicated in the 2007 Nuffield Council report, the UK NDNAD Strategy Board has received an increasing number of requests to conduct research relevant to the 'prevention and detection of crime', which, broadly conceived, could include many different types of research and not only that which is related to the operation of the database itself. The pressure group GeneWatch UK in its investigations in 2006 identified uses of the database by public and commercial bodies to carry out research on developing new methods of familial searching and genetic tests to attempt to determine the racial/ethnic identity of perpetrators (GeneWatch UK 2006). As both GeneWatch UK and the authors of the Nuffield Council report observe, there is at present a lack of public information about the research that has been allowed or disallowed, and, as such, there are justifiable fears about the kinds of possible research that could be carried out both now and in the future.

In the context of police databases, biological citizenship looks quite different. This is not a context in which we can talk about self-organising groups of active citizens but one in which those convicted of crimes, those charged but not subsequently prosecuted and those arrested but not subsequently charged are compelled to provide a sample to be retained indefinitely on a police database. If we think of this as a form of genetic citizenship or biological citizenship, then it clearly has very different features than the celebrated cases of the self-organising citizens who form support and advocacy organizations, for whom biological knowledge is a source of their self-definition.

This is not about self-definition but definition by the state; a social sorting into the suspect and the non-suspect for the operational purposes of policing. While this is not based on the identification of specific biological characteristics that are seen to have some social relevance – the putative genes for criminal behaviour would be an example of that – it is nonetheless a situation where state ownership and control of biological information imposes a certain status upon individuals, about whom the state makes certain claims even if this is rendered as a largely socially invisible automated process of database searching. This process can, however, sometimes become dramatically visible when positive matches are made and the guilty are identified, tried and incarcerated (or, more rarely, those otherwise thought or found guilty are freed).

The development of police DNA databases also highlights that biological citizenship should be understood not simply in terms of activities 'from below'. Rose and Novas (2005) emphasise individual self-governance in the sense that individuals 'discipline' themselves to accord with societal norms. At the same time, however, they acknowledge that strategies for 'making up' biological citizens are employed from above by the state, institutions of medicine or welfare, the police or private industry. These strategies involve the creation of categories by which individuals are then known by these actors (e.g. the 'genetic carrier', the 'arrestee', or the 'DNA donor'), creating boundaries between those who are treated in different ways. Outside the domain of biomedicine, the state is active in 'making up' certain kinds of citizen through its own technologies of social categorisation, control and disciplining (Hacking 1986). The police DNA database, as one of these technologies, could deter individuals from committing offences or pursuing 'criminal careers' because they know that it has expanded to include even relatively minor offences, and that their bioinformation will be used to identify and prosecute them.

Looking at police DNA databases through the lens of biological citizenship highlights the underexplored theme of obligation in the account of Rose and Novas (2005). Of course, Rose and Novas see that the acts of 'calculation and choice' in which biological citizens engage with their own health are also to some extent matters of obligation. Individuals are obliged to be informed, to know about their current health and predispositions and then to act on this knowledge. Police databases sometimes involve similar dynamics of obligation – people who are asked by the police to volunteer elimination samples are, of course, free to refuse but this itself may cast suspicion upon them. Therefore, the voluntary nature of these exercises is always in doubt. From the perspective of biological citizenship, we might read inclusion in the police database as the removal of choice and self-responsibility, as reducing individuals to the 'bare life' of being a data subject on a police database. We also see the implicit articulation of the boundaries of community – those included in the databases are rendered as 'statistical suspects' (Cole and Lynch 2006) in every single crime that takes place in the UK from which DNA evidence can be gathered. The boundaries are drawn not between the criminal and the non-criminal but between those who have had contact with the police as suspects or offenders and those who have not (as yet, it might be said). The British parliamentarian Jenny Willot (2008) remarked that these individuals are 'considered innocent under British law, but considered potentially guilty by the Home Office. By retaining that DNA, the state is saying: "Well, you might not have been convicted, but we think you may commit an offence in future and we want to make sure we can catch you when you do". That is not acceptable.'

The retention of profiles (and samples) not only of convicts and suspects but also of arrestees for relatively small ('any recordable') offences has a particular significance in relation to minority ethnic groups and the history of institutionalised racism within police forces in the UK. Civic society organisations such as Liberty have argued that racial discrimination in the criminal justice system will result in 'a disproportionate number of innocent people from ethnic minorities' to be included in the NDNAD in the UK (Liberty 2002). The BBC reported in 2007 that approximately 40% of black male Britons are included in the database as opposed to 9% of white men and 13% of Asian men (BBC News 2007). While these classifications might not be comparable to those used in the 2001 census, they are in stark contrast to the finding in that census that 2% of the persons who completed the ethnic group question on the census form elected to categorise themselves as Black or Black British in England and Wales, 4% assigning themselves to an Asian British category and 92.1% classifying themselves as white (Office of National Statistics 2008).

It might be thought naive to expect a database used for forensic purposes to be governed by the same arrangements as a database for medical research. It is usual for those convicted of criminal offences to be deprived of some human rights, proportional to the seriousness of the offence and, except in the most serious cases, for a limited period. Children who commit an offence are not treated in the same way as adults in the criminal justice system, so there is particular concern about stigmatisation and premature labelling of young offenders by their inclusion in the NDNAD. Other European databases retain profiles for a limited range of offences and differing time periods but do not include young children and do not use the profiles for research. Without doubt, in the UK, people arrested but never charged with an offence, or charged and later acquitted, may in practice be regarded with continued suspicion by police and the public. The NDNAD, as a consequence of its inclusion and retention policies, legitimises such suspicion and extends it to families through familial matching, as discussed in Chapter 2.

CONCLUSIONS

This chapter draws on three influential analyses concerned with developments in the biomedical arena to address the socio-technical configuration of the NDNAD in the UK and to explore the convergences and divergences in the ways in which forensic and biomedical databases have developed. Taking each of these analyses, in turn, it is argued that this police DNA database is amenable to being usefully analysed from each of these perspectives. As in the case of biomedical databases such as UK Biobank, certain 'futures' were imagined and promised by the investment in and subsequent development of the UK NDNAD, such as enhancing policing, speeding up and securing a greater number of convictions, and detering future offenders. Second, drawing on what has been written about the politics of legitimation of biobanks, the different trajectories followed by UK Biobank and the NDNAD can be clearly seen. Initially, the latter did not experience particular difficulties with its formation and acceptance. Now, as with large-scale genetic projects in the medical arena, it has needed to engage in a politics of legitimation by enrolling new epistemic communities (particularly from medicine) into its governance to deal with perceived public anxieties about the increasing numbers and categories – especially the innocent – of those registered on the database. From the perspective of recent discussions of 'biological citizenship', the policing context presents a counterpoint to the medical one. Genetic databases operate differently to medical ones and entail a form of citizenship that stands in opposition to the 'biological citizenship' celebrated by Rose and Novas.

So what should we conclude about the conceptual frameworks and tools that have been developed and debated by social scientists concerned with biomedicine and their applicability or explanatory power in relation to the policing context? One argument might be that it is inappropriate to carry over conceptual models from one arena to the other, because to do so transfers assumptions from medical databases to discussions about police forensic DNA databases, when they are not the same. There is merit to this argument. Equally, it can be said that as the NDNAD has expanded to include offenders of less-serious crimes and those only arrested and not charged or convicted, so it changed from being a forensic database to a 'policing' database, with greater social impact and of greater public concern. In this sense, perhaps, it is more open to analysis from a greater number of theoretical perspectives than might otherwise be the case had its rules of inclusion and governance been differently fashioned.

In summary, this chapter has sought to make a contribution to the question we posed at the start about how to theorise across contexts of opposing practices to identify the intersections and discontinuities that characterise the use of genetic databases in policing and biomedicine. The analyses discussed are useful for illuminating the different trajectories of genetic databases in these contexts and highlighting their similarities and differences. They provide innovative and productive insights into the forensic arena, highlighting the many social, ethical and regulatory issues entailed by the growth and governance of police DNA databases in many countries, not only in the UK. Looking through the lens of other conceptual frameworks such as 'biological citizenship' helps to throw into relief the different configurations that genetic technologies and their governance take in relation to opposing imperatives of compulsion and voluntary consent in the name of different outcomes – health and wealth and law and order.

REFERENCES

Anon. (2003). Editorial: Have the police hijacked our DNA? *Lancet*, 362, 927.
Anon. (2009). DNA of one-year-old baby stored on national database. *Telegraph*, 10 March http://www.telegraph.co.uk/news/newstopics/politics/4966168/DNA-of-one-year-old-baby-stored-on-national-database.html (accessed 12 August 2009).
BBC News (2007). All UK must be on DNA database. http://news.bbc.co.uk/2/hi/uk_news/6979138.stm (accessed 30 September 2008).
Brown, N., A. Kraft and Martin, P. (2006). The promissory pasts of blood stem cells. *BioSocieties*, 1, 329–348.
Brewer-Hartley, J. (1999). Straw's DNA plans are deeply flawed. *Guardian*, 31 July.

Busby, H. (2006). Biobanks, bioethics and concepts of donated blood in the UK. *Sociology of Health and Illness*, 28, 850–865.

Children off National DNA Database http://www.cond.org.uk/ (accessed 23 March 2009).

Cole, S.A. and Lynch, M. (2006). The social and legal construction of suspects. *Annual Review of Law and Social Science*, 2, 39–60.

Cragg Ross Dawson (2000). *Public Perceptions of the Collection of Human Biological Samples*. Lancaster, UK: Centre for the Study of Environmental Change.

European Court of Human Rights (2008). *Applications 30562/04 and 30566/04*. Strasbourg: Council of Europe http://www.bailii.org/eu/cases/ECHR/2008/1581.html (accessed 12 August 2009).

Fitzgerald, R. (2008). Biological citizenship at the periphery: parenting children with genetic disorders. [In *Life Sciences Governance: Civic Transitions and Trajectories*, eds. R. Hindmarsh and R. Du Plessis.] *New Genetics and Society*, 27 (Special Issue), 251–266.

GeneWatch U K (2006). *Using the Police National DNA Database: Under Adequate Control?* Buxton, UK: GeneWatch UK.

GeneWatch U K (2007). *Nuffield DNA Consultation*. Buxton, UK: GeneWatch UK http://www.genewatch.org/sub-548276 (accessed 23 February 2010).

GeneWatch U K (2008). *Would 114 murderers Have Walked Away if Innocent People's Records were Removed from the National DNA Database?* Buxton, UK: GeneWatch UK http://www.genewatch.org/uploads/f03c6d66a9b354535738483c1c3d49e4/brown.pdf (accessed 6 October 2008).

Hacking, I (1986). Making up people. In *Reconstructing Individualism: Autonomy, Individuality, and the Self in Western Thought*, eds. T. Heller, M. Sosna and D. Wellbery. Stanford, CT: Stanford University Press.

Hedgecoe, A and Martin, P. (2003). The drugs don't work. *Social Studies of Science*, 33, 327–364.

Hillier, M. (2008). Under-Secretary in the Home Office in reply to Gordon Prentice, MP. *Hansard*, 21 April, Column 1029 http://www.publications.parliament.uk/pa/cm200708/cmhansrd/cm080421/debtext/80421-0001.htm (accessed 6 October 2008).

Home Office (2005a). *DNA Expansion Programme 2000–2005: Reporting Achievement*. London: The Stationery Office http://police.homeoffice.gov.uk/publications/operational-policing/DNAExpansion.pdf (accessed 6 October 2008).

Home Office (2005b). *The National DNA Database Annual Report 2004–2005*. London: The Stationery Office http://www.homeoffice.gov.uk/documents/NDNAD_AR_04_0512835.pdf?view=Binary (accessed 3 March 2010).

Home Office (2008). *Using Science to Fight Crime: The National DNA Database*. London: The Stationery Office http://www.homeoffice.gov.uk/science-research/using-science/dna-database/ (accessed 6 October 2008).

House of Lords Science and Technology Committee (2001). *Human Genetic Databases: Challenges and Opportunities*. [Fourth Report Session 2000–01.] London: The Stationery Office http://www.publications.parliament.uk/pa/ld200001/ldselect/ldsctech/57/5701.htm (accessed 6 October 2008).

Human Genetics Commission (2000). *Public Attitudes to Genetic Information. People's Panel Quantitative Study conducted for the Human Genetics Commission*. London: Human Genetics Commission http://www.hgc.gov.uk/UploadDocs/DocPub/Document/morigeneticattitudes.pdf (accessed 6 October 2008).

Human Genetics Commission (2002). *Inside information: Balancing Interests in the Use of Personal Genetic Data*.London: Human Genetics Commission.

Human Genetics Commission (2008). *A Citizen's Inquiry into the Forensic use of DNA and the National DNA Database*. Blackburn, UK: Vis-à-Vis Research Consultancy http://

www.genomicsnetwork.ac.uk/media/citizens%27_inquiry_   mainfindings.pdf (accessed 6 October 2008).

Kaye, J. and Martin, P. (2001). Safeguards for research using large scale DNA collections. *British Medical Journal*, 321, 1146–1149.

Levitt, M. (2007). Forensic databases: benefits and ethical and social costs. *British Medical Bulletin*, 83, 235–248.

Levitt, M. and Tomasini, F. (2006). Bar-coded children: an exploration of issues around the inclusion of children on the England and Wales National DNA database. *Genomics Society and Policy*, 2, 41–56.

Levitt, M. and Weldon, S. (2005). A well placed trust? Public perceptions of the governance of DNA databases. *Critical Public Health*, 15, 311–321.

Liberty (2002). *Databasing the DNA of Innocent People: Why it Offers Problems not Solutions.* [Press release.] London: Liberty http://www.liberty-human-rights.org.uk/news-and-events/1-press-releases/2002/databasing-the-dna-of-innocent-people-why-it.shtml (accessed 6 October 2008).

Lynch, M. and McNally, R. (2009).DNA, biolegality, and changing conceptions of suspects. In *Genetics and Society: Mapping the New Genetic Era*, eds. P. Atkinson P. Glasner and M. Locke. London: Routledge, pp. 283–301.

Margo, J. and Stevens, A. (2008). *Make Me a Criminal: Preventing Youth Crime.* London: Institute for Public Policy Research, pp. 1–92 http://www.ippr.org.uk/publicationsandreports/publication.asp?id=587 (accessed 30 March 2009).

Moreira, T. and Palladino, P. (2005). Between truth and hope: on Parkinson's disease, neurotransplantation and the production of the 'self'. *History of the Human Sciences*, 18, 55–82.

Nightingale, P. and Martin, P. (2004). The myth of the biotech revolution. *TRENDS in Biotechnology*, 22, 564–569.

Novas, C. and Rose, N. (2000). Genetic risk and the birth of the somatic individual. *Economy and Society*, 29, 485–513.

Nuffield Council on Bioethics (2007). *The Forensic Use of Bioinformation: Ethical Issues.* London: Nuffield Council on Bioethics www.nuffiedldbioethics.org/go/our-work/bioinformationuse/publication_441.html (accessed January 2009).

Office of National Statistics (2008). *Ethnicity: Population Size.* London: Office of National Statistics http://www.statistics.gov.uk/CCI/nugget.asp?ID=273&Pos=4&ColRank=2&Rank=1000 (accessed 6 October 2008).

Parliamentary Office of Science and Technology (2006). The national DNA database, *Postnote*, 258, 1–4.

People, Science and Policy (2002). *BioBank UK: A Question of Trust: A Consultation Exploring and Addressing Questions of Public Trust.* [Report prepared for the Medical Research Council and Wellcome Trust.] London: People Science and Policy.

Petersen, A. (2006). Securing our genetic health: Engendering trust in UK Biobank, *Sociology of Health and Illness*, 27, 271–292.

Petryna, A. (2002). *Life Exposed: Biological Citizens After Chernobyl.* Princeton, NJ: Princeton University Press.

Rabinow, P. (1996). Artificiality and enlightenment: from sociobiology to biosociality. *Essays on the Anthropology of Reason*, Ch.5. Princeton, NJ: Princeton University Press.

Rose, N. and Novas, C. (2005). Biological citizenship. In *Global Anthropology*, eds. O. Aihwa and S. Collier. Oxford: Blackwell.

Salter, B. and Jones, M. (2002). Regulating human genetics: the changing politics of biotechnology governance in the European Union. *Health Risk and Society*, 4, 325–340.

Salter, B. and Jones, M. (2005). Biobanks and bioethics: the politics of legitima-tion, *European Journal of Public Policy*, 12, 710–732.

Tutton, R. (2007). Banking expectations: the promises and problems of biobanks. *Personalized Medicine*, 4, 463–469.

Tutton, R. and Corrigan, O. (eds.) (2004). *Genetic Databases: Socio-ethical Issues in the Collection and Use of DNA*. London: Routledge.

UK Biobank Ethics Consultation Workshop (2002). *Report for Wellcome Trust and Department of Health*. London: Wellcome Trust http://www.ukbiobank.ac.uk/docs/ethics_work.pdf (accessed 6 October 2008).

UK National DNA Database (2008). *First Annual Report of the Ethics Group: National DNA Database April 2008*. London: The Stationery Office http://police.homeoffice.gov.uk/publications/operational-policing/NDNAD_Ethics_ Group_Annual_Report?view=Binary (accessed 6 October 2008).

Waldby, C (2002).Stem cells, tissue cultures and the production of biovalue. *Health*, 6, 305–323.

Weldon, S. (2004). Public consent or 'scientific citizenship?; In *Genetic Databases: Socio-ethical Issues in the Collection and Use of DNA*, eds. R. Tutton and O. Corrigan. London: Routledge.

Williams, G. (2005). Bioethics and large-scale biobanking: individualistic ethics and collective projects, *Genomics, Society and Policy*, 1, 50–66.

Williams, R. and Johnson, P. (2008). *Genetic Policing: The Use of DNA in Criminal Investigations*. Cullompton, UK: Willan.

Willot, J. (2008). DNA database (removal of samples). *Hansard*,11 June 2008, Column 310 http://www.publications.parliament.uk/pa/cm200708/cmhansrd/cm080611/debtext/80611-0004.htm (accessed 6 October 2008).

CASE

S and Marper v. the United Kingdom (2008). A summary of the judgment is avail-able at http://cmiskp.echr.coe.int/tkp197/view.asp?action=html&documentId=843937&portal=hbkm&source=externalbydocnumber&table=F69A27FD8 FB8 6142BF01C1166DEA398649 (accessed January 2009).

6

DNA profiling versus fingerprint evidence: more of the same?

INTRODUCTION

While other chapters in this book address the role of national political cultures in the governance of DNA databases, in this chapter we examine how other forensic systems have provided models for the organisation of such databases. We will argue that many, but by no means all, aspects of DNA profiling followed patterns established historically by earlier techniques. In particular, we focus on fingerprint identification, the technique we view as most closely analogous to DNA profiling on several levels. Both fingerprinting and DNA profiling seek to identify particular bodies as sources of crime scene traces by examining correspondences between those traces and reference samples taken from persons in police custody. Both techniques proved useful enough from a social control perspective to warrant large government investments in developing databases of records indexed according to bodily markers. Finally, both fingerprinting and DNA profiling have enjoyed primacy as 'gold standards' in an imagined hierarchy of forensic techniques. Indeed, even today when DNA profiling is sometimes viewed as having supplanted fingerprinting, one recent report has noted that, 'the more humble fingerprint retains its status as the most commonly used method of identification and is a cornerstone of forensic crime scene investigation' (Nuffield Council on Bioethics 2007: 15).

Profiling with DNA is widely heralded as a novel and distinctively *scientific* technique for analysing criminal evidence that is having revolutionary impact on criminal justice systems throughout the world. In popular media, exemplified by American television programmes such as *Crime Scene Investigation* (*CSI*), a whole array of forensic sciences rides the coat-tails of DNA to claim unquestioned integrity and certainty. In more scholarly legal and forensic writings, DNA profiling is treated as

Genetic Suspects: Global Governance of Forensic DNA Profiling and Databasing, ed. Richard Hindmarsh and Barbara Prainsack. Published by Cambridge University Press. Copyright © Cambridge University Press 2010.

exceptional, and its factual status is often contrasted with the doubtful credibility of forensic comparison evidence (such as hair, fibre and bite-mark analysis) as well as ordinary forms of legal evidence such as eyewitness testimony and confessions. Indeed, DNA evidence is often cited as a *test* of other forms of evidence – a factual baseline against which to calculate error rates (Saks and Koehler 2005) – and a source of legal leverage with which to overturn old convictions (Scheck *et al.* 2000).

This was not always so. In fact, not so long ago, in the late 1980s and early 1990s, DNA 'fingerprinting' (as it often was called at the time) was considered less certain than its namesake, latent fingerprint identification. Overwhelmingly, in the criminal courts, fingerprint evidence was accepted without question. Defendants often confessed when confronted with fingerprint evidence, and police organisations amassed extensive files (what would now be called databases) of fingerprints collected from suspects, convicts and crime scenes. Courts in the USA routinely accepted fingerprint examiners' reports as *facts* rather than expert *opinions* (Mnookin 2001). Until recently, practitioners in the UK and in many other countries required a given number of points (corresponding ridge characteristics) in order to declare a fingerprint match in court. However, points have no clear relation to probabilities, and latent print examiners in North America, and later the UK, moved away from that system. Consequently, examiners could simply declare that, for example, a suspect's finger made the latent prints lifted from the door of a getaway car – they did not, and were not required to, qualify their declarative judgments with probability figures. Although the international fingerprint community made strenuous efforts to secure expert 'scientific' status in the eyes of the law, no extensive knowledge of statistics, population genetics or laboratory technique was, or is, deployed in routine fingerprint practice (Grieve 1990). Nevertheless, early in the twentieth century, fingerprinting became established as a highly credible, and rarely questioned, forensic method worldwide (Cole 2001).

When DNA profiling techniques were first introduced into criminal investigations in the UK in 1986, and shortly afterwards in the USA, Australia and other nations, proponents drew an explicit analogy with the established and highly credible practice of fingerprinting. One way in which this connection was drawn was in grounding claims of the supposed 'uniqueness' of the combination of biological markers in a profile solely on a visual appearance of great complexity and variability. Sir Alec Jeffries, the inventor of DNA profiling, claimed that the earliest method used in the UK (the multilocus probe (MLP) technique) could 'produce somatically stable DNA "fingerprints" which are completely specific to an individual (or to his or her identical twin)...' (Jeffreys et al. 1985). Similar statements were made in the USA by forensic scientists testifying for the prosecution in early cases involving DNA evidence (Aronson 2007: 46).

Despite such claims, there seemed good reason at the time to treat DNA profiling as less powerful than fingerprinting. Critics of the new technique highlighted key differences: DNA matches were probabilistic, not absolute, identifiers, and the probability estimates were based on questionable genetic and statistical assumptions; the collection and analysis of DNA samples was novel, complicated and error prone, and not at all like the fingerprint examiner's craft; and interpreting DNA evidence was not simply a pattern-matching operation. At first, such differences seemed to lower the credibility of DNA evidence, relative to fingerprint evidence, but within a few years an 'inversion of credibility' occurred (Lynch 2004; Lynch *et al.* 2008). The resolution of controversies over the above issues was cited as evidence that DNA had been more thoroughly vetted than fingerprinting ever had been. Probabilistic error rates, the calculation of which had been a topic of fractious argument and an apparent source of disadvantage in comparison with the non-quantified judgments given by latent fingerprint examiners, became a source of strength – a measure of precision and a criterion that some critics cited to distinguish 'scientific' from 'non-scientific' evidence. Not only did DNA gain credibility in courts and within science (often, the acronym 'DNA' was used as a shorthand reference for a changing array of techniques that sampled and ordered selected fragments of DNA for identification purposes), it provided a basis for invidious comparisons with other forms of evidence, especially fingerprinting.

The leverage provided by the ascendancy of DNA profiling has spurred efforts to evaluate and reform forensic science (Saks and Koehler 2005). Post-conviction DNA testing has led to the exoneration of more than 200 inmates in the USA, including more than a dozen who had been on death row, thus giving new force to criticisms of the death penalty (Rosen 2003). For many of those exonerations, forensic evidence had been implicated in their wrongful convictions (Garrett and Neufeld 2009). In addition, because of the fact that current DNA profiling techniques are readily digitised and stored on searchable databases, DNA profiling has been the basis for the initiation and rapid growth of large national databases throughout Europe, the USA, Australia and elsewhere.

While we applaud many of these developments, we also notice that excited talk of the novelty, precision and certainty afforded by DNA evidence tends to obscure the extent to which uses and interpretations of the new techniques are continuous with those of earlier modes of criminal identification and older archives of individual biometric information. We explore such connections in this chapter because they offer reminders of earlier hopes and investments in 'sciences' of individual identification; hopes and investments that were disappointed or downgraded when the systems of governance in which they were embedded penetrated the laboratory and the archive to infuse 'data' with the judgments, skewed attentions and bureaucratic prerogatives of criminal justice administration.

As social control technologies subject to error and abuse, criminal databases are themselves the means as well as the objects of governance. Older systems of bioinformation provide historical antecedents, and to some extent legal precedents, for examining how DNA databases act as tools of governance that themselves require governing. The relationship between older and newer forensic systems is dynamic and interactive, as fingerprinting and fingerprint databases have changed in recent years, in part through a reaction to the model and challenge offered by DNA. An examination of such historical interactions gives us reason to question the extent to which 'DNA' is an essentially different, scientific, and even infallible, technique (Thompson 2008).

Particularly in the UK and USA, the argument that DNA profiling is infallible has been used to justify seemingly inexorable expansions of so-called 'criminal' databases. We argue that fingerprint evidence also was, and to some extent still is, heralded as infallible, and that its presumed objective status influenced the way it was (or was not) subjected to state regulation. In part because of the ascendancy of DNA profiling, fingerprinting has undergone more critical scrutiny in recent years, with some tentative moves undertaken toward regulating examiners' practices and judgments. If, as we argue, DNA evidence is bound to the history and practices of criminal justice, then we should be less eager to let go of the protections and restrictions that have developed through that history in so-called liberal democracies.

HISTORY OF FINGERPRINT DATABASES

Although criminal records, in the form of registers of criminals' names, have existed since the Middle Ages and perhaps even earlier (Groebner

2007), *archives* of criminal records, or in contemporary terms *databases*, must be counted as a nineteenth century development. These archives attempted to type, classify and index bodily traits and traces in a way that enabled identification and re-identification of the same individuals (Sekula 1986). They were associated with the rise in governmentality – the biopolitical disciplines for discovering the nature of body and population, together with 'scientific' regimes of administration and self-management – described by Foucault (1991) and others (Lemke 2001). Criminal justice officials compiled lists of names, registers of physical descriptions (features such as tattoos and other distinctive marks) and collections of photographs. As Sekula (1986) has noted, the Parisian police official Alphonse Bertillon introduced what we now recognise as the crucial feature of the modern criminal identification database: a means of *indexing* records according to somatic measurements and marks, rather than testimonial information such as names and addresses, or actuarial information on age, national origin, gender, and so on. This, in turn, enabled the *retrieval* of information from the archive, instead of depending upon the veracity of reported information.

Bertillon's records contained three primary sources of information: (1) physical descriptions, such as ear, nose and lip types; (2) 'peculiar marks', such as a birthmarks, scars and tattoos; and (3) anthropometric measurements of 'bony lengths' of the prisoner's body. This last measure formed the system for indexing records. Bertillon's achievement lay not so much in the somatic markers he chose but rather in how he conceptualised the information. Easily quantified anthropometric measurements provided a way to narrow a search to a small number of records. This approach would soon be adapted for a different somatic marker, more familiar to us today: the fingerprint. In late nineteenth century colonial India, British Government officials had begun collecting impressions of individuals' fingertips – fingerprints – using ink. Without an indexing system, however, officials could do little more than compare an individual's fingerprints with a print on a record that was presumed to derive from the same person. If an individual's identity was unknown or doubted, there was little alternative to flipping through the entire archive in search of a corresponding record (if there was one). Adapting Bertillon's approach, police officials in India and Argentina nearly simultaneously devised methods for ordering fingerprint records according to the pattern types of the ten fingertips. Like Bertillon, therefore, they devised an entirely somatic criminal identification database, free of reliance on self-reported information.

Many of the earliest databases were not for criminal identification, but for civil purposes. The earliest fingerprint archive in India was for pensioners. Subsequent archives held the fingerprint records of contractors, labourers, individuals sitting examinations and medical records (Sengoopta 2002: 151–154; Singha 2008). Fingerprints were used to track immigrants in the USA and labourers in South Africa (Breckenridge 2008). Eventually, however, criminal identification databases emerged as the dominant application of fingerprints. By the 1920s, with even the French, the staunchest advocates of anthropometry, having acceded to fingerprints, most of the world's criminal justice systems had established criminal identification databases based on fingerprints (Cole 2001).

Role of race/diagnostics

Fingerprints appealed to colonial authorities in India who complained of the difficulty of distinguishing natives from one another. The bodily 'signatures' furnished by fingerprints seemed ideal for recording the identities and movements of illiterates, whom the authorities deemed 'cunning and deceptive'. Though race was salient from the outset, fingerprint databases also were explored with a more precise interest in the subject. No sooner did the fingerprint emerge as a marker of somatic identity than researchers immediately turned their attention to using it to index heredity, race, ethnicity, gender and propensity for disease and criminality. Fingerprint samples were taken from different ethnic and tribal groups around the world, from prisons, asylums and from epileptics. Racial differences in the frequency of gross pattern types were identified, and these differences led to proposed evolutionary hierarchies of pattern types in which conflicting theories arose about which pattern type was 'most' evolved. Fingerprint patterns were used occasionally in paternity cases, and researchers speculated that insurance companies would mine fingerprint patterns for actuarial data (Cole 2001: 110).

This diagnostic research, however, though it never completely died out, was largely pushed to the margins, as databases expanded and the idea became established that the fingerprint was solely a mark of individual, not group, identity. Race was still 'in' databases, as in the case of large databases that, remarkably, were subdivided by race (Cole 2007a). But in these cases 'race' was purely a visual assessment; no attempt was made to correlate visually perceived race with, say, frequencies of fingerprint pattern types.

Expansion of fingerprint databases

As the twentieth century progressed, criminal identification databases expanded as populations grew and ever-pettier crimes were designated as warranting the creation of a fingerprint record. An advantage of fingerprinting over anthropometry (the Bertillon system described above) would have always been its lower front-end costs. Taking fingerprints requires little training whereas recording anthropometric measurements required extensive training and discipline, and mandating the creation of fingerprint records or a print taken from a suspect for petty crime was a relatively low cost anti-crime measure. The expansion of such databases to pettier crimes (what today would be called high-volume offences) had an important implication in that it classified ever-larger portions of the population not only as criminals but also as recidivists. Expanding criminal identification databases allowed governments to detect repeat offences at much higher rates, thus, greatly increasing the number of people labelled repeat offenders and justifying more severe punishment for them. This labelling fell disproportionately on those with convictions for petty crimes (Cole 2001: 153–159).

In addition, civil uses of fingerprint databases continued to grow. In the USA, fingerprints were taken from military personnel, civil servants, immigrants and employees of New Deal organisations such as the Works Progress Administration (Cole 2001: 247). These expansions necessitated the development of ever-more elaborate schemes for indexing larger collections of fingerprints according to ridge patterns (whorls, loops, etc.).

Inclusion of individuals in fingerprint databases was viewed as relatively benign from a privacy standpoint. Courts in the USA, for example, have largely treated the taking of fingerprints from persons placed under arrest as routine and unremarkable (Kaye 2001: 485–489). They have declined to view such actions as invasions of personal privacy, and in 1969, the US Supreme Court ruled that taking fingerprints did not violate Fourth Amendment protection against unreasonable search and seizure (Kaye 2006). Indeed, US courts have generally ruled against requests to expunge fingerprint records of individuals who were arrested but acquitted or never convicted, reasoning that the government's interest in 'promoting effective law enforcement' outweighed any violation of privacy. A few courts, however, have taken the opposite position, ruling that failure to expunge such records constituted a privacy violation (Spivey 2005). The reputed uniqueness and objectivity of fingerprints seemed to exempt examiners' practices and judgments from the critical scrutiny and concerns about prejudice that attend other forms of evidence.

There were, however, limits to fingerprint database expansion. Schemes for so-called 'universal' fingerprint databases, advocated most enthusiastically by Argentine fingerprint pioneers, were not successful. In Argentina in the 1910s, efforts to develop the world's first universal fingerprint database were deemed unconstitutional amid great public hostility to the plan (Ruggiero 2001: 193; Rodriguez 2004). In the USA, three bills to create universal fingerprint databases were defeated in the late 1930s and early 1940s, despite the support of J. Edgar Hoover and the Federal Bureau of Investigation (FBI) (American Civil Liberties Union 1938; Cole 2001: 249). Even today, only a few nations have pushed forward with efforts to include non-suspect citizens in databases, with Chile, Malaysia and South Africa perhaps furthest along (Breckenridge 2008). Therefore, generally speaking, fingerprint databases were subject to relatively light governance, with the important proviso that they did not implicate most non-suspect citizens. Fingerprint databases exemplified the notion that archiving somatic information from individuals convicted of, or even arrested for, crimes would be relatively uncontroversial, but that the archiving of such data from non-suspect citizens would be subject to heightened scrutiny. Decades later, this precedent had important implications for the development of genetic databases.

FORENSIC FINGERPRINTING

Of course, fingerprints were not used solely for databases. In the late nineteenth century, after police in many nations had begun to archive fingerprint records, they realised a very different kind of fingerprint could be used – one left accidentally at a crime scene, rather than one taken deliberately from an individual in custody. Latent prints, as they usually were called, could be compared with fingerprint records to investigate particular crimes. Fingerprints thus became something quite different from identifying records in an archival index: they became *forensic evidence*.

Once again, authorities in the USA, UK and other nations exercised a light touch in regulating how fingerprint evidence was used. Few standards were developed for training and providing credentials for examiners, collecting and comparing evidence or characterising the value of the evidence. Although fingerprint examiners formed professional associations, the American organisation, the International Association for Identification (IAI), for example, decried its inability to police the profession given the ability of the courts to certify their own experts. Indeed, in the USA the profession called upon the state to regulate the practice. This did not occur, and by the turn of the twenty-first century it was noted that crime laboratories were subject to less government regulation than similar scientific laboratories (Jonakait 1991; Giannelli 1997).

Courts, likewise, exercised little control over forensic experts or evidence. They sometimes allowed experts to testify even when deemed unqualified by their peers (Cole 2001: 210–211). The courts also did not regulate the assessment of the evidence; instead, they allowed fingerprint experts to testify with unqualified certainty that the defendant *was* the source of particular latent print. Courts, at that time, did not conceive of forensic evidence as inherently probabilistic and did not require it to be described in a probabilistic manner. At the same time, the special epistemic status of fingerprint evidence seemed to exempt it from regulation. It was as though examiners' conclusions about matching evidence were mere vehicles for 'facts' that spoke for themselves. This association between objective status and (lack of) regulation eventually applied to 'DNA fingerprinting' as well, although with some significant differences: courts required DNA evidence to be presented in probabilistic form, and it was subject to contentious disputes that were resolved through a series of technical, legal and administrative changes (Lynch *et al.* 2008: 157ff.).

GOVERNING DNA

The following discussion will draw an analytic distinction between the governance of DNA databases (focusing on such issues as the scope of inclusion in such databases) and the governance of forensic conclusions (focusing on such issues as the policing of how DNA 'matches' should be characterised by an expert witness to a fact finder, such as a judge or jury). In both cases, it will be argued that the relatively light governance of fingerprinting has shaped the governance of DNA profiling. However, DNA did not just follow along the lines set by fingerprinting, it also established a model that fed back into evaluations and proposals to standardise the older practice.

Governing forensic DNA profiling

The development of DNA profiling techniques raised the question of how DNA 'matches' were to be declared and communicated in courts of law. As noted above, profiles developed with the early MLP technique were likened to fingerprints, and were presumed unique to individuals. This presumption made some sense because of the large number of bands arrayed in an MLP profile (representative of numerous genomic loci in which a given sequence of base pairs occurred). However, the very complexity of the bar-code-like pattern maximised the likelihood of artefacts and mismatched bands, which would need to be explained or explained away, and no precise way existed to quantify match probabilities.

Another technique, known as single-locus probe (SLP), initially adopted in the USA and later in the UK, became the standard for DNA profiling in the late 1980s. Profiles obtained in this way were more precisely quantified than MLP profiles, because they targeted discrete loci with alleles (alternative variants of a gene) whose frequency in human populations could be estimated. The relatively small number of loci in early SLP profiles also meant the technique was *less* discriminating. Analysts could not simply assume the uniqueness of the patterns produced by SLP techniques. Instead, courts required that they report the results of a 'two-step process': a finding of consistency (a 'match') followed by a calculation of the estimated probability of finding the particular combination of loci within a specified population (Aitken 2005). A common expression for this is the random match probability (RMP): the probability that a person chosen at random from the relevant population group or subgroup (e.g. northern European Caucasian men) would have the specific combination of alleles in a given DNA profile. However, just how to calculate this number led to extensive debates about the assumptions that supported such calculations.

Debates in the courts and in scientific journals led to advisory commission efforts to specify rules for governing how evidence matches should be reported. The case of *People* v. *Castro* (1989), which was among the first in which defence attorneys retained qualified experts capable of scrutinising the processes used by the experts retained by the government (many of whom were academics or employees of private laboratories), demonstrated that DNA laboratories were deviating from their own rules governing findings of consistency (Lander 1992; Mnookin 2006; Aronson 2007). The ensuing controversy led to efforts to generate rules governing findings of consistency and estimates of rarity, such as match windows, the product rule and the ceiling principle. These rules were variously promulgated by ad hoc agreements among experts (in *Castro*), legal rulings and, especially, in two reports issued by the US National Research Council (NRC) (1992, 1996). The NRC reports, in particular, appeared to lay down clear guidelines for calculating and declaring match probabilities: rules vouched for by the authority of a prestigious scientific institution. However, as we shall see; both the supposed clarity of these rules and their authority to govern would turn out to be at least partially illusory.

Because contemporary DNA profiles are the output of a largely automated process and the calculation of RMPs seems straightforward and transparent (governed by formulae), one might assume that the process of generating conclusions for 'public' (i.e. legal fact finder) consumption would be far more governable than, say, fingerprinting, in which the 'instrument' is a human examiner and RMPs are not available. Instead, the process of generating conclusions about forensic DNA evidence continues to be unruly and shaped by long-standing practices in forensic science. For example, efforts to generate rules governing the interpretation of DNA evidence – such as the insistence that examination of the unknown sample be completed before proceeding to examine the known one, in order to avoid biasing the examination – have gained momentum among experts only recently and have yet to penetrate routine practice (Krane *et al.* 2008).

Forensic DNA testimony thus continues to be susceptible to issues concerning exaggeration of the probative value of the evidence; a problem also found in fingerprinting and other forensic assays, with their common claims to have excluded all other possible donors in the universe (Cole 2007b). For example, the 'prosecutor's fallacy' – a common logical fallacy that was identified early in the history of forensic DNA profiling (Lynch *et al.* 2008) – continues to be salient today. This fallacy (sometimes called the fallacy of the transposed conditional) occurs when the probability of the evidence given a specific *hypothesis* – that the defendant is innocent – is confused with the probability of the hypothesis itself – the defendant *is* innocent. For example, a finding that evidence from a defendant is consistent with a profile that is expected to appear in 1 out of every 10 000 people in the population is erroneously characterised as a finding that the likelihood that the defendant is *not* the source of the profile is only 1 in 10 000 (Thompson and Schumann 1987).

A recent report by the Nuffield Council devotes a great deal of attention to the prosecutor's fallacy, and it advises UK courts to follow the appeal court ruling in R v. *Doheny and Adams* (1997), which found the prosecutor's fallacy rendered the trial unfair (Nuffield Council on Bioethics 2007: xix). In the USA, however, the prosecutor's fallacy still appears common. A recent case marked the first time that a US court freed a convict because the prosecutor's fallacy misled the jury as to the probability of innocence, though that outcome may now be reversed (*McDaniel* v. *Brown* 2010). But, as William Thompson notes (quoted in Rubin and Felch 2008), the singularity of the case probably speaks less to the rarity of the prosecutor's fallacy than it does to the rarity of defence attorneys' ability to challenge it successfully.

Prosecutors in the USA still attempt to present the strongest probabilities available, despite the NRC's recommendations to err on the side of caution (National Research Council 1992, 1996), and contrary to the practice in the UK of reporting a figure of one in a billion whenever calculations indicate a smaller RMP (e.g. one in five trillion). Beginning around 2005, a series of cases adjudicated how the results of database trawls should be presented. In those cases, the defendants cited a disagreement among statisticians over how the RMP of a DNA 'match' generated from a cold hit (a match generated through database searches with, as yet, no corroborating evidence) should be adjusted to take into account the number of profiles on the database relative to the population at large (Balding 1997; Donnelly and Friedman 1999; Devlin 2006). In these cases, the US courts concluded that the disagreements among the statisticians concerned issues of interpretation, not science, and they permitted the smallest (that is, weightiest) probabilities to be presented (*United States* v. *Jenkins* 2005; *People* v. *Nelson* 2008). In another case, the court drew an explicit analogy with fingerprinting, reasoning that a DNA database trawl was no different from one in which a suspect was identified through a cold search of a fingerprint database. In either case, the court argued, the database is simply an investigative tool; the 'real' evidence is generated when the suspect's profile is compared with the crime profile: 'The fact [that the] appellant was first identified as a possible suspect based on a database search simply does not matter' (*People* v. *Johnson* 2006).

The court ruling in *People* v. *Johnson* ignored the argument made by statisticians - including statisticians who otherwise disagree with one another about how to deal with cold hits - that the means through which the evidence is produced *must* matter in evaluating the probability of the evidence: the investigator or fact finder cannot simply 'forget' how the evidence was generated.

Moreover, the governance of DNA evidence might reverberate back onto fingerprinting. Defendants could argue that the cold database search issue also pertains to fingerprint cold hits. In one particularly notorious case, a latent print from the 2004 Madrid terrorist bombing was widely searched for in fingerprint databases worldwide. The US FBI eventually attributed the print to an Oregon attorney named Brandon Mayfield and imprisoned him for 10 days as a material witness. There was no evidence that Mayfield had left the country or communicated with anyone linked to the bombing, but there was circumstantial evidence that, when interpreted in light of the latent print match, seemed to make him suspicious (for instance, he had converted to Islam and had represented Muslim defendants). Mayfield was eventually exonerated and the latent print attribution was deemed erroneous (Cole and Lynch 2006).

Some propose doing away with troublesome probabilities altogether and simply making what are called 'source attributions', statements that the defendant *is* the source of a particular crime scene trace without any reference to probabilities (Budowle *et al.* 2000). Such proposals, which have generated controversy among DNA experts (Buckleton 2005), clearly draw on fingerprint practice to simply assert in testimony that the defendant *is* the source of the crime scene trace, rather than inviting all the messy contingencies attendant upon seeking to ask lay people to comprehend probabilistic evidence and to delve into possible sources of error, and even fraud. It illustrates how expert testimony about DNA profile matches can be modelled after fingerprint evidence (thus affirming the conventional way the latter has been presented in court), although certainly there is a significant difference between making source attributions in which extremely low calculated probabilities have been 'black-boxed' (calculated, but no longer expressed explicitly – as in DNA profiling) and making source attributions for which probability calculations simply are not made (as in latent print identification).

Efforts to 'black-box' the probabilities underlying the presentation of DNA evidence, by glossing over seemingly academic differences of interpretation and rounding off low RMPs to zero and then characterizing them as 'source attributions', depend on the assumption that the data underlying the generation of probabilities are beyond dispute. (And this is to leave aside further complications having to do with the, usually unmeasured, probability of practical error in the handling, analysis and interpretation of DNA samples.) However, another recent development threatens to destabilise even that issue with a dispute

over the frequency estimates and assumptions of statistical independence that underlie the probability debates. In 2004, Bicka Barlow, a
San Francisco defence attorney uncovered a report of a search of the
Arizona state database that revealed an unexpectedly high number of
coincidental matches between samples at 9 or more of the 13 loci in
the profiles. Coincidental matches between so many loci should be
exceedingly rare (on the order of, say, 1 in 100 trillion), and the
Arizona database at the time was relatively small, numbering around
65 000. The FBI spokespersons disputed the significance of these findings, by noting that the Arizona search compared all 65 000 profiles
with each other – a procedure that differs from running a single profile
(e.g. from a crime scene investigation) against the 65 000 on the database. The first procedure produces a huge matrix of comparisons
(around two billion), thus compounding the possibility of finding
coincidental matches (Felch and Dolan 2008). Still, in the Arizona
case, and in later runs on other state databases, the number of coincidental matches was high enough to at least raise doubts about long-
standing assumptions about the statistical independence of different
alleles, though they have not, as yet, definitively led to refutation of
those assumptions (Mueller 2008).

    In the Arizona case, debates about the governance of genetic
databases reverberated into debates about the governance of forensic
conclusions. The statistical independence issue can be resolved specifically by determining the degree of blood relatedness of the individuals
whose profiles compose the Arizona database, and more generally by
making the anonymised profiles contained in large databases available
for research by non-government scientists (Murphy 2007: 783; Mueller
2008; Thompson 2008). However, the FBI reprimanded the Arizona
laboratory for releasing the data. In Illinois and Maryland, where, in
contrast to Arizona, the courts ordered searches of state databases for
coincidental matches at high numbers of loci, the FBI threatened to
disconnect the state laboratories from CODIS, the national DNA database, if they complied with those court orders (Felch and Dolan 2008).
Paradoxically, the FBI invoked the privacy rights of convicted criminals, who (still mostly) make up the database, in order to justify their
threat. Thus, controversies over the governance of databases impelled
the FBI to take uncharacteristic positions, such as defending criminals'
privacy rights, in order to keep the lid on the black-box of forensic
database matches.

    Just as governing databases appears to depend on unproblematic
characterisations of matching forensic evidence, technological change

threatens to destabilise the black-box. Since the mid 1990s, it has become commonplace to treat DNA evidence as a gold standard, and to cite the impressive probability estimates associated with current techniques to support claims that coincidental matches are essentially impossible, except in cases of identical twins and very close relatives. However, new techniques such as mitochondrial DNA analysis and low-copy number DNA analysis pose further challenges for interpreting matches. Low-copy number DNA analysis is a method for developing DNA profiles from samples containing miniscule amounts of DNA-containing cellular material. At present, this technique has been refused by courts in some cases (*The Queen* v. *Sean Hoey* 2007). Even with more settled techniques, such as the CODIS system with its 13 loci, or the current multiplex STR system used for the national DNA Database (NDNAD) of England and Wales, scrutiny in specific cases often shows that the probabilities associated with those systems are based on idealized estimates. Crime samples often are less than perfect. Crime scene profiles often are partial and samples sometimes are mixed. In the near future, such cases will doubtless present further demands for the governance of conclusions – the standards and regulations applying to evidential reports, including match probabilities, attributions of source and expressions of certainty or doubt.

Reverberations for fingerprinting

At the same time, developments and debates concerning the governance of forensic DNA conclusions have had a boomerang effect on the governance of fingerprint conclusions. The characterisation of DNA conclusions in terms of a two-step process – a finding of consistency followed by an estimate of the probability of finding the combination of features in a relevant population – has had an impact on the rest of forensic science. In particular, there are numerous forms of comparison evidence – toolmarks, bitemarks, hair, fibres and, perhaps most significantly, fingerprints – that are potentially amenable to the same approach, except for the fact that there is not currently, and may never be, a credible way to calculate RMP figures for evidence matches using such techniques. Therefore, the process of governing DNA profile evidence leads to the question of why the same approach should not be adopted for fingerprints, and, indeed, some research is currently underway to develop probabilistic (particularly Bayesian) indices for fingerprint evidence (Egli *et al.* 2006; Meuwly 2006; Neumann *et al.*

2006, 2007). This research, however, is not yet complete and, more importantly, has not yet penetrated actual forensic practice, where the estimation of rarity remains intuitive and the reporting of matches makes no use of quantified probability estimates (Champod and Evett 2001).

Efforts to extend DNA's governance model for forensic conclusions using other techniques have met with mixed success. The Nuffield Council on Bioethics report (2007), despite engaging in a nuanced discussion of the potential pitfalls that characterise DNA matches, fails to acknowledge that precisely the same issues pertain to fingerprint matches, with the only difference being that fewer data are available. Instead, the Nuffield Council proposed an 'opinionization' solution for fingerprints, in which characterising matches as 'opinions' (as opposed to 'facts') forgives ignorance about the rarity of the matching features (Cole 2008).

In the USA, meanwhile, the NRC finally convened a long awaited panel on forensic identification evidence. Fingerprints made up a large portion of the Committee's agenda (10 of 57 presentations before the Committee). In light of attempts by previous NRC Committees to resolve controversies about the use of forensic DNA evidence (as well as voice analysis, polygraphy, comparative bullet lead analysis and ballistics), many commentators expected that the current panel might do for fingerprints and other trace identification techniques what earlier panels did for DNA evidence (Cooley and Oberfield 2007; Hamilton and Cohen 2008 (quoting Scheck)). In its presentation to the Committee, however, the IAI, the leading professional organisation of latent print examiners in the USA and perhaps the world, expressed staunch resistance to the two-step DNA model of governing forensic conclusions. In obvious reference to arguments that fingerprint matches, like DNA matches, require rarity estimates, a letter from the IAI to the Committee stated that '[i]t is the business model, not the science based model, which needs to be fostered for the remaining forensic sciences' (International Association for Identification 2007). In other words, what the IAI seemed to envy about DNA evidence was its success at procuring a level of funding sufficient to create an infrastructure that supported its credibility: a disciplinary regime including training, protocols and material standards. It is as though the IAI had taken instructions from Latour (1987) on how to add weight and solidity to evidence by surrounding it with a network of standards, measures and routine practices. The NRC Report concluded that claims that the error rate of latent print identification was zero were 'not plausible', that there was 'limited' information as to the accuracy of the technique, that the method of latent print analysis had not been validated and that the available evidence could not support claims of source attribution (National Research Council 2009).

Governing DNA databases

As discussed elsewhere in this volume, DNA databases through steady expansion are beginning to seem less like an entirely novel development in forensic science. Instead, they seem to be largely following the path set by fingerprint databases (which continue to be much larger than DNA databases and used in a higher number of criminal investigations). The example of fingerprint databases has served as a resource for both sides of debate about genetic database expansion. Interestingly, both sides take for granted the acceptability of fingerprint databases. Proponents of genetic databases have drawn on the relatively uncontroversial nature of fingerprint databases to argue that genetic databases are 'no different' from fingerprint databases. Since fingerprint databases have not been the subject of political opposition, these proponents reason that DNA databases should not be opposed either. Opponents of genetic databases, meanwhile, seek to break this chain of analogical reasoning. They argue that DNA databases *are* different and that the acceptance of fingerprint databases does not necessarily imply the acceptability of genetic databases. The basis for this difference is the supposed diagnostic power of genes, as opposed to fingerprints. Although the markers used in DNA profiling systems are for non-coding sequences (with the exception of one that denotes sex), some markers are highly correlated with racial or ethnic categories, and when bodily samples are retained from which DNA profiles are developed, it remains possible to conduct research on a broader range of genetic relationships. Concerns about these possibilities are variants of a view that is known as genetic exceptionalism, the idea that there is something special, or even unique, about DNA databases. Such a view is also evident when proponents of DNA databases entertain hopes of unprecedented crime control and promote expansive schemes (including 'universal' databasing), while forgetting that such hopes and schemes once characterised fingerprint databases (see above: Expansion of fingerprint databases).

While proposals for universal national DNA databases may have superficial appeal, so far they have generated little more than token rhetorical support among legislators and forensic organisations. While efforts to compile criminal databases from samples taken from stigmatised 'others' appear to be popular with voters, proposals to place legislators' and their constituents' own genetic material on the database do not appear to have much political support, in part because of practical problems and expenses associated with the routine laboratory and clerical work necessary to process samples and enter information. Previous expansions of criminal databases in the USA and UK have created persistent backlogs, confusion from wrongly coded clerical information and (at least initially) police resistance to collecting DNA samples associated with petty crimes. For these reasons and more, proposals for universal DNA databases remain hypothetical, and neither the US nor the UK Government shows any inclination to expand its database anywhere close to the limits of its population. In this sense, it seems likely that DNA will replicate the balance struck with regard to fingerprint databases, in which it was deemed acceptable to store biometric data about criminalised 'others,' including arrestees and immigrants, but not of 'innocent' citizens who fell outside those categories.

The recent decision by the European Court of Human Rights, finding that the indefinite retention of DNA and fingerprint samples violated human rights, may slow the worldwide trend towards arrestee databases, but it allows room for limited retention of samples given adequate justification (*S and Marper* v. *United Kingdom* 2008). Consistent with our thesis that the governance of any given forensic technique is shaped by the governance of *other* forensic techniques, in this judgment, DNA practically brought fingerprinting down with it. The court declined to distinguish between DNA profiles and fingerprints. In the same breath that it ruled genetic arrestee databases werea violation of 'the right to respect for private life', the court also ruled that hitherto unquestioned fingerprint arrestee databases likewise violated human rights.

CONCLUSIONS

In several ways, therefore, procedures for governing DNA data have not been discovered anew but have followed patterns established by earlier forensic and biometric technologies, particularly fingerprinting. Despite genetic exceptionalism and calls for universal databases, DNA databases appear to be following in the path of fingerprint databases, to include information from citizens (and non-citizens) with arrest records or other modes of stigmatisation but not from individuals free of such stigmas (the NDNAD of England and Wales is a partial, and controversial, exception to this rule, though the judgment in *S and Marper* v. *the United Kingdom* may curb the exception somewhat). Similarly, although current DNA profiling techniques enable the production of precise and impressive probability estimates, there have been consistent efforts to return to the sorts of simple, non-quantified declarations so successfully exploited by fingerprint examiners. Such tendencies were visible early in the history of DNA typing, when 'DNA fingerprinting' (the early MLP technique) was first used in the UK, and then later when random match probabilities grew so small that some forensic scientists in the USA felt justified in promoting DNA 'matches' as a basis for declaring the individual source of the evidence.

The notions that DNA is an infallible technology and that DNA matches enable definitive declarations of fact have been crucial for this technology, just as they were in the past for establishing and perpetuating fingerprint identification (Cole 2005; Saks 2007). These claims are highly persuasive for the general public and for actors in the criminal justice system, including defendants and convicts (Prainsack and Kitzberger 2009). What makes the situation paradoxical is the fact that the claims of infallibility associated with fingerprinting have been challenged, in part because of the ascendency of DNA to the status of 'gold standard'. By exonerating convicts, by providing a model of statistically defensible identification evidence and by making more persuasive claims to 'high science', DNA helped to undermine fingerprinting's position in the hierarchy of forensic techniques and its claims to infallibility. Yet, the impact of this inversion, far from encouraging scepticism about such infallibility claims in general, has encouraged many analysts to transfer such claims to the next 'infallible' forensic technique.

While the governance of new genetic identification technology is surely shaped by the larger historical context as well as local contingencies and the exigencies of the technology itself, we suggest that it is also shaped by the biometric technologies that preceded it. In many ways, DNA is following a well-worn path laid down principally by fingerprinting but also by other technologies. Yet the path is not linear. Rather, at times it seems to loop back upon itself, as when developments in the governance of DNA profiling reshape the governance of fingerprinting and other older biometric techniques.

REFERENCES

Aitken, C. (2005). The evaluation and interpretation of scientific evidence. In *Proceedings of a Colloquium on Forensic Science: The Nexus of Science and the Law*. Washington, DC: National Academies Press http://www.nasonline.org/site/ PageServer?pagename=sackler_forensic_presentations (accessed 23 February 2010).

American Civil Liberties Union (1938). *Thumbs Down! The Fingerprint Menace to Civil Liberties*. New York: American Civil Liberties Union.

Aronson, J. (2007). *Genetic Witness: Science, Law, and Controversy in the Making of DNA Profiling*. New Brunswick, NJ: Rutgers University Press.

Balding, D. (1997). Errors and misunderstandings in the second NRC report. *Jurimetrics*, 37, 469–476.

Breckenridge, K. (2008). The biometric obsession: Trans-Atlantic progressivism and the making of the South African state. In *Identi-net Conference: The Documentation of Individual Identity: Historical, Comparative and Transnational Perspectives since 1500*, Oxford, September http://identinet.org.uk/people/ members/#breckenridge (accessed 6 March 2010).

Buckleton, J. (2005). Population genetic models. In *Forensic DNA Evidence Interpretation*, eds. J. Buckleton, C. Triggs and S. Walsh. Boca Raton, FL: CRC Press, pp. 65–122.

Budowle, B., Chakraborty, R., Carmody, G. *et al.* (2000). Source attribution of a forensic DNA profile. *Forensic Science Communications* 2, July http://www.fbi. gov/hq/lab/fsc/backissu/july2000/source.htm (accessed 24 February 2010).

Champod, C. and Evett, I. (2001). A probabilistic approach to fingerprint evidence. *Journal of Forensic Identification*, 51, 101–122.

Cole, S.A. (2001). *Suspect Identities: A History of Fingerprinting and Criminal Identification*. Cambridge, MA: Harvard University Press.

Cole, S.A. (2005). More than zero: accounting for error in latent fingerprint identification. *Journal of Criminal Law and Criminology*, 95, 985–1078.

Cole, S.A. (2007a). Twins, Twain, Galton and Gilman: fingerprinting, individualization, brotherhood and race in *Pudd'nhead Wilson*. *Configurations*, 15, 227–265.

Cole, S.A. (2007b). Where the rubber meets the road: thinking about expert evidence as expert testimony. *Villanova Law Review*, 52, 803–842.

Cole, S.A. (2008). The 'opinionization' of fingerprint evidence. *BioSocieties*, 3, 105–113.

Cole, S.A. and Lynch, M. (2006). The social and legal construction of suspects. *Annual Review of Law and Social Science*, 2, 39–60.

Cooley, C. and Oberfield, G. (2007). Increasing forensic evidence's reliability and minimizing wrongful convictions: Applying Daubert isn't the only problem. *Tulsa Law Review*, 43, 285–380.

Devlin, K. (2006). *Statisticians Not Wanted*. Washington, DC: Mathematical Association of America.

Donnelly, P. and Friedman, R. (1999). DNA database searches and the legal consumption of scientific evidence. *Michigan Law Review*, 97, 931–984.

Egli, N., Champod, C. and Margot, P. (2006). Evidence evaluation in fingerprint comparison and automated fingerprint identification systems: modeling with finger variability. *Forensic Science International*, 167, 189–195.

Felch, J. and Dolan, M. (2008). How reliable is DNA in identifying suspects? *Los Angeles Times*, 19 July.

Foucault, M. (1991). Governmentality. [Translated by R. Braidotti and revised by C. Gordon.] In *The Foucault Effect: Studies in Governmentality*, eds. G. Burchell,

C. Gordon and P. Miller. Chicago, IL: University of Chicago Press, pp. 87–104.

Garrett, B. L. and Neufeld, P. (2009). Improper forensic science and wrongful convictions. *Virginia Law Review*, 95, 1–97.

Giannelli, P. (1997). The abuse of scientific evidence in criminal cases: the need for independent crime laboratories. *Virginia Journal of Social Policy and the Law*, 4, 439–478.

Grieve, D. (1990). The identification process: traditions in training. *Journal of Forensic Identification*, 40, 195–213.

Groebner, V. (2007). *Who Are You? Identification, Deception, and Surveillance in Early Modern Europe*. [Trans. M. Kyburz and J. Peck.] New York: Zone Books.

Hamilton, B. and Cohen, S. (2008). 'Clueless' crime labs. *New York Post*, 21 September.

International Association for Identification (2007). *IAI Positions and Recommendations*. Mendota Heights, MN: International Association for Identification http://www.theiai.org/ (accessed 24 February 2010).

Jeffreys, A., Wilson, V. and Thein, S. L. (1985). Individual-specific 'fingerprints' of human DNA. *Nature*, 316: 76–78.

Jonakait, R. (1991). Forensic science: the need for regulation. *Harvard Journal of Law and Technology*, 4, 109–191.

Kaye, D. (2001). The constitutionality of DNA sampling on arrest. *Cornell Journal of Law and Public Policy*, 10, 455–508.

Kaye, D. (2006). Who needs special needs? On the constitutionality of collecting DNA and other biometric data from arrestees. *Journal of Law, Medicine and Ethics*, 34, 188–98.

Krane, D., Ford, S., Gilder, J. *et al.* (2008). Sequential unmasking: a means of minimizing observer effects in forensic DNA interpretation. *Journal of Forensic Sciences*, 53, 1006–1007.

Lander, E. (1992). DNA fingerprinting: Science, law, and the ultimate identifier. In *The Code of Codes: Scientific and Social Issues in the Human Genome Project*, eds. D. Kevles and L. Hood. Cambridge, MA.: Harvard University Press, pp. 191–210.

Latour, B. (1987). *Science in Action: How to Follow Scientists and Engineers through Society*. Cambridge, MA: Harvard University Press.

Lemke, T. (2001). The birth of bio-politics: Michel Foucault's lectures at the College de France on neo-liberal governmentality. *Economy and Society*, 30, 190–207.

Lynch, M. (2004). 'Science above all else': the inversion of credibility between forensic DNA profiling and fingerprint evidence. In *Expertise in Regulation and Law*, ed. G. Edmond. Aldershot, UK: Ashgate, pp. 121–135.

Lynch, M., Cole, S. A., McNally, R. and Jordan, K. (2008). *Truth Machine: The Contentious History of DNA Fingerprinting*. Chicago, IL: University of Chicago Press.

Meuwly, D. (2006). Forensic individualisation from biometric data. *Science and Justice*, 46, 205–213.

Mnookin, J. (2001). Fingerprint evidence in an age of DNA profiling. *Brooklyn Law Review*, 67, 13–70.

Mnookin, J. (2006). *People v. Castro*: challenging the forensic use of DNA evidence. In *Evidence Stories*, ed. R. Lempert. New York: Foundation Press, pp. 207–238.

Mueller, L. (2008). Can simple population genetic models reconcile partial match frequencies observed in large forensic databases? *Journal of Genetics*, 87, 101–108.

Murphy, E. (2007). The new forensics: criminal justice, false certainty, and the second generation of scientific evidence. *California Law Review*, 95, 721–797.

National Research Council (1992). *DNA Technology in Forensic Science*. Washington, DC: National Academies Press.

National Research Council (1996). *The Evaluation of Forensic DNA Evidence*. Washington, DC: National Academies Press.

National Research Council (2009). *Strengthening Forensic Science in the United States: A Path Forward*. Washington, DC: National Academies Press.

Neumann, C., Champod, C., Puch-Solis, R. *et al.* (2006). Computation of likelihood ratios in fingerprint identification for configurations of three minutiae. *Journal of Forensic Sciences*, 51, 1–12.

Neumann, C., Champod, C., Puch-Solis, R. *et al.* (2007). Computation of likelihood ratios in fingerprint identification for configurations of any number of minutiae. *Journal of Forensic Sciences*, 52, 54–64.

Nuffield Council on Bioethics (2007). *The Forensic Use of Bioinformation: Ethical Issues*. London: Nuffield Council on Bioethics www.nuffieldbioethics.org/go/ourwork/bioinformationuse/publication_441.html (accessed January 2009).

Prainsack, B. and Kitzberger, M. (2009). DNA behind bars: Other ways of knowing forensic DNA technologies. *Social Studies of Science*, 39, 51–79.

Rodriguez, J. (2004). South Atlantic crossings: fingerprints, science, and the state in turn-of-the-century Argentina. *American Historical Review*, 109, 387–416.

Rosen, R. (2003). Innocence and death. *North Carolina Law Review*, 82, 61–113.

Rubin, J. and Felch, J. (2008). Man convicted in sex assault should be freed or retried, court rules. *Los Angeles Times*, 6 May.

Ruggiero, K. (2001). Fingerprinting and the Argentine plan for universal identification in the late nineteenth and early twentieth centuries. In *Documenting Individual Identity*, eds. J. Caplan and J. Torpey. Princeton, NJ: Princeton University Pres, pp. 184–196.

Saks, M. (2007). Protecting factfinders from being overly misled, while still admitting weakly supported forensic science into evidence. *Tulsa Law Review*, 43, 609–626.

Saks, M. and Koehler, J. (2005). The coming paradigm shift in forensic identification science. *Science*, 309, 892–895.

Scheck, B., Neufeld P. and Dwyer J. (2000). *Actual Innocence: Five Days to Execution, and Other Dispatches from the Wrongly Convicted*. New York: Doubleday.

Sekula, A. (1986). The body and the archive. *October 39*, Winter, 3–64.

Sengoopta, C. (2002). *Imprint of the Raj: How Fingerprinting Was Born in Colonial India*. London: Macmillan.

Singha, R. (2008). Settle, mobilize, verify: identification practices in colonial India. *Studies in History*, 16, 151–198.

Spivey, G. (2005). Right of exonerated arrestee to have fingerprints photographs, or other criminal identification or arrest records expunged or restricted. *American Law Reports*, 3, 900.

Thompson, W. C. (2008). The potential for error in forensic DNA testing. *GeneWatch*, November-December, 5–8.

Thompson, W. and Schumann, E. (1987). Interpretation of statistical evidence in criminal trials. *Law and Human behavior*, 11, 167–187.

CASES

McDaniel v. *Brown* (2010). 130 S.Ct. 665 (U.S.)

People v. *Castro* (1989). 545 N.Y.S.2d 985 (NY Sup. Ct. Bronx Cty).

People v. *Johnson* (2006). 139 Cal.App.4th 1135 (Cal.App. 5 Dist.).

People v. *Nelson* (2008). 185 P.3d 49 (Cal.).

R v. *Doheny and Adams* (1997). 1 Court of Appeal R 369, C.A.

S and Marper v. *the United Kingdom* (2008). A summary of the judgment is available from http://cmiskp.echr.coe.int/tkp197/view.asp?action=html&documentId =843937&portal=hbkm&source=externalbydocnumber&table=F69A27FD8 FB86142BF01C1166DEA398649 (accessed January 2009).

The Queen v. *Sean Hoey* (2007). NICC 49 (Crown Court Sitting in Northern Ireland).

United States v. *Jenkins* (2005). 887 A.2d 1013 (D.C. Cir.).

Section 2 National contexts of forensic DNA technologies and key issues

7

DNA databases and the forensic imaginary

INTRODUCTION

Technologies promising to provide reliable knowledge of individual human subjects are the constant companions of efforts to control personal and collective action in all societies. There is an extensive scholarly literature on the ways in which the historically and socially conditioned 'interiors' of such subjects, variously formulated as subjectivity, self-identity or ipse, have been constructed and interrogated in recent and contemporary social formations (see especially Foucault (1977, 1979), Martin *et al.* (1988) and Rose (1985, 1990, 1996, 1999)). Less well studied, at least until recently, have been the related efforts of a range of commercial, industrial and state agencies (especially security and criminal justice agencies) to acquire and use knowledge of the uniquely differentiated and securely fixed individualising characteristics of their employees, customers and citizens. Methods for reliably grasping and securing what passes for 'individual identity', 'individuality' or 'idem' are desiderata for any such agency seeking to attribute or evaluate claims of individual identifiability amongst the population members with which it deals. There is an increasingly large number of institutional actors who seek assurance that each individual with whom they deal can be differentiated from all others, and that there are ways of anchoring the changing biographies of such individuals to some ineradicable and unchanging bodily substrate.

It is widely asserted that a concern with the 'legibility' of such individualised identities is a key feature of modern state formations, and also that this concern has become intensified in many societies in the first decade of the twenty-first century, especially those that define themselves as being at risk from uncontrolled migration or from terrorist violence. The desire to visualise stable and ineradicable biological

Genetic Suspects: Global Governance of Forensic DNA Profiling and Databasing, ed. Richard Hindmarsh and Barbara Prainsack. Published by Cambridge University Press. Copyright © Cambridge University Press 2010.

correlates of individuality, and to preserve such visualised bio-identities for subsequent examination and comparison is central to all contemporary approaches to the exercise of authority over persons. In the words of Aas (2006: 144), 'bodies, fused with the latest technologies, are proving to be vital to contemporary governance'.

Perhaps in response to the contemporary intensification and proliferation of interests in bio-identification, the current forms and longer history of such efforts are being increasingly documented and interrogated. Studies include accounts of the rise and proliferation of a variety of such 'machineries of identification' (e.g. Caplan and Torpey 2001), analyses of methods to support the regulation of movement within and between nation states (e.g. Torpey 1998, 2000) as well as more general accounts of the expansion of contemporary regimes of surveillance (e.g. Lyon 2001; Williams and Johnson 2004a, 2004b; Zuriek and Salter 2005; Lyon 2006). A central resource for many of these efforts has been Michel Foucault's account of a series of developing practices that make up this broader 'political economy of bodies', in which the material body itself has been central to the development of both disciplinary and governmental discourses and practices (e.g. Foucault 1977, 2003, 2004). While Foucault suggests that the most important aspect of bodily examination is the introduction of methods for observing, knowing and recording individuality, he wrote little about how such methods were used to render the individuation of bodies though specific techniques of bio-identification such as those of anthropometry or fingerprinting.

This chapter, and others in this book, discusses the development of forensic DNA profiling as a particularly conspicuous instance of one such method. The focus here is on the particular trajectory of scientific and operational innovations that occurred in England and Wales as the first criminal jurisdiction in which a national DNA database was established. First, the historical sources of the contemporary enthusiasm for the use of DNA profiling to support the control of crime through increasing the effectiveness of criminal investigations is outlined. The chapter then goes on to describe the ways in which scientific advances were harnessed to legislative changes and financial support to enable the creation of a DNA database that currently holds more than five million profiles.

Later, the chapter considers a range of issues that have arisen – in the UK and elsewhere – as the enthusiasm for forensic DNA profiling and databasing as resources for governing individuals and populations has been supplemented by questions about how such technologies should themselves be governed. In particular, the focus is on those features of their deployment that have raised significant operational, legal and ethical challenges, as well as the forms of regulation, oversight and accountability that may have to be put in place to meet these challenges.

POLICING AND BIO-IDENTIFICATION

Attempts to individuate, document and categorise bodies formed a major feature of the emergent policing practices of the nineteenth century as an element in the wider conglomerate of its 'technology of governmentality' (Ericson and Haggerty 1997: 53). Caplan and Torpey (2001: 9) have written about the ways in which modern policing itself has been 'the source of repeated efforts to rationalize and standardise practices of identification and the systems for storage and retrieval of the expanding documentation this generated'. However, as Simon Cole (2001) reminds us, official methods for working on and with the body to support social control have a longer history; for instance recording the ways that bodies have been marked by branding, tattooing or amputation. Yet while the marking of individuals to identify them as criminals may differentiate such individuals, it does not provide a method of unique identification. The practice of subjecting a body to examination in order to differentiate it from all others is founded in a range of techniques that culminate in the 'sciences' of identification. There are themes and variations amongst these sciences, but the objective is constant: to locate seemingly unique aspects of individual bodies – of bio-individuality – and to find ways of linking a representation of that uniqueness to a record of other officially established facts about persons.

However, while many of these technologies focus on methods for verifying the identity claims of individuals who present themselves at a variety of spatial, judicial, penological and administrative 'borders', another parallel set of endeavours have been concerned to generate knowledge of the presence of such individuals at specific places in the past– in particular, at 'scenes of crime'.[1] This investigatory

---

[1] While the term 'scene of crime' is found in common usage, police investigatory discourse prefers the term 'crime scene' and specifies this as referring to a complex variety of locations, persons and objects, including 'a piece of land or part of a street; a building, or a room within a building; the houses, vehicles, vessel and other property of a suspect, witness, or victim; stolen or recovered property; the body, personal possessions and clothing of a suspect, witness or victim; ambulances or other vehicles used to convey victims or offenders to hospital premises, police stations or mortuaries' (Association of Chief Police Officers 2005: 12).

complement to efforts to corroborate the identity of present suspects seeks to support the investigation of crime by the practical application of methods of individuation (largely but not wholly by reference to fixed physical characteristics) in ways that also reflect an interest in the divisibility and transfer of matter (including biological matter). What is called here the 'forensic imaginary' is dominated by a commitment to two principles, where their joint application to the process of crime scene examination within the practice of criminal investigations is expected to hugely improve efforts to detect crime and identify suspects. In turn, these achievements are expected to lead to the reduction of crime and a corresponding increase in public safety (see Chapter 10).

The first principle, that 'individuation' (or the unique identifiability) of any object is always possible, is an incorrigible and emphatic assertion of the forensic imaginary. The principle is protected from refutation by faith in the development of knowledge. Whenever two objects seem incapable of being differentiated here and now, this principle supports the view that the two objects are capable of being differentiated or individualised whenever the means become available to do so. It may be necessary to 'look more carefully', 'test more sensitively' or 'measure more accurately'. It is tempting to characterise the principle of unique identifiability as 'proto-scientific' if this term can be used without irony. Certainly it does not seem to be derived from any particular set of disciplinary propositions but instead from an earlier and more widespread assertion that 'no two things in nature are exactly the same'. But whatever its origins, the assertion is so central to forensic practice that, for some authorities, it constitutes its essence as 'the science of individualization' (Nickell and Fischer 1999: 3).

The second principle can be summarised as the proposition that 'exchange always happens'. This can be understood as a compressed expression of the claim that whenever individuals make physical contact with other individuals, or enter and leave defined physical spaces, a (usually unintentional) transfer of varying amounts of matter – inert and vital material – takes place between the parties and places involved. Some refer to this proposition as Locard's law, Locard's truth or Locard's doctrine, while others prefer to call it Locard's theory or even Locard's principle. In fact, Edmund Locard himself did not originate a proposition with any of the exactitude implied by these designations, but the closest he seems to have come to it is in a statement in 1923 (Inman and Rudin 2001: 93):

No one can act (can commit a crime) with the force (intensity) that the criminal act requires without leaving behind numerous signs (marks) of it; either the wrong-doer (felon, malefactor, offender) has left signs at the scene of the crime, or on the other hand, has taken away with him – on his person (body) or clothes – indications of where he has been or what he has done.

It is this forensic imaginary – of the actual or promised ability increasingly to recover individualisable traces of biological and other materials transferred between persons and objects at crime scenes and to use these traces as evidence to support criminal prosecutions – that has powered the rise of forensic science from the nineteeenth century onwards. This trajectory has been documented in studies of the growing uses of forensic science in general within criminal investigations (e.g. Eco and Sebeok 1983; Spufford and Uglow 1996; Valier 1998, 2001) and in accounts of the trajectories of several specific forensic technologies (e.g. Lambourne 1984; Cole 2001; Sengoopta 2003).[2] Some of the technologies that were developed to contribute to the realisation of the forensic imaginary, such as anthropometry, proved unable to deliver on the promises of their inventors, while others, like fingerprinting, have been strongly embraced by police and the courts and arguably constitute an established success.[3] The successful ones – like fingerprinting – have allowed investigators to capture past actions through the artefactualisation and informatisation of the residual presences of individuals at crime scenes.

Typically, this imaginary has been carried in 'images, stories and legends' (Taylor 2004: 23) and shaped by hopes, worries, desires and a range of other emotional energies, and it has contributed hugely to the willingness of governments to fund forensic science developments and ambitions. Nevertheless, at the same time, the inevitable failure to deliver the full range of expected benefits can make for difficulty in retaining confidence in the capacity of forensic science institutions to contribute to criminal justice in the way previously imagined.

[2] There is some overlap here between forensic technologies used to investigate crimes and the use of available closed circuit television imagery both to initiate police intervention in ongoing events and to support the investigation of prior events. The differences between these are explored by Johnson and Williams (2004).

[3] It is worth noting here that even the seemingly most secure technology of bio-identification can come under attack. There is an emerging literature which documents significant problems in fingerprint comparison, at least as far as latent print comparison is concerned. See, especially, Cole (2001).

Accordingly, the current and future shapes of forensic science and the investigations that rely on it are heavily determined by the attentiveness of key actors to the promises and hopes as well as to the demands of the realities with which such expectations can be compared.[4]

THE TRAJECTORY OF DNA PROFILING

The quick and strong embrace of forensic DNA profiling by key state actors in the UK and elsewhere is best understood as the most recent instance of the forensic imaginary at work. Its technical trajectory is best understood as the rapid and total dislodgement of an earlier reliance on blood groups for the bio-identification and bio-differentiation of individuals involved – as suspects, victims or 'persons of interest' – in criminal investigations. Between 1900 and the early 1980s, the 'class identification' rather than individuation of human subjects was based on ABO blood groups, according to which human blood cells fell into four inherited antigenic groups: A, B, AB and O. Subsequently, additional red blood systems were discovered along with other polymorphic markers based on serum proteins and enzymes. A number of these were regularly used by investigators and courts to confirm or challenge the involvement of individuals in crime, but recent years have seen a marked shift from the analysis and use of the protein products of DNA (in blood, semen and other bodily fluids) to the analysis and use of nuclear DNA itself.[5]

Since the early 1980s, molecular biologists have identified three main types of genetic markers: those based on repeat sequences in

[4] The social and policy significance of imaginaries, along with similar notions of 'expectations', 'anticipations' etc., has been discussed elsewhere, and sometimes with specific reference to individual forensic domains (e.g. Gerlach 2004; Taylor 2004; Borup *et al.* 2006; also Chapter 5). There are many policy actors with strong aspirations to alleviate a range of fears concerning security, safety, crime control and the management of 'risky' individuals by the demonstrably effective use of current and emerging technologies able to capture and interrogate a range of material and informational attributes of individuals and their actions.

[5] Many of the cases identified as miscarriages by Innocence Projects (www.innocen ceproject.org/) in the USA and elsewhere have been resolved by using the individualising capacity of DNA profiling to overpower what were regarded earlier as definitively incriminating results obtained from the blood group analysis of biological samples, particularly, but not exclusively, of semen samples recovered from crimes involving sexual assault. Such opportunities exist only when original crime scene samples can be located in a form and condition suitable for DNA processing.

DNA, those based on systems with sex-specific transmission and those based on alternative sequences found at particular nucleotide sites in the genome. Variants of the first type provided the basis for DNA profiling and databasing. Studies in the mid 1980s (Gill *et al.* 1985; Jeffreys *et al.* 1985) established that forensic samples from potentially crime-relevant objects could contain sufficient quantities and quality of DNA for profiling. These objects include blood, semen, saliva, hair, dandruff, skin, vaginal and nasal secretions, sweat and urine.

The possibility of deriving DNA from the unintentionally abandoned biological matter left by criminal suspects at crime scenes generated immediate and huge interest amongst investigators in the UK, the USA and elsewhere. The UK Forensic Science Service (FSS), along with other state and commercial organisations in the UK and the USA, quickly initiated research programmes that focused on techniques for the production of DNA profiles from crime scenes capable of being reliably compared with profiles derived from samples taken directly from criminal suspects by the police. Prior to the implementation of polymerase chain reaction (PCR) methods of nucleic acid analysis in the 1990s, these initial uses of DNA profiling (based on multiple and single locus probes) were largely confined to reactive forensic casework. In this modality, laboratories directly compared DNA profiles obtained from biological material left at crime scenes with those taken from individuals already in police custody who were suspected of involvement in the specific criminal offence under investigation. However, the subsequent ability to construct digital representations of profiles, following on from the implementation of PCR methods in the 1990s, meant that such profiles could be stored in continuously searchable computerised databases. In addition, a continuous and successive series of laboratory improvements enabled the reliable extraction of genetic material from a wider range of samples in varying conditions and has meant that forensic laboratories can more easily generate usable DNA profiles. Sometimes (as in cold case reviews) such DNA profiling may succeed when other forms of forensic or witness evidence has proved insufficient or unreliable in helping to bring offenders to justice for crimes committed some years earlier.

Several features of DNA profiling make it particularly attractive in support of criminal investigations. The most obvious is that, unlike any other instances of bio-identification, DNA is a truly universal feature of all living (and dead) human beings. The use of DNA to construct documents of identity introduces a significant change in the relationship between the physical body and the resulting record of identity: a

DNA profile is not constructed from an impression of the body; it is not created by measuring external bodily features, and it is not a document of physical appearance. While all previous identification methods used by investigators manipulated *aspects* of the body into a standardised form of information, the analysis of DNA departs from a very different premise. Fundamentally, as van der Ploeg (2005: 96) argues, 'there is no clear point where bodily matter first *becomes* information. The "essence" of the stuff of DNA, both the reason for its scientific isolation in the first place, and, in its watered down version, its forensic significance, is precisely that it *is* information.'

In contrast to the processes in all the other practices, from anthropometry, to fingerprinting, to iris recognition, this assumed property of DNA allows investigators to capture, store and use not just the representation or documentation of the body but bodily matter itself. This potential to collect a 'part' of the body, and not simply an impression of it, from which a representation of individual identity can be constructed is a significant development in criminal investigation. More than any other technological development in human identification, then, the practice of DNA profiling renders the human body as a system of standardised and repeatable properties. It goes, as David Lyon (2001) puts it, 'under the skin' to capture the very essence of the body itself (see also Chapter 13), bypassing the need to measure any external surface or to engage with the outward aspects of human corporeality. The durability of DNA in the body matter that can be collected (blood, hair, semen) means that investigators now have access to a powerful form of body data at scenes of crime.

In the UK, this technological advance, together with widening police powers to take and retain biological samples from individuals, made possible a vastly expanded role for DNA profiling. In particular, police investigators became able to apply this technology inceptively rather than reactively. In other words, it increasingly came to be used to shape enquiries by identifying potential suspects from the start rather than being used later to lend authoritative support to the incrimination or exoneration of otherwise nominated suspects. This development of proactive uses of bio-identification was made possible not only by the increasing size of the national DNA Database of England and Wales (NDNAD) but also through a continuous and suc- · cessive series of laboratory improvements, which enabled the reliable extraction of genetic material from a wider range of samples. When forensic laboratories can more easily, more quickly and more cheaply generate usable DNA profiles, then investigators increasingly are able to make use of such bio-identifications to generate suspects for crimes that might otherwise be impossible (or far too expensive) to detect. Many of these features underlie the strong support given to DNA profiling by those concerned to realise the promise of the wider forensic imaginary. However, its very focus on, and utilisation of, molecular attributes has been the reason that a wider group of scholars and stakeholders have sought to question its uses and users in ways that have no previous parallel in the history of forensic bio-individuation (as this volume makes clear).

CONTESTING FORENSIC DNA DATABASING

The rapid rise of forensic DNA databasing, first in the UK and subsequently elsewhere in the world, has met with a mixed reception from a variety of observers. On the one hand, there are those who argue that DNA databases enhance the public good for the reasons described above. In particular, that they have the potential to make speedy and robust suspected offender identifications through automated profile comparisons in centralised criminal justice databases; further they provide the ability to confidently eliminate innocent suspects from investigations and increase the likelihood of generating reliable and persuasive evidence for use in court. In more general terms, they may also provide a deterrent effect for potential criminal offenders, and increase public confidence in policing and in the wider judicial process. On the other hand, there are those who argue that police DNA databasing threatens the bodily integrity of citizens who are subject to the forced and non-consensual sampling of their genetic material; further, that they violate privacy rights by allowing the storage and use of tissue samples, storage which itself creates the potential for the future misuse of such samples held in state and privately owned laboratories.

Both of these positions are regularly encountered in debates over the establishment and expansion of any forensic DNA database. Accordingly, database legitimacy requires legislative authorisation, judicial endorsement, funding and a governance framework that secures stakeholder and public confidence in the supervision and accountability of its scientific contributors and police users. The operational success of such a database depends on the inclusiveness of its coverage and on the degree to which police investigators respond to the intelligence opportunities it provides. However, both of these features – inclusiveness and utilisation – are subject to a number of both factual and normative questions. Uncertainties about the proven effectiveness of technologically enhanced methods of criminal investigation necessarily overlap with reservations about the acceptability and consequences of the collection, retention and use of personal data from a widening category of individuals. This overlap means that, however much police actors seek increases in the uses of bio-identification in support of criminal investigations, fundamental practical and moral dilemmas remain at the centre of all such efforts to collect, retain and use personal information about those who are the perpetrators, victims or witnesses of criminal actions. An examination of the design and operation of the governance of DNA databases provides a way of unearthing how different social aggregations have resolved (or failed to resolve) these dilemmas.

Three separate but interrelated social and epistemic domains are the usual focus of DNA database governance.[6] The first of these is a juridico-scientific domain in which profile construction procedures, profile matching methods and subsequent police operational procedures have to have sufficient integrity to satisfy legal requirements and resist adversarial – including scientific – challenge. The second is an operational domain in which the economy, efficiency and effectiveness of police uses of a forensic DNA database can be authoritatively established and assessed through various forms of audit appraisals. The third is an ethico-political domain in which attention is focused on the necessity for the oversight and independent scrutiny of the total operation and uses of any such database, the social implications of the retention of DNA from various categories of persons and the permissible analysis and uses of both biological material and digital profiles.

There are now many instances of such databases, and there are varied ways in which they have been constructed and governed within different criminal jurisdictions. The next section of this chapter focuses on the recent history of NDNAD, the first such national DNA database, to consider how policy makers and policing strategists in this jurisdiction have struggled with these matters.

---

[6] The legislative framework within which such databases operate is not the subject of this chapter, although it should be obvious that this framework forms a stable legal background against which a variety of softer regulatory processes are brought to bear to control the details of database workings.

THE NATIONAL DNA DATABASE OF ENGLAND AND WALES

There is no single model for the organisation of public sector services in the UK; instead it is recognised that there exists 'a diversity of organisations providing and delivering public services with constitutions, funding arrangements and operational procedures appropriate to the work they do' (House of Lords Select Committee on Science and Technology 1997: 7). This diversity reflects the character of the contemporary public sector as a network of public and private bodies operating within market or quasi-market social environments rather than as elements in a complex but unitary state bureaucracy. However, despite important differences among the various bodies within this network, a degree of uniformity in the ways that they are governed has been attempted through the application of a common framework of 'public accountability'. The following three subsections of this chapter outline the ways in which this requirement has been addressed in the case of the NDNAD.

The juridico-scientific domain

The routine management and uses of the NDNAD are subject to a legal framework that both enables and constrains how DNA profile data are managed as well as how they may legitimately be used. Both the system of database governance and the scope of allowable uses of DNA for operational purposes are the product of the legislative arm of government to which the police are ultimately responsible. A series of statutory instruments and government circulars provides the essential limits of legitimate action (e.g. Home Office Circular 25/2001, updating Home Office Circular 16/95). However, a wide range of organisational issues is involved in the exercise of these responsibilities, the most important of which are described below.

The first of these responsibilities is standard setting, and here there are a number of ways in which the NDNAD Custodian sets standards for the operation of the database to 'ensure the reliability, compatibility and legality' (Forensic Science Service 2000) of all data held on it. First, the Custodian advises the NDNAD Board on the DNA data that are to be used for the construction of profiles to be held on the database along with the minimum standards that have to be met for each separate profile to be loaded and searched. Second, the Custodian has the duty to establish 'appropriate protocols, procedures and standards of performance' (Forensic Science Service 2000) for database entries, information derived from them and of the reports provided to relevant users. Third, the Custodian sets standards for the specification of all collection kits that may be used by police forces to take samples from individual suspects or volunteers and from scenes of crime. Finally, the Custodian advises the Board on the suitability of laboratories wishing to become suppliers of data to the NDNAD. In this last instance, standards are set by an external body, the United Kingdom Accreditation Service (UKAS), which assesses and, where appropriate, accredits laboratories seeking to supply profiles to the database.[7]

In addition to setting the scientific and procedural standards to be maintained by suppliers of information to the database, the Custodian is also responsible for monitoring the performance of suppliers against those standards. The quality assurance programme includes assessment of both declared samples (where samples are submitted to the laboratory for use in criminal trials) and undeclared samples (where samples are submitted by individual police forces as originating from criminal suspects). All instances of profiles supplied that are 'subsequently found to be in error' (Forensic Science Service 2000) are recorded by the Custodian, who also facilitates the checking of all near-miss matches (matches on all but one allele). Successful completion of proficiency tests by all staff involved in DNA analysis in each supplier laboratory is a condition of continued accreditation.

The NDNAD Custodian has a duty to establish and maintain arrangements for the safe and accurate transfer of data between profile suppliers and the NDNAD as well as to oversee the accuracy, storage, management and deletion of profiles and the demographic data associated with them. Since March 2000, the Information Commissioner has been responsible for enforcing the provisions of the UK Data Protection Act 1998 and for ensuring that data controllers (those who decide how and why 'personal data' are processed within their organisation) comply with its provisions. The Secretary of State and Association of Chief Police Officers (ACPO) are designated as data controllers of the NDNAD, and the Custodian, as data handler, is required

---

[7] The ISO/IEC 17025 standards (*General Requirements for the Competence of Testing* and *Calibration Laboratories and Supplementary Requirements for Accreditation in the Field of Forensic Science*) along with the additional requirements stipulated by the Custodian and set out in the FSS document *The National DNA Database Standards of Performance*. The details of the UKAS assessment are specified by reference to further UKAS documents and international standards (including NIS46, NIS96, ISO25, ISO9000 and ASO9001).

to ensure that access to, and use of, all records on the NDNAD are compliant with the Act.

In summary, the concerns of this juridico-scientific domain are largely focused on the establishment, preservation and enhancement of the reliability of the NDNAD as an aid to criminal investigations in general. The effectiveness of its uses by investigators in individual police forces raises different questions. These are described here as constitutients of an operational domain, and they are considered in the next section.

The operational domain

By far the most important aspect of the NDNAD's operation is that the loading, searching and matching of profiles is carried out as speedily as possible so that information about matches reaches the police in a timely way. There is agreement between the NDNAD Board and the ACPO that speculative searches of newly loaded profiles against existing profiles be made on a daily basis so that any match identified is relayed to the relevant authority. Reports of matches are then issued as soon as possible to the forces who supplied the profiles (crime scene and/or subject samples). In addition, there is provision for the 'one-off speculative searching' of scene-of-crime samples that have provided insufficient allelic data to be permanently added to the NDNAD. Despite not being loaded, such profiles can be re-checked against data-based profiles 'at agreed specified intervals' (Forensic Science Service 2000).

Measurement of the effectiveness and efficiency of NDNAD performance has been modelled on extensive efforts by several government bodies to audit the delivery of forensic science and its usefulness to policing. The late 1980s and early 1990s witnessed a series of positive evaluations of the potential of forensic science to contribute to the effectiveness of criminal investigations and prosecutions. In particular, both the House of Commons Home Affairs Committee (1989) and the House of Lords Select Committee on Science and Technology (1993) made strong arguments for increasing existing budgetary provisions for forensic science in support of crime investigation. These arguments rested on claims of cost-effectiveness, especially when compared with the costs of other investigative strategies, but neither the accountancy firm Touche Ross, who reviewed forensic science support for the Home Office (Touche Ross 1987), nor the House of Lords Select Committee were precise about the possible level of efficiency gains or the general framework for measuring the contribution of forensic science to crime investigation and prosecution.

The ethico-political domain

At the beginning of this century, ACPO stated (in their Memorandum to the House of Lords Select Committee on Science and Technology (2003: 115) that:

> Police fully recognise the sensitivity of maintaining DNA data on individuals on the National DNA Database and we accept the need for high standards of probity/integrity at all stages of the process. That includes the need for DNA profiles to be removed from the Database whenever a person is acquitted in a case for which a DNA suspect sample has been taken or that case is discontinued for whatever reason. Police also acknowledge the concern that people have about genetic information held on the National DNA Database being misused for purposes other than those for which it is originally gathered and stored.[8]

However, this memorandum identified no particular measures designed to address the acknowledged 'concern', and further elaboration of issues of public interest in this document was foreclosed by the rhetorical claim that all 'social, ethical, legal and economic implications of the National DNA Database should be viewed in the light of its enormous success in helping to prevent and detect crime'.

Elsewhere in the world, various bodies largely independent of operational policing and criminal justice administration have been set up to consider the more general social and ethical implications of the growth of forensic DNA profiling and databasing. These bodies have sometimes been formed as commissions in advance of the establishment of national or subnational forensic DNA databases (most recently, in the Republic of Ireland) and sometimes to provide oversight of their subsequent development and uses. When the 1993 Royal Commission recommended the establishment of the NDNAD, although it provided no detailed discussion of its governance, it did suggest the establishment of an 'independent body' with the remit to oversee forensic science.

---

[8] Readers may note that the acknowledgement of the need to remove the profiles of persons acquitted or against whom cases were discontinued quickly vanished following legislative changes in 2001 and 2003.

In fact, no such independent body has yet been established in England and Wales. Instead, the instruments of database governance have largely remained in the hands of a very small network of forensic DNA database stakeholders, in particular members of the scientific agencies that provide profiles and interpret the results of profile comparison, the database custodian and representatives of police users. From its origins in 1995, the main governing body comprised a 'User Board' which was jointly chaired by members of ACPO and the FSS. Relationships between parties, and expectations of their actions, were governed only by a *Memorandum of Understanding* (Forensic Science Service 2000), a form of agreement less binding than a legal contract but one that suited the quasi-clan nature of the relationship between those who were recognised as stakeholders in the NDNAD.[9]

In subsequent years, the User Board became the NDNAD Board, and its membership widened slightly, but only to include more police actors and Home Office officials. In 2003, the strategic focus of the NDNAD Board was sharpened by passing to a subsidiary DNA Operations Group responsibility for practical operational issues relating to the use of DNA profiling and matching. However, the main interests of both of these bodies remained focused on the juridico-scientific and operational domains described above. In particular they focused on issues of 'evidential rigour', 'standard setting and quality assurance' and 'efficiency, effectiveness and cost-effectiveness' as the central preoccupations of their governance regime.

A number of actors have commented critically on this governance arrangement and its outcomes, in particular on the way that it has failed to deal with the wider ethical and social issues that might be expected to be the subject of debate when biological samples are taken from widening categories of individuals (usually without consent), analysed and used to support criminal justice objectives. For example, the House of Lords Select Committee on Science and Technology (2001: para 7.66) recommended that 'the Government should establish an independent body, including lay membership, to oversee the workings of the National DNA Database, to put beyond doubt that individuals' data are being properly used and protected'. A year later, the Human Genetics Commission (2002: 153) made several suggestions of alternative ways to make possible independent participation in the current arrangements for governing the NDNAD and to increase the transparency of its operation: '[A]t the very least, the Home Office and ACPO

---

[9] See Ouchi (1980) for a discussion of 'clan'-based organisations.

should establish an independent body, which would include lay membership, to have oversight over the work of the National DNA Database custodian and the profile suppliers'.

This absence of independent scrutiny contrasts markedly with the governance arrangements for medical DNA databases in England and Wales. It is also significantly different from the arrangements for the oversight of forensic DNA databases found in some other legal jurisdictions. In such contexts, it is generally agreed that oversight bodies are chaired by a respected public figure and that a board will include a substantial representation of individuals who are unconnected with database custodianship or use, but at least some of whom have professional knowledge of relevant scientific and criminal justice matters.

In 2003, a Government-commissioned review of the Forensic Science Service considered – as a minor element of its work – a response to these observations (Home Office 2003). Its author, Robert McFarland, concluded (p. 3) that,

> the present custodian arrangements [of the NDNAD] needed to be made more independent, more transparent and more accountable. .... Equally the Review accepts that there is an overriding public interest in maintaining the effectiveness and operational efficiency of the NDNAD. The Review is recommending that the NDNAD Custodianship is removed from the FSS. It is proposing that the NDNAD database becomes the responsibility of the NDNAD Board reconstituted into a (public sector classified) Company Limited by Guarantee (CLG), with an independent chairman but with a majority of the membership nominated by ACPO.

In the six years since the publication of McFarland's report, a new governance model for the NDNAD has begun to emerge, but changes in the organisation of relevant Home Office departments and related policing agencies (as well as other political priorities) have slowed the speed of its development. A new NDNAD Strategy Board has been established with membership including representatives from the Association of Police Authorities, the Home Office, ACPO (and the Scottish equivalent, ACPOS) and two members from the Human Genetics Commission (one of whom is described as the 'principal advisor to the Board on ethical matters'). An independent Ethics Group 'drawn from a wider audience' was established in 2008 and its first report was published in 2009. However, the NDNAD Board has discretion over whether or not to act on any advice provided on ethical matters by this Ethics Group, and the power of the group has yet to be tested. Finally, the relevance for the future governance of the NDNAD by the establishment of the new post of Forensic Regulator in England and Wales has yet to be determined.

In a range of public statements and reports, the UK Government Ministers and Chairs of the NDNAD Board have consistently declared a commitment to openness, transparency and accountability in the operation of the NDNAD, including its use for research. They have also recognised the importance of these values for the maintenance of public confidence in police uses of forensic genetics. The degree to which discussions of changes in its governance structure themselves remain open, transparent, accountable and inclusive provides an interesting test of the depth of these commitments, and in this respect their record has been rather mixed.

At the time of writing, however, the UK Government is preparing a further White Paper on Forensic Science in England and Wales, and it seems possible that this White Paper will include proposals for further changes in NDNAD governance. In part, this White Paper is a response to the recent decision of the European Court of Human Rights in the case *S and Marper* v. *the United Kingdom* (2008). In this judgment, the Court ruled that the current UK legislation, which allows the retention of DNA samples and profiles from those not convicted of a crime, violates their rights under Article 8 of the *Convention for the Protection of Human Rights* (Council of Europe 1950). Accordingly, there is an expectation that the law in England and Wales will have to be changed in response to this judgment. Arrangements will also have to be put in place to manage the European Court of Human Rights requirement to delete the profiles, and destroy the samples, taken from more than 800 000 individuals who fall into the category of those whose genetic material should not have been retained.

CONCLUSIONS

The UK's leading position in the development of forensic DNA profiling and databasing is largely attributable to the fact that a number of interconnected state, academic, scientific and commercial institutions have enthusiastically supported and facilitated a range of government investments in genetic profiling and databasing. The forensic imaginary underpinning this enthusiasm has been stimulated and shaped by relatively narrow understandings of the uses of bio-identification in criminal investigations and the promise of such uses to achieve a reduction in reported crime through increasing detections and deterring offenders and others from future criminal activity. While the establishment and expansion of such databases are generally celebrated by police actors, little research has been done on how policing in particular, and the administration of criminal justice in general, are re-shaped by the use of (and innovations in) the expanding technologies that make these collections possible.

For some commentators, investment in forensic databases in which individuals' bodily attributes are recorded and which can subsequently be consulted by a variety of policing agencies, contributes to a refashioning of the social relationship between law enforcement officials, citizens and 'suspects' (e.g. Amey et al. 1996; Barton and Evans 1999; Heaton 2000; John and Maguire 2003). Likewise, Innes (2003: 74) argues that capturing identities in searchable databases makes those recorded individuals permanent potential suspects and, as a result, 'in the process of targeting these individuals the organization is more likely to generate further intelligence on them, thus justifying their selection as targets both retrospectively and prospectively'. Similar claims have been made by others who have written about the 'chilling effect' of the expanding uses of genetic profiling and databasing.

Nevertheless, it seems unlikely that criminal justice agencies will lose their enthusiasm for the establishment and expansion of such databases. In such circumstances, the maintenance of public confidence, or trust, in the police uses of such databases is heavily dependent on the openness and transparency of their operation. It is important that individuals and agencies independent of law enforcement have sufficient knowledge about the detailed workings of such databases to arrive at authoritative judgments of the claims made by those who manage them. Such claims include those concerning the confidentiality of records, assurances given to victims and others who give consent to the use and retention of their profiles and samples, the security of the database and the results of various scientific quality assurance trials.

Despite its 15-year life span, only recently have discussions of the governance arrangements of the UK NDNAD begun to address these issues. Here, and elsewhere in the world, a range of individual and institutional actors have continued to imagine and promote established and innovative ways in which police investigators can or could deploy forensic technology, including DNA profiling. However, it is very rare for any of these actors to acknowledge the rather varied historical trajectories taken by these innovations.[10] For the moment, DNA profiling seems to have been the most conspicuously successful of them, but exactly how successful, and with what wider social effects, remain open to question.[11] Finally, there is also a pressing need to document and consider the varied ways in which efforts to govern the uses of these technologies reflect and reconstitute understandings of the social and political identities of the subjects whose bio-identities they seek to capture and profile.

ACKNOWLEDGEMENTS

This chapter draws on work carried out with Paul Johnson and Paul Martin on the police uses of genetic profiling. Further details can be found in Johnson *et al.* (2003), Johnson and Williams (2004a, 2004b), Williams and Johnson (2004a, 2004b, 2005) and Williams *et al.* (2004). All of this work has been generously supported by two Wellcome Trust grants (GR 067513 and GR 073520)

REFERENCES

Aas, K. (2006). The body does not lie; identity, risk and trust in technoculture. *Crime Media Culture*, 2, 143–158.

Amey, P, Hale, C. and Uglow, S. (1996). *Proactive Policing*. Edinburgh: Scottish Central Research Unit.

Association of Chief Police Officers (2005). *DNA Good Practice Manual*. London: Association of Chief Police Officers.

Barton, A. and Evans, R. (1999). *Proactive Policing on Merseyside Police*. London: The Stationery Office.

Borup, M., Brown, N., Konrad, K. *et al.* (2006). The sociology of expectations in science and technology. *Technology Analysis and Strategic Management*, 18, 285–298.

Caddy, B., Taylor, G. and Linacre, A. (2008). *A Review of the Science of Low Template DNA Analysis*. London: The Stationery Office.

Caplan, J. and Torpey, J. (2001). *Documenting Individual Identity: The Development of State Practices in the Modern World*. Princeton, NJ: Princeton University Press.

---

[10] In the case of fingerprinting, for example, critical work by Cole (2001, 2004), Dror and Charlton (2006) and others have been largely ignored by the expert community of fingerprint experts and have had no discernable effect on the willingness of courts to allow fingerprint evidence as expert evidence.

[11] The judicial challenge to low template DNA has been answered by the work of Caddy *et al.* (2008), commissioned by the new Forensic Regulator for England & Wales. However, the much more devastating effects on the NDNAD retention regime of the European Court of Human Rights Judgment in the case of *S and Marper* are only now beginning to emerge (Home Office 2009).

Cole, S.A. (2001). *Suspect Identities: A History of Fingerprinting and Criminal Identification.* Cambridge MA: Harvard University Press.

Cole, S.A. (2004). Fingerprint identification and the criminal justice system: historical lessons for the DNA Debate. In *DNA and the Criminal Justice System: The Technology of Justice*, ed. D. Lazer. Cambridge, MA: MIT Press, pp. 63–90.

Council of Europe (1950). *Convention for the Protection of Human Rights and Fundamental Freedoms.* Strasbourg: Council of Europe http://conventions.coe.int/Treaty/Commun/QueVoulezVous.asp?NT=005&CL=ENG (accessed February 2010).

Dror, I. and Charlton, D. (2006). Why experts make errors. *Journal of Forensic Identification*, 56, 600–616.

Eco, U. and Sebeok, T. (1983). *The Sign of Three: Dupin, Holmes, Peirce.* Bloomington IL: Indiana University Press.

Ericson, R. and Haggerty, K. (1997). *Policing the Risk Society.* Oxford: Oxford University Press.

Forensic Science Service (2000). *Memorandum of Understanding between the Association of Chief Police Officers and the Custodian of the National DNA Database.* Birmingham: Forensic Science Service.

Foucault, M. (1977). *Discipline and Punish: the Birth of the Prison.* London: Allen Lane.

Foucault, M. (1979). *The History of Sexuality*, Vol. 1: *An Introduction.* London: Allen Lane.

Foucault, M. (2003). *Society Must Be Defended.* London: Picador

Foucault, M. (2004). *Security, Territory, Population.* London: Palgrave Macmillan.

Gerlach, N. (2004). *The Genetic Imaginary: DNA in the Canadian Criminal Justice System.* Toronto: University of Toronto Press.

Gill, P., Jeffreys, A. and Werrett, D. (1985). Forensic application of DNA 'fingerprints'. *Nature*, 318, 577–579.

Heaton, R. (2000). The prospects for intelligence-led policing: some historical and quantitative considerations. *Policing and Society*, 9, 337–356.

Home Office (2003). *Executive Summary of McFarland Review of the Forensic Science Service.* London: The Stationery Office.

Home Office (2009). *Keeping the Right People on the DNA Database: Science and Public Protection.* London: The Stationery Office.

House of Commons Home Affairs Committee (1989). *Report on 1988–1989 Session.* London: HMSO.

House of Lords Select Committee on Science and Technology (1993). *Fifth Report on Forensic Science.* London: HMSO.

House of Lords Select Committee on Science and Technology (1997). *The Governance of Public Bodies.* London: HMSO.

House of Lords Select Committee on Science and Technology (2001). *Human Genetic Databases: Challenges and Opportunities.* London: The Stationery Office.

House of Lords Select Committee on Science and Technology (2003).*Committee Report.* London: The Stationery Office, p. 115.

Human Genetics Commission (2002). *Inside Information: Balancing Interests in the Use of Personal Genetic Data.* London: The Stationery Office.

Inman, K. and Rudin, N. (2001). *Principles and Practice of Criminalistics: The Profession of Forensic Science.* London: CRC Press.

Innes, M. (2003). 'Signal crimes': Detective work, mass media and constructing collective memory. In *Criminal Visions: Media Representations of Crime and Justice*, ed. P. Mason. Cullompton, UK: Willan, pp. 51–69.

Jeffreys, A., Wilson, V. and Thein, S. (1985). Individual-specific 'fingerprints' of human DNA. *Nature*, 316, 76–79.

John, T. and Maguire, M. (2003). Rolling out the National Intelligence Model: key challenges. In *Essays in Problem-oriented Policing*, eds. K. Bullock and N. Tilley. Cullompton, UK: Willan, pp. 38-68.

Johnson, P. and Williams, R. (2004a). DNA and crime investigation: Scotland and the UK National DNA Database. *Scottish Journal of Criminal Justice*, 10, 71-84.

Johnson, P. and Williams, R (2004b). Post-conviction DNA testing: the UK's first exoneration case? *Science and Justice*, 4, 77-82.

Johnson, P., Martin, P. and Williams, R. (2003). Genetics and forensics: a sociological history of the National DNA Database. *Science Studies*, 16: 22-37.

Lambourne, G. (1984). *The Fingerprint Story*. London: Harrap.

Lyon, D. (2001). *Surveillance Society: Monitoring Everyday Life*. Buckingham, UK: Open University Press.

Lyon, D. (ed.) (2006). *Theorizing Surveillance: The Panopticon and Beyond*. Cullompton, UK: Willan.

Martin, L., Gutman, H. and Hutton, P. (eds.) (1988). *Technologies of the Self*. London: Tavistock.

Nickell, J. and Fischer, J. (1999). *Crime Science: Methods of Forensic Detection*. Lexington KY: University of Kentucky Press.

Ouchi, W. (1980). Markets, bureaucracies and clans. *Administrative Science Quarterly*, 25, 129-141.

Rose, N. (1985). *The Psychological Complex: Psychology, Politics and Society in England 1869-1939*. London: Routledge & Kegan Paul.

Rose, N. (1990). *Governing the Soul: The Shaping of the Private Self*. London: Routledge & Kegan Paul.

Rose, N. (1996). *Inventing Ourselves: Psychology, Power and Personhood*. Cambridge, UK: Cambridge University Press.

Rose, N. (1999). *Powers of Freedom*. Cambridge, UK: Cambridge University Press.

Sengoopta, C. (2003). *Imprint of the Raj: How Fingerprinting Was Born in Colonial India*. London: Macmillan.

Spufford, F. and Uglow, J. (eds.) (1996). *Cultural Babbage: Technology, Time and Invention*. London: Faber.

Taylor, C. (2004). *Modern Social Imaginaries*. Durham, NC: Duke University Press.

Torpey, J. (1998). Coming and going: on the state monopolization of the legitimate 'means of movement'. *Sociological Theory*, 16, 239-259.

Torpey, J. (2000). *The Invention of the Passport: Surveillance, Citizenship and the State*. Cambridge, UK: Cambridge University Press.

Touche Ross (1987). *Review of Scientific Support for the Police*. London: The Stationery Office.

Valier, C. (1998). True crime stories: scientific methods of criminal investigation, criminology and historiography. *British Journal of Criminology*, 38, 88-105.

Valier, C. (2001). Criminal detection and the weight of the past: critical notes on Foucault, subjectivity and preventative control. *Theoretical Criminology*, 5, 425-443.

van der Ploeg, I. (2005). *The Machine-readable Body: Essays on Biometrics and the Informatization of the Body*. Maastricht: Shaker Publishing.

Williams, R. and Johnson, P. (2004a). Circuits of surveillance. *Surveillance and Society*, 2, 1-14.

Williams, R. and Johnson, P. (2004b). 'Wonderment and dread': representations of DNA in ethical disputes about forensic DNA databases.*New Genetics and Society*, 23, 205-222.

Williams, R. and Johnson, P. (2005). Inclusiveness, effectiveness and intrusiveness: Issues in the developing uses of DNA profiling in support of criminal investigations. *Journal of Law Medicine and Ethics*, 33, 545-558.

Williams, R., Johnson. P. and Martin, P. (2004). *Genetic Information and Crime Investigation*. London: Wellcome Trust.

Zureik, E. and Salter, M. (2005). *Global Surveillance and Policing*. Cullompton, UK: Willan.

CASE

S and Marper v. *the United Kingdom* (2008). A summary of the judgment is available from http://cmiskp.echr.coe.int/tkp197/view.asp?action=html&documentId=843937&portal=hbkm & source=externalbydoc number&table=F69A27FD8 FB86142BF01C1166DEA398649.

BARBARA PRAINSACK

8

Partners in crime: the use of forensic DNA technologies in Austria

INTRODUCTION: AUSTRIA'S 'FIRST SERIAL KILLER' AND THE ESTABLISHMENT OF THE FORENSIC DNA DATABASE

In spring 2008, the small Alpine republic of Austria made it into world headlines in the reporting on a bizarre and disconcerting criminal activity. On 27 April, police had discovered a cellar dungeon in which 42-year-old Elisabeth F. had been abused and held captive by her father for 24 years. During that period she bore seven children, of whom six survived. Upon being rescued by the police, some of Elisabeth's children saw sunlight for the first time in their lives. The entire world was in shock.

Many Austrians remembered that this was not the first time they had made headlines with a horror story. Johann ('Jack') Unterweger, the illegitimate son of a Viennese prostitute and a member of the US Armed Forces, 'gave Austria its first serial killer', as an Austrian weblog put it.[1] At age 24, Unterweger was sentenced to life in prison after murdering an 18-year-old German woman who, as he later stated, reminded him of his mother (Gepp 2007). He spent his life in prison writing short stories and he also authored a book, an autobiography titled *Purgatory: A Journey to Jail* (1983), which in a twist had also rendered him the darling of the local celebrity scene. Released from prison in 1990 – after having served only 16 years of his life sentence as a result of devoted lobbying efforts on the part of celebrity friends – Unterweger was celebrated as a model case of rehabilitation. Soon, however, he was to return to his old passion of murdering women.

---

[1] This can only be maintained, of course, if political mass murders (which are not defined as serial killings) are not counted.

Genetic Suspects: Global Governance of Forensic DNA Profiling and Databasing, ed. Richard Hindmarsh and Barbara Prainsack. Published by Cambridge University Press. Copyright © Cambridge University Press 2010.

Over two years, he strangled at least nine women in Austria, the Czech Republic and in the USA, mostly prostitutes. In 1992, he was arrested in Miami, Florida, and extradited to Austria. On 29 June 1994, he received his second life sentence for nine counts of murder (Leake 2007). In the same night, he hanged himself in his cell with the string of his track bottoms.

Apart from 'giving' Austria its first serial killer', the Unterweger case marked a milestone in another respect: It was the first case in Austrian history in which decisive evidence for the conviction was provided by DNA analysis. Unterweger had hidden the bodies of his Austrian victims in the *Wienerwald*, the woods in the vicinity of the city of Vienna; they were discovered months after being killed. A hair of one of the victims found in Unterweger's car, as well as a fibre of his scarf on one of the corpses, eventually gave him away.

Less than three years later, on 1 October 1997, the Austrian Forensic DNA Database became operational as one of the first world-wide, with Europe's first forensic DNA database having become operational about two years earlier in England and Wales (see Chapter 7).[2] One year after its establishment, police announced that 149 crimes had been solved on the basis of profiles stored in the database; by 2004, that number had grown to over 1400 and continued to rise to 9973 in 2008 (European Network of Forensic Science Institutes 2009). Measured by the ratio of the number of subject samples stored in the database to the total population, the Austrian forensic DNA database is among the largest in Europe and even in the world (Prainsack 2008a) (Table 8.1).

This chapter starts by outlining the relevant legal provisions as well as actual practices pertaining to forensic DNA profiling and data-basing in Austria and discusses the circumstances that led to the establishment of the forensic DNA database. Drawing upon empirical research carried out in 2006 and 2007, it provides an overview of understandings and practices of forensic DNA databasing on the part of individuals in law enforcement, on the one hand, and the understandings and responses of convicted criminal offenders on the other. The final section reflects on the impact that DNA profiling and data-basing has had on crime commission and crime investigation. This

---

[2] The National DNA Database (NDNAD) in England and Wales was officially established in April 1995. As Van Kamp and Dierickx (2007: 9) point out, however, the method of forensic DNA profiling had been used to solve crimes in the UK much earlier, namely from 1985 onwards; the first person being convicted on such evidence being Colin Pitchfork (Walker and Cram 1990; Sanders 2000).

Table 8.1. *Comparison between the largest forensic DNA databases in Europe (June 2009)*[a]

	No. subject profiles[b]	No. crime scene profiles[c]	Ratio subject profiles to crime scene profiles	Total population (millions)	Ratio subject profiles to total population	Ranking by subject profile to total population ratio
UK	4 458 340	329 482	14:1	53.7	1:12	1
Scotland[d]	202 236	9 987	27:1	5.1	1:22	2
Austria	116 258	32 032	4:1	8.1	1:69	3
France	856 911	110 000	3:1	59.3	1:69	3
Switzerland	104 625	21 278	5:1	7.3	1:70	4
Germany	611 867	145 122	4:1	82.4	1:134	5

[a] Figures for profiles and populations are approximate.
[b] Subject profiles are profiles obtained from cheek swabs from suspects, convicts, victims, volunteers, etc.
[c] Crime scene profiles have been obtained from stains found at crime scenes.
[d] The Scottish database is linked to the UK database in the sense that Scotland exports profiles (but not samples) to the UK database.
Sources: Nuffield Council 2007; European Network of Forensic Science Institutes 2009; personal communication.

discussion is particularly relevant to one of the core issues of this volume, namely, how new technologies reinforce and/or alter prevailing configurations of power and governance.

First, let us take a look at the legal and regulatory framework in which the Austrian forensic DNA database is embedded.

LEGAL AND REGULATORY BACKGROUND OF THE
AUSTRIAN FORENSIC DNA DATABASE

The custodian of the Austrian forensic DNA database is the Federal Ministry of the Interior (*Bundesministerium fuer Inneres*). Its legal basis is the Security Police Act (*Sicherheitspolizeigesetz*, SPG, BGBl I Nr 56/2006). Of particular relevance are §67–75. These sections stipulate that DNA can be obtained in the context of forensic identification only if the person in question is suspected of having committed a dangerous assault (*gefährlicher Angriff*),[3] and if, with regard to the deed,[4] or the personality[5] of the suspect, it can be expected that the person will leave, in the course of committing further dangerous assaults, further traces that enable his or her identification (*Wiedererkennung*) on the basis of the retrieved genetic information (SPG §67.1).

The law also stipulates that the contractor who performs the molecular genetic examination must not receive any information about the identity of the person (*erkennunsdienstliche Identitätsdaten des Betroffenen*) (SPG §67.2). Laboratories may examine only parts of the DNA that serve the purpose of identification, which rules out phenotypic profiling (see Chapter 2). In addition, in all cases where authorities are obliged to delete the *data* stored in the database, the physical *samples* must be destroyed as well (SPG §67.3). The conditions under which the data stored in the database must be deleted are listed in §73.1: first, when the originator of the DNA has reached 80 years of age, if s/he has not undergone any examination for forensic identification purposes in the previous five years; second, if the originator of the DNA had been a minor at the time of the provision of the material for forensic identification and if s/he was not examined for forensic identification purposes in the previous three years; third, five years after the death of the originator; and, fourth, if the originator of the DNA is no longer suspected of having committed a dangerous assault. In these cases, data deletion on the authorities' own initiative is mandatory unless the authorities' retention of the data is necessary because of concrete circumstances indicating that the person will commit further dangerous assaults (two more cases apply only in specific circumstances, see §73.1, points 5–6). The subsequent paragraph, §74, pertains to

---

[3] A 'dangerous assault' is defined in SPG §16[2] as a threat posed to a legal good through the unlawful commission of a crime which is punishable by a court and which has been committed with intent, and not only upon the request of a another person (*Beteiligten*). Furthermore, the criteria of a 'dangerous assault' are met only if (1) the crime is punishable according to the Penal Code [few exceptions]; or (2) it is punishable according to the Prohibition [of Nazi activities] Law (*Verbotsgesetz*); or (3) it is punishable according to the Illegal Substances Law (*Suchtmittelgesetz*), unless the possession or acquisition of the illegal substance is only for the purpose of the suspect's consumption.

[4] By means of internal guidelines (*Deliktkatalog per Erlass*) the Federal Criminal Police Office authorised a list of crimes in the context of which the retrieval of DNA (by means of buccal swabs) should be standard procedure; these are considered to be categories of crime with a particularly high risk of recidivism, such as property and sex crimes, crimes involving illegal substances, violent crimes.

[5] This condition is met if the suspect already has a criminal record.

the deletion of stored DNA profiles on request of the originator. This is the case if the original suspicion against the originator no longer exists, or if the deed has been found to be not unlawful.[6]

Other notable features of governance include §72, which stipulates that data obtained for forensic identification purposes (*erkennungsdienstliche Daten*) may be provided to Austrian universities as well as to federal ministries for evaluation in the context of academic and scientific research. However, the identification of particular individuals must not result from such research.

In addition, Austrian law does not specify a minimum age of persons from whom DNA can be taken for forensic identification purposes. In practice, however, police officers typically refrain from taking DNA of individuals younger than 10 years. Another feature, which has drawn severe criticism from privacy watchdog groups (including *Arge Daten* Privacy Service, which criticised the potential extension of police power) was the legalisation in 2004 of DNA dragnets through revision of the Code of Criminal Procedure (*Strafprozessordnung*; effective from January 2008; see, in particular, §123.2). Austrian police representatives countered the criticism by referring to their general policy of restraint, in that DNA dragnets would be used only if deemed to be absolutely necessary (that is, in high-profile cases in which no other means of investigation could be expected to render any clues), and only in crimes for which the law prescribes prison sentences of five years or more. Any dragnet would need to be requested by a public prosecutor and approved by a court. So far, no dragnet has been carried out in Austria.

In general, those representatives of the police and the Ministry of Interior whom I interviewed were generally aware of the benefits of both tight regulation and monitoring of forensic DNA technologies. Public support for the existing laws, as well as for investigative police work in general, was recognised as a necessary precondition of the smooth functioning of their work.

---

[6] In practice, §73.1 accounts for the vast majority of all data deletions and sample destructions. Most of these are through the expiry of the time limits for profiles storage. Section 74, which grants the former suspect the right to request the deletion of the data, is applied in only a small number of cases annually. Most of these cases pertain to suspects who escape conviction through diminished responsibility at the time of the commission of the deed. Suspects receive an information sheet explicating this right at the time of the provision of the DNA sample.

FORENSIC DNA PROFILING AND DATABASING

Having looked at the relevant laws, let us now examine how forensic DNA profiling and databasing is carried out in practice.

PRACTICES OF LAW ENFORCEMENT AUTHORITIES

Every arrest procedure routinely entails the collection of verbal data, photographs and dactyloscopic data in the form of whole palm prints from the suspect. Although Austrian law does not restrict the range of crimes in which DNA may be taken from suspects, in practice (and based on a decree (*Erlass*) of the Ministry of Interior), DNA is only taken on suspicion of a limited number of crimes (see above). Verbal data comprises all personal details as reported to the authorities by the arrested individual, including details displayed on legal documents and identity cards. Second, photographs are taken of the face and body of the person as well as so-called 'particular characteristics', such as tattoos. Third, dactyloscopic data are obtained with the help of an automatic 10-finger and palm flat-screen scanner, which feeds the data directly into the database(s). Dactyloscopic data, as police representatives informed me, are the best tool for forensic identification purposes at the time of arrest, because they are 'unique for every individual and relatively cheap to obtain and process'.

Approximately 25 000 cases of evidence taking, as described above, are carried out nationwide per year. Less than half include the retrieval of a DNA sample. In cases where a DNA sample is taken, the officer first checks whether the arrestee's profile is already stored in the database, with the help of finger and palm prints. While arrestees could attempt to hide their identity by giving a false name, the dactyloscopic data would betray them. If the finger and palm prints produced a match in the database it could be assumed that the person was not a first-time arrestee and might have a DNA profile stored as well. Subject profiles entered into the DNA database stem from two groups of people: suspects (from them, DNA may be taken by force, typically from the forehead or from the neck[7]) or individuals under no suspicion who

---

[7] DNA samples obtained by force are normally not derived from blood but taken in a situation in which one or two officers restrain the suspect and another takes a skin swab from the forehead (if the head of the suspect is being held back) or the neck (if the neck of the suspect is fixed in a forward position).

might have left their DNA at the crime scene unintentionally but for legitimate reasons ('persons of happenstance', *Gelegenheitspersonen*). Typically, these are partners, family members, flat-mates and neighbours, cleaning personnel, but also victims. Their DNA profiles are used only for elimination purposes: If their DNA profiles match with profiles from crime scene traces, these matches will be known not to refer to the perpetrator (unless a 'person of happenstance' later turns into a suspect, of course). DNA profiles obtained from such 'persons of happenstance' are not entered into the database, which means they are not used for routine searches against profiles from crime scene traces (so-called 'speculative searches').[8]

Theoretically, elimination samples from 'persons of happenstance' could be obtained by force. However, according to my informants, this does not happen very often in practice. This is because, if a 'person of happenstance' is reluctant to provide a buccal swab voluntarily, they are then informed that they could be labelled a suspect; in that case, their DNA profile would be uploaded to the database and used for speculative searches in the future. If they provided a sample voluntarily they would remain a 'person of happenstance'.

Turning to the transportation of DNA samples to the contracting laboratories, certain barcodes are used for suspects/convicts and then other barcodes for each of the following groups: volunteers ('persons of happenstance'); victims; profiles of unidentified corpses; profiles of missing persons if there are concrete indicators for a crime, suicide, or an accident; and, finally, profiles of police officers. Different levels of comparison (routine or ad hoc) also apply to profiles held in these different sections of the database.

To reduce the risk of contamination of crime scene evidence with suspect DNA, crime laboratories are legally required to process crime scene traces and subject samples on different premises. The DNA analyses are carried out at three different laboratories, all part of academic research institutions: the Departments of Forensic Medicine at the University of Innsbruck, the Medical University of Vienna and the University of Salzburg. Contracts with these institutions stipulate a maximum number of crime scene and subject samples to be analysed; at present, the limit is at approximately 4000 crime scene traces and 12 000 subject samples per year. (Excluded from this

---

[8] An exception are profiles of police officers who routinely work at crime scenes; their profiles are stored at a separate database and routinely compared with crime scene traces for purposes of elimination.

limitation are DNA samples related to the investigation of particularly serious crimes, for which the courts cover costs.) If the actual number of samples submitted for analysis is below the limit in any given year, then the Ministry receives a 'traces credit' (*Spurenguthaben*), which can be carried forward to the following year. Because subject samples are typically uncontaminated with other substances (retrieving and storing DNA from a buccal swab is a relatively straightforward task), they are much cheaper to process than DNA samples retrieved from crime scenes under difficult technical circumstances from various surfaces and in varying quantities. The average cost of the creation of a DNA profile based on a crime scene sample is currently about €255, while a profile based on a buccal swab (the typical 'subject profile') lies at about €90.

The strict limit on the number of crime scene samples which can be submitted for analysis forces crime scene investigators to divide DNA traces secured at crime scenes into two groups: the first comprises samples deemed essential to be submitted for analysis; the second group will be stored and perhaps analysed later if in the course of the investigation it turns out that a particular sample might bear clues to the solution of the case. For example, if a murder has happened on the third floor of an apartment building, crime scene investigators might secure cigarette butts from the front of the building. For reasons of cost containment, these cigarette butts will not be submitted for DNA analysis but go into storage. If, for example, a witness indicated that s/he had seen a stranger smoking a cigarette in front of the building prior to the crime, the cigarette butts would then be submitted to a forensic laboratory for DNA analysis. Besides these tactical considerations, decisions on what traces to analyse and which ones to store are made also according to the quality of the sample. Blood, saliva and sperm, for example, are considered 'good' media from which usable DNA is likely to be extracted, whereas other media such as sweat are more problematic. Consequently, for analysis, evidence in the form of 'good' media usually has priority.

The DNA profiles established in the laboratory consist of 11 so-called 'system values' (*Systemwerte*), which are results from 10 genetic loci plus the result from the analysis of the sex chromosome (see Chapter 2). If the quality of the sample allows it, the profile derived from it is fed automatically into the database. Reinhard Schmid, the Director of the Department of Forensic Identification at the Austrian Federal Criminal Police Office, asserted that the automatisation of this process renders the Austrian database relatively immune to mischaracterisations of profiles, as the greatest potential for mistakes is inherent in manual transfer of results into the database (as is routinely done in Germany, for example): 'Human beings are prone to typing mistakes' (interview Schmid). In case a profile matches another one in the database, the Ministry requests a so-called 'confirmatory analysis' (*Bestätigungsanalyse*) from the relevant laboratory. (In such cases, a second DNA sample, which had been submitted together with the first but not yet analysed, is used.)

In sum, the practices of forensic DNA profiling and databasing in Austrian reality are, of course, clearly less glitzy and fast paced than usually portrayed in television thrillers. In addition, my informants seemed far less enthusiastic about high-technology solutions in crime investigation than their television 'colleagues'. The arduous, time-consuming and often monotonous nature of their work was often emphasised. But what do people in law enforcement generally think about forensic DNA technologies? Are they happy with the legislative and regulatory situation? Do they agree with media coverage that DNA profiling and databasing have 'revolutionised' crime investigation and crime prevention?

The views of law enforcers: 'when you have a hit, this is when the real work starts'

When asked whether he would prefer a situation such as in England and Wales, where DNA profiles and samples can be retained even if the suspect has been acquitted (seeChapter 7), the Director of the Department of Forensic Identification at the Austrian Federal Criminal Police Office answered that in his opinion there was no need to include profiles of individuals without a criminal record. According to Dr Schmid, the Austrian system works very efficiently in the sense that individuals identified for DNA sample retrieval upon arrest are rarely found innocent later. The low number of complaints, and even of applications for deletion of data and samples, he says, proves his point: 'We have very few deletions [due to a suspect's being found not guilty], and hardly any complaints. Except, maybe, the occasional murderer [who complains]. But then, this probably only proves the success of our approach.'

In contrast, when I asked one of my informants on the DNA team of the Criminal Police Office of Vienna (*Landeskriminalamt Wien*), whether he was in favour of a more permissive legislative framework pertaining to retention of profiles and samples, he answered affirmatively: 'In the context of DVI [disaster victim identification], it would be handy [to have a similar legal framework for forensic DNA storage as in England and Wales]. ... And for criminal investigation, this would certainly help to deter people [that is, to have as many people in the database as possible].'

My informants also told me that the Ministry of Interior had encountered much resistance from stakeholders, especially from the prison population, when it first took steps to establish a forensic DNA database. With many prisoners refusing to have their DNA taken, initial success rates based on database hits were modest: When it came to the number of crime scene profiles matching subject profiles, 'we barely made it to a two-digit hit rate'.

But then, a crucial breakthrough based on forensic DNA evidence helped their cause, reminiscent of the Unterweger case in 1994 that catalysed public support for the establishment of the forensic DNA database. Between 1988 and 1990, a series of cruel murders of girls and young women in Favoriten – a district of Vienna – engrossed the nation. The murders of Christina Beranek, Alexandra Schriefl and Nicole Strau confronted the police with a riddle. It was assumed that the same person had committed all three murders, but the investigations were unproductive. A decade later, in September 2000, a DNA sample was taken as part of a routine arrest procedure from a man involved in a violent fight. The man's DNA sample instantly matched a profile from the crime scene of murder victim Alexandra Schriefl. The suspect, Herbert P., was tried and eventually convicted and sentenced to 18 years in prison. Shortly thereafter, in October 2001, 33-year-old Michael P. was arrested after his DNA profile was found to match evidence from the crime scene of the murder of eight-year-old Nicole Strau in December 1990. Michael P., who had been the boyfriend of the victim's aunt at the time of the murder, was always a suspect but had refused to provide a DNA sample for comparison with the evidence secured at the crime scene. Only when arrested in connection with a series of burglaries in 2001, when a DNA sample was routinely taken, could he be linked to the crime scene? Michael P. was charged with raping and murdering the girl, and eventually convicted as well. The extensive media attention aided the forensic DNA team in their quest to expand police competencies around forensic DNA databases. After the breakthroughs in connection with these murders in Favoriten, one of my informants explained: 'We got approval for almost anything'.

The law enforcers interviewed agreed that the biggest impact that forensic DNA databases has had in the field of criminal investigation is that they enable 'cold hits' – that is, they can establish a link not previously possible between crime scene traces and subjects. This ability to generate a virtual link between otherwise unconnected individuals or places is an important characteristic of 'second-generation' evidence (Murphy 2007). Apart from cold hits, my inform-ants in law enforcement did not consider DNA evidence as particularly powerful or 'unique' in any way. In contrast to television representa-tions of crime scene investigators, who notoriously refer to DNA as a new 'language of truth' (for critical discussions of such claims, see Lynch 2003; Dahl 2008), police representatives were critical of the overuse of DNA evidence. In some cases, they said, 'good old' police methods such as interrogation or the analysis of footprints and fibres were much more meaningful. As an Austrian police official put it: 'When you have a hit, this is when the real work starts … It is not like on TV where you have a hit, and then the case is closed.' So, I asked, do all crime investigators share this view? 'Yes', I was told, 'this is well known on the part of law enforcement, and the more specialised an investigator or officer, the less likely she or he is to be affected by the "*CSI* effect" – the belief that DNA evidence is a privileged way to the truth.[9] … It is mainly the offenders who think that DNA profiling is scientifically infallible and always perfectly accurate.' But to what extent do offenders hold this same view?

What offenders say: DNA behind bars

What do criminal offenders think about DNA evidence? Frequent refer-ences are made to this question in conversations with representatives of law enforcement authorities and in the literature (e.g. Dahl 2008), especially about how criminals try to avoid leaving DNA traces and how much probative value they think DNA evidence has. This triggered my interest in exploring this question in more depth, including what implications this might have for governance of forensic DNA data-bases. With permission from the Austrian Ministry of Justice, I approached administrators of two large prisons, requesting a list of potential informants based on the following two criteria: first, the crime that led to the current imprisonment dated back no longer than 1997 (the year in which the DNA database in Austria became

---

[9] The empirical relevance of the '*CSI* effect' has recently been challenged by Schweitzer and Sachs (2007). See also Hewson and Goodman-Delahunty (2008); Hughes and Magers (2007); Houck (2006); Cole and Dioso (2005).

operational), and, second, the individual had a prison sentence of more than 18 months. The length of the prison sentence was included as a requirement to ensure that the crimes were sufficiently serious to motivate a search for DNA trace evidence. During the recruitment process, prisoners were asked to participate in a social science research project on crime scene traces, particularly DNA traces. Twenty-six volunteered for interviews; in most cases, I had access to case files on the circumstances leading to the person's imprisonment.

The 26 volunteers were all male, as interviews were carried out in male-only prisons. This population was typical of Austria for my criteria, as less than 5% of all prisoners serving sentences of more than 18 months are female.[10] Respondents were between 20 and 60 years of age at the time of the interviews (median, 32 years). The most frequent crimes committed were murder or attempted murder, battery, rape, robbery (both armed and unarmed) and sexual abuse of minors (for more details, see Table 1 in Prainsack and Kitzberger 2009). Prison sentences ranged from 18 months to life (for some individuals, the date of release was unclear because of their participation in special therapy programmes that end when a psychological or psychiatric expert evaluates the prisoner's progress positively; §21.2 of the Austrian Penal Code). In just under half, DNA evidence played a significant role in the investigation and/or trial.

In the interviews, carried out in closed rooms without the presence of a prison guard and which lasted 70–80 minutes on average, I avoided starting the conversation with questions about DNA evidence. Instead, I was interested in hearing my informants' life stories first, and then the stories related to the crime that had led to imprisonment. If DNA profiling, or the DNA database, did not come up in my informants' accounts, then I asked about this in the second half of the interview. Had they ever heard about DNA profiling and the forensic database; in what context had they ever talked about it to anybody, and did they have any personal experience with it? While most prisoners stated '[I] do not know much about it at all', some had considerable knowledge about how to best avoid leaving DNA traces. All, however, held strong opinions regarding 'good' and 'bad' uses of forensic DNA databases.

---

[10] On 1 September 2007, 209 (4.2%) out of a total of 4983 individuals serving prison sentences of more than 18 months were female (figures obtained by a database search of *Integrierte Vollzugsverwaltung*, courtesy of the Enforcement of Sentences Office (*Vollzugsdirektion*) of the Republic of Austria; see Prainsack and Kitzberger (2009).

Knowledge and opinions on forensic DNA technologies revolved around three core themes: first, DNA evidence during the commission of the crime, and during criminal investigation; second, the values and risks of DNA profiling and databasing; and third, the effects of forensic DNA profiling and databasing on my informants' own identities and lives.

DNA evidence during commission of crime and criminal investigation

A striking finding of the study was that, on the whole, DNA was not seen as a particularly 'dangerous' kind of evidence in the prisoners' stories. Generally, my informants attributed success and failure in their operations to human behaviour: They thought that criminals sufficiently skilled and disciplined (e.g. not being too greedy) would not get caught, while beginners – or those who chose the wrong accomplices – were prone to get caught (Prainsack and Kitzberger 2009). However, weight was given to DNA evidence once it had entered the stage. Offenders then considered their case to be lost. It emerged that part of the professional knowledge of many informants was that one should deny charges as long as possible but one should confess once DNA evidence was found. An example was Ygor, convicted of rape (which he admitted to committing).

YGOR: I denied everything for almost three days, up to the point when the criminal investigator told me that they had found DNA traces of me and the victim – and he said: 'One thing I can assure you of: if you confess now you'll get three years, but if you don't, then it will be six to ten'. And then I thought: now it's time.

INTERVIEWER: But – if I may say so – it wasn't extremely smart on your part to deny charges in the beginning, because you knew that you had not used a condom, right?

YGOR: That's true, but I had not expected [that they would find DNA]. Denial has always been the best method. Deny it until it's too late! [Thus] then I thought, before it's all too late, I'll confess.

Also interesting was that, besides thinking that one should admit one's crime as soon as DNA evidence has been found, informants also thought that upon being confronted with DNA evidence one should even admit crimes *not* committed. Gert, for example, charged with rape but insisting he had not done it, confessed immediately when he was told that incriminating DNA evidence had been found; he said that he was 'not as dumb as not to know' that DNA evidence was 'always correct'. In attempting to explain how DNA evidence had turned up at the scene of a crime he said he had never committed, Gert did not challenge the scientific accuracy of forensic DNA profiling but rather believed that the victim had 'set him up' by planting Gert's DNA at the crime scene (Prainsack and Kitzberger 2009). Some expressed their belief in the science by stating that the results of DNA profiling were, as Walter put it, '1000% true'.

It might seem paradoxical, then, that some prisoners seemed relatively careless when it came to avoiding leaving traces. Many stated they 'did not think' about traces at all when committing their crime. This was particularly the case in drug-related crimes (such as burglaries committed in an intoxicated state), or in cases of first-time offenders in the context of violent crimes within the family or sex crimes (Prainsack and Kitzberger 2009). While the careless attitude of offenders could partly be explained by ignorance, intoxication or the overwhelming urge or need that they felt when committing their crime, my impression was that some also wanted to get caught and sometimes even longed for punishment. Walter, for example, described the moment his sentence (for rape and several counts of attempted rape) was announced: 'At two o'clock I went back to the courtroom. The light was switched off, and in front of me the judge had lit two candles, and this is when I thought: "Now it's time, now the death sentence will be announced".' I had no other explanation for Walter's mentioning the death sentence (the death penalty was abolished in Austria in 1968) other than his deep conviction that he had done something so horrible that he deserved to die.

Another reason for an offender's 'carefree' attitude could be the conviction that everything would go very smoothly, as in Christoph's case (he had kidnapped a teenage girl and demanded ransom from her aunt):

INTERVIEWER: Did you *ever* think about traces throughout the commission of your crime?

CHRISTOPH: No. Never in my life would I have thought that they would catch me. Not at all. I was so sure. What the heck? They had neither a DNA [sic] of me nor a fingerprint. Where would they find it?

DNA profiling and databasing is 'a good thing'

Perhaps somewhat surprisingly at first, most informants saw it as 'a good thing' that forensic DNA technologies existed. The main reasons given for this view were (a) that it was good for catching the truly bad guys (such as child rapists and murderers), and (b) that DNA profiling could also prove innocence. Concomitantly, many informants also mentioned a third factor, the possibility of misuse of the technology (mainly by planting evidence but also by carrying out a 'sloppy' investigation under media pressure for a conviction). Finally, they thought that the existence of forensic DNA profiling and databasing made the life of certain criminals harder. Let us take a look at some emblematic statements for those central themes.

Zeno, a so-called 'poly-criminal' person convicted of various offences against the Addictive Drugs Law, including battery and attempted armed robbery, had had his DNA profile stored in the national forensic DNA database for many years. Nevertheless, he was of the opinion that DNA profiling was a positive development:

> The DNA [sic] has solved many crimes … If somebody did something 20 years ago, before [DNA profiling] existed, and you can catch him now…, I do think it's a good thing. Why should he be free? It is supposed to be like this [that he gets caught], he's sick and for those things [DNA profiling] is good, I think. For child abuse and things like that, by all means.

The prisoners' statements on this topic sometimes suggested a longing for order and protection; Dorian, for example (serving 15 years for murder), said that 'if society didn't catch the criminals then everything would sink into chaos'.

The need for protection, in various contexts, was a prominent theme among prisoners in general. With regard to DNA evidence, close to 20% of my informants mentioned that, as well as proving guilt, it could also prove innocence – such as Karl, serving 17 years for murder and attempted murder:

> I personally think it's a good thing. One can prove innocence as well as guilt. I think it's a good thing when done correctly. And of the latter I am convinced. I think it's a good thing that it exists because one can solve cases which had previously been impossible [to solve]. Where would it get us, if somebody knew he could do something horrible and with 100% certainty he won't get caught? I think it's a good thing.

Interestingly, however, Karl's conviction that forensic DNA profiling 'is done correctly' seems to be overshadowed by distrust less than a minute later when I next asked whether he thought that 'all people should be entered into the database at birth':

Basically yes. But the condition is, in what country do we live? And how is it handled? It is a strong instrument that can be abused. If the wrong person is in the wrong place and moves the wrong 'handle', then you just need the DNA traces – and he's guilty. Boom, boom, let's bust him . . . . Basically, if it were guaranteed that it is always handled correctly then I'd be in favour [of it]. [Take DNA] from everybody. Absolutely.

Prior to his conviction, Karl had been a successful 'old school' entrepreneur in the prostitution business, which he saw as an honourable occupation – this partly explains his emphasis on honour and trust. While most informants shared his concern about potential abuse of DNA profiling, Karl was one of the very few who could see anything potentially positive in the establishment of a population-wide forensic DNA database. This is what Sigi said (serving 18 months for battery and coercion) when I discussed the same idea with him:

SIGI: . . . That would be as if you'd take traces [sic!] from my three year old son, only because maybe sometime something might happen – that's stupid.
INTERVIEWER: So you're clearly opposed [to this idea]?
SIGI: Yes, for sure. That would be a police state. That would be like in America. I don't think it's OK.
INTERVIEWER: In your opinion, we're not yet a police state, is this correct?
SIGI: Well, they do it [police surveillance] in a very smart way. With mobile phones, and so on. They locate you. Through your mobile phone you're traceable everywhere. Surveillance is everywhere, all the cameras, the computers. They do it in a smart way. People are usually not aware of how much the police can see in no time.

while Sigi's statement reflected a common conviction among his peers, namely that only people who had actually 'done something' should have their profile stored (as Oliver argued, 'They are not in the database for no reason, after all!'), a somewhat unexpected insight came from Quentin. Quentin initially voiced support for a population-wide forensic DNA database (which, similarly to Karl, he reversed later in the interview):

INTERVIEWER: [What would you say if] everybody had to provide DNA at birth?
QUENTIN: Give DNA and that's it. Then you've got everybody's perpetrator profile [sic!]. And then we'll see how fast the crime rate will go down. You can see it in prisons. Things used to happen in the past that do not happen anymore.

INTERVIEWER: What things, for example?

QUENTIN: Sexual assault. That doesn't happen anymore. Because everybody knows that the DNA is behind it [sic]. I prefer to watch TV and I want my peace and quiet and thus the others are left in peace as well.

Quentin is currently serving eight years for robbery and attempted theft. Being a chronic offender, he, like some other informants, is now keen on going straight, as forensic DNA technologies are making the lives of criminals harder (Prainsack and Kitzberger 2009). As Vincent (serving eight years for various offenses against the Addictive Drugs Law, against the Law on Firearms, and for severe battery and robbery) outlined:

> Fingerprints, that is – well, [every offender is aware of it and simply] puts on gloves; but that somebody drops a cigarette somewhere, or loses a hair, this happens much more easily. I can put on gloves and nothing can happen [in terms of leaving traces]. . . . But the other thing [DNA] isn't in your hands 100%. For a criminal, the DNA [stuff] is crazy . . . if you didn't shave your head bald, or if you had a tiny little hair on your pullover, or a bit of liquid, that's enough. It has become very difficult, crime, it has become extremely difficult. For the opponents of crime it's great, but for us, it's total shit.

Vincent concluded that the availability of forensic DNA technologies was tipping the balance in his professional cost–benefit analysis (Prainsack and Kitzberger 2009) in such a way that his only option was to end his criminal career. Vincent brings us to the last theme in this context: Does the existence of forensic DNA technologies deter other convicted offenders from re-offending as well, and if not – what other effects does it have on their lives?

Effects of forensic DNA profiling and databasing on offenders' identities and lives

As I argued elsewhere (Prainsack and Kitzberger 2009), an effect of the widespread use of forensic DNA profiling and databasing was that my informants' identities as delinquents were 'deepened'. Jürgen, for example, said that when both fingerprints and saliva were taken at the time of his arrest for battery, he minded the saliva more 'because it comes from the inside'. Jürgen seems to feel that while fingerprints are something 'superficial' (easy to avoid, and on the 'outside' of the body), DNA comes from 'inside' of him, and thus presumably from a place closer to the core of his self (see also Prainsack 2008a). However, while the knowledge that something from 'inside' them is stored in a national forensic database might deepen and intensify their feeling of stigmatisation as criminal offenders, we should not forget that criminality has always had a very physical dimension: 'Signs' of deviant behaviour, or even a deviant character, have always been ascribed to particular features of the human body. Examples are the shapes of the face, types of body shapes and even fingerprint patterns (Cole 2000; Prainsack 2008b). The offender's body is always an ambivalent affair: It is both a tool for the commission of the crime and a liability in the sense that it could give its owner away. As we have seen in the quotes from prisoners who said that forensic DNA technologies have made the lives of criminals harder because it is virtually impossible not to lose bodily substances at a crime scene, the body enters the risk–benefit analysis, which many offenders carry out before they commit crimes.

Another effect of forensic DNA technologies is, as mentioned above, that it has rendered most types of crime a more risky business. According to my informants, this can be expected to affect mostly chronic offenders who are not deviant enough to live, as Paul called it, 'outside of the borders of society'. Those people will have to consider going legitimate or move into a different field of crime. According to Paul, if one manages to escape leading a 'normal' life with a family, a steady circle of friends and a permanent address, one is relatively safe. In addition, highly specialised and professional criminals, according to my informants, are not really affected by forensic DNA technologies, either:

YGOR: I am fully convinced that if somebody is a capable guy, a good contract assassin, he won't be caught because of any traces.

INTERVIEWER: Why not?

YGOR: Because he doesn't leave any.

INTERVIEWER: How does he do that?

YGOR: Well, if he's good, then he knows how not to. [Police] will never find a corpse or anything else.

INTERVIEWER: But in theory, they could find something...

YGOR: Yes, OK, they'll find a DNA trace in the apartment of the victim, but what will they do with it?

INTERVIEWER: Because it's not seen as a crime scene?

YGOR: Exactly.

Others simply mentioned that forensic DNA technologies – and particularly the fact that their DNA profile was stored in a centralised database – had significant effects, without being specific about what those effects might be. They articulated a diffuse feeling of unease and fear rather than specific concerns. Richard, for example, who at the time of the interview was serving a five year prison sentence for severe sexual abuse and rape of minors, said that his DNA profile would be 'sent away and stored in a computer: And when I [go to] the US then I'll be sent back'. As a rule of thumb, however, knowledge about the practical consequences of forensic DNA profiling tended to be more detailed and concrete the longer the criminal record of my informant.

CONCLUSIONS

The first part of this chapter provided an overview of the history and the legal framework of forensic DNA profiling and databasing in Austria. In that context, the second part, drawing upon fieldwork carried out in Austria in 2006 and 2007, examined different understandings of and attitudes towards forensic DNA technologies on the part of law enforcement officers and prisoners. Comparing these two central groups in the DNA debate – namely those who employ forensic DNA technologies and those on whom they are employed – we can discern one crucial difference and several similarities. The difference is in the probative value, and more generally, the power ascribed to forensic DNA technologies. Whereas on the side of law enforcers I encountered rather nuanced understandings of what forensic DNA profiling can and cannot do, my prisoner informants tended to regard forensic DNA profiling as infallible and true. In cases of wrongful convictions, prisoners tended to blame human action or human judgment, and never regarded the technology as potentially fallible.

A somewhat surprising finding is that prisoner informants held views relatively similar to those of law enforcers in their overall assessment of the benefits of forensic DNA profiling and databasing. Both law enforcers and prisoners thought that forensic DNA profiling was a good thing (with divergent reasons). Similarly, neither law enforcers nor prisoners supported the idea of a population-wide forensic DNA database. Finally, both law enforcers and prisoners emphasised human action, and not technological 'agency', to account for professional success or failure. For police officials the meaningfulness of forensic DNA technologies clearly lies in the social and professional contexts of their use. When used judiciously, forensic DNA technologies were seen as immensely helpful tools in crime investigation. The technological tools by themselves, however, were not seen as sufficient to guarantee success. On the part of the prisoners, the increasing prevalence of forensic DNA technologies was primarily regarded as making the lives of criminals harder. None of my informants challenged the supposed infallibility of the technology of DNA profiling.

This comparison of the views and understandings of law enforcers and prisoners was necessarily compromised by the fact that fewer interviews were conducted with law enforcers than with prisoners. In addition, all the informants in law enforcement were from the same geographical area. However, it was not the aim of this chapter to provide a linear and generalisable comparison between attitudes held in these groups. Instead, I sought to draw attention to the extent to which the particular objectives in the functional areas of criminal investigation and crime prevention, on the one hand, and in the area of crime commission and rehabilitation, on the other, shape understandings of the science underlying these technologies. Whereas one could have expected both groups to relate to the science underpinning forensic DNA technologies in similar ways, and at the same time view the benefits of these technologies differently, the exact opposite was the case: prisoners never doubted the scientific accuracy and soundness of forensic DNA technologies, while law enforcers were quite aware that no science or technology could ever be infallible.

This brings us back to the initial question about how DNA technologies in this field reinforce or shift prevailing configurations of power and governance. I argue that the most apparent effect of forensic DNA technologies in this respect is to increase and enhance tendencies and values that are already inscribed in the criminal justice system. Put differently, rather than creating radically new practices or identities, the increasing importance of DNA technologies in crime scene work, in court and in our individual and collective crime imageries renders existing trends more entrenched. For example, as has been argued elsewhere (Duster 2006; see Chapter 4), the demographic structure of DNA profiles in most police databases overrepresents ethnic minorities and men and thereby perpetuates a problematic feature of the criminal justice system and reinscribes it ever deeper into the system. The most important effect of forensic DNA technologies that is apparent in this case study is the widening of the power gap between societal elites and underprivileged groups; between those who have access to institutional knowledge and those who typically do not; or those who can successfully navigate their way through society and those who cannot. In my case study, those who use forensic DNA technologies belong to the first group, and those on whom they are used belong to the second. The first group understands the science behind forensic DNA profiling and, therefore, knows its limits; the second group does not even dare to try to understand or question the science, which, like much other elite knowledge, is out of reach. It is out of reach not only, and perhaps not even primarily, because they seem prevented from obtaining it by the societal elite. Although virtually every prisoner could have obtained an understanding of the scientific underpinnings of DNA profiling from books and textbooks, none of my informants attempted to take that step. It was the perception of powerlessness, and of being at the bottom – or outside of – society that rendered the gap too difficult to cross. In that sense, forensic DNA profiling, with its supposedly sophisticated science, serves as a tool for governing and self-governance at the same time: It makes people, and their bodies, 'confess', to 'tell the truth'. These very same people use the technology to anchor their own position at the other end of knowledge.

REFERENCES

Cole, S. A. (2000). *Suspect Identities: A History of Fingerprinting and Criminal Identification*. Cambridge, MA: Harvard University Press.

Cole, S. A. and Dioso, R. (2005). Law and the Lab. *Wall Street Journal*, 13 May, W13.

Dahl, J. (2008). Another side of the story: defence lawyers' views on DNA evidence. In *Technologies of Insecurities: The Surveillance of Everyday Life*, eds. K. Aas, H. Gundhus and H. Lomell. London: Routledge-Cavendish, pp. 219-237.

Duster, T. (2006). Explaining differential trust of DNA forensic technology: grounded assessment or inexplicable paranoia? *Journal of Law, Medicine and Ethics*, 34, 293-300.

European Network of Forensic Science Institutes (2009). *ENFSI Survey on DNA Databases in Europe*. The Hague: European Network of Forensic Science Institutes www.enfsi.eu/page.php?uid=98 (accessed June 2009).

Gepp, J. (2007). Der geliebte Psychopath. [*The Beloved Psychopath.*] *Falter 48*, 28 November. www.falter.at/web/print/detail.php?id=598 (accessed June 2009).

Hewson, L. and Goodman-Delahunty, J. (2008). Using multimedia to support jury understanding of DNA profiling evidence. *Australian Journal of Forensic Sciences*, 40, 55-64.

Houck, M. (2006). CSI: reality. *Scientific American*, 295, 84-89.

Hughes, T. and Magers, M. (2007). The perceived impacts of crime scene investigation shows on the administration of justice. *Journal of Criminal Justice and Popular Culture*, 14, 259-276.

Leake, J. (2007). *The Vienna Woods Killer: A Writer's Double Life*. London: Granta Books.

Lynch, M. (2003). God's signature: DNA profiling, the new gold standard in forensic science. *Endeavour*, 27, 93-97.

Murphy, E. (2007). The new forensics: criminal justice, false certainty, and the second generation of scientific evidence. *California Law Review*, 95, 721-797

Nuffield Council on Bioethics (2007). *The Forensic Use of Bioinformation: Ethical Issues*. London: Nuffield Council on Bioethics www.nuffieldbioethics.org/go/our work/bioinformationuse/publication_441.html (accessed January 2009).

Prainsack, B. (2008a). Forum on the Nuffield report *The Forensic Use of Bioinformation: Ethical Issues*: An Austrian perspective. *BioSocieties*, 3, 92–97.

Prainsack, B. 2008b. Über Bio(a)soziale und uns, die wir über sie schreiben. [On bio (a)social people and us, who write about them.] In *Gefährliche Menschenbilder. Biowissenschaften, Gesellschaft und Kriminalität* [*Dangerous human images. Biosciences, Society, and Criminality*], eds. L. Böllinger, S. Krasmann and Pilgram, A. Baden-Baden: Nomos, pp. 82–96.

Prainsack, B. and Kitzberger, M. (2009). DNA behind bars: other ways of knowing forensic DNA technologies. *Social Studies of Science*, 39, 51–79.

Sanders, J. (2000). *Forensic Casebook of Crime*. London: True Crime Library/Forum Press.

Schweizer, N. and Saks, M. (2007). The *CSI* effect: popular fiction about forensic science affects public expectations about real forensic science. *Jurimetrics*, 47, 357–364.

Unterweger, J. (1983). *Fegefeuer oder Die Reise ins Zuchthaus*, auflage 2 [*Purgatory, or The Journey to Jail*, 2nd edn]. Augsburg: Maro Verlag.

Van Kamp, N. and Dierickx, K. (2007). *European Ethical-Legal Papers*, No. 9: *National Forensic DNA Databases in the EU*. Leuven: Centre for Biomedical Ethics and Law.:.

Walker, C and Cram, I. (1990). DNA profiling and police powers. *Criminal Law Review*, 479–493.

VICTOR TOOM

9

Inquisitorial forensic DNA profiling in the Netherlands and the expansion of the forensic genetic body

INTRODUCTION

During the 1990s in the Netherlands, DNA profiling became established as a mechanism to provide legal evidence for severe, violent crimes, including sexual assault, manslaughter and murder. That development was in accord with what Williams and Johnson (2008: 1) have observed in many jurisdictions regarding the transformation of DNA profiling into an important tool in processes of crime investigation and usage in so-called 'volume crimes'. Drawing upon the Dutch situation, this chapter first describes how Dutch DNA profiling became governed through legal measures and the inquisitorial orientation of the Dutch legal system. Second, the trajectory – the lines of development – of Dutch DNA profiling practices, is described, outlining who and what has been involved in DNA profiling. This account will provide insight into, first, the strategies employed by various stakeholders to achieve the current situation, where DNA profiling is deployed extensively and routinely in volume crimes, and, second, DNA profiling applied in the process of crime investigations. Hence, the analysis contributes to the understanding of how current DNA profiling practices were realised in a country – the Netherlands – that has what can be referred to as an 'inquisitorial legal orientation', where judges function as impartial fact finders.[1] Finally, some implications

---

[1] In an inquisitorial legal system it is the suspect who is the object of the process of finding juridical truth. The Office of Public Prosecution leads the process of criminal investigation, makes the decision on bringing legal cases and suspects to court and prosecutes; judges actively search for truth during court proceedings and impose sanctions. Van Kampen (1998) has a further elaboration on the differences between 'inquisitorial' and 'adversarial' legal systems.

Genetic Suspects: Global Governance of Forensic DNA Profiling and Databasing, ed. Richard Hindmarsh and Barbara Prainsack. Published by Cambridge University Press. Copyright © Cambridge University Press 2010.

for current directions in the governance of Dutch forensic DNA profiling practices are outlined.

DNA PROFILING IN THE NETHERLANDS: LEGAL SYSTEM AND LAWS

Most continental European legal systems, including that of the Netherlands, are organised according to an inquisitorial principle. Van Kampen (1998: 48) compared the Dutch and American legal systems regarding admissibility of forensic evidence. She found that the Dutch inquisitorial legal system rests on the assumption that different members of the system working on legal cases – judges, prosecutors, police officers and experts – act impartially, fairly and expediently. Whereas 'adversarial' forensic evidence, as examined in US courts, is sceptically regarded and questioned intensively by lawyers and prosecutors; inquisitorial forensic evidence is attached to a practice that puts trust in experts, legal professionals and institutes.

Trust in Dutch forensic evidence has been achieved through the institutionalising of dealings with forensic technologies on various levels. First, the Netherlands Forensic Institute (NFI) is an agency of the Ministry of Justice and is under the Directorate-General for Law Enforcement. This means that the Minister of Justice is responsible for the NFI as an impartial, fair and expedient organisation. Second, the role and functions of expert witnesses and the NFI are circumscribed in the Code of Criminal Procedure. The Code prescribes, among other things, that expert witnesses have to take an oath that they will competently perform their tasks, such as drafting a written testimony or being in 'good conscience' when they appear in court as expert witnesses. After taking the oath, an expert witness is installed permanently. Third, expert evidence usually takes the form of documents, for example of DNA evidence, autopsies or drug analyses, which are supposed to be unbiased and neutral; however, expert witnesses themselves hardly ever appear in Dutch courts and are thus hardly ever questioned or cross-examined (for exceptions see Bal (2005) and M'charek (2005)).

However, DNA evidence has a somewhat special position within Dutch forensic practices. In the early 1990s, the Minister of Justice decided that DNA profiling should be governed through special DNA profiling legislation to form a fourth mode of institutionalising forensic evidence. In September 1994, the Forensic DNA Profiling Act (Government of the Netherlands 1993) came into force. This law mainly sought to regulate two aspects of DNA profiling. First, it became a legal possibility to issue a compulsory 'body search' (the legal term) of individuals suspected of having committed severe and violent crimes – such as murder, homicide and sexual assault – to obtain blood (on medical grounds, saliva or hair roots could also be obtained) for DNA profiling. Second, the law laid down various measures to ensure the reliability of DNA evidence. It stipulated, for example, that only DNA profiles produced in a laboratory with accreditation according to international standards (ISO 17025) may be used as DNA evidence in court.

In November 2001, an amendment was added (Government of the Netherlands 2001), which offered an important widening of the scope and applicability of DNA profiling. Most importantly, individuals suspected of having committed more minor crimes, such as theft, break-ins and mistreatment or being a (severe) public nuisance – so-called volume crimes – could be body-searched to obtain saliva for DNA profiling. Another amendment was issued in 2003 (Government of the Netherlands 2003). This amendment, the Law on External Visible Personal Characteristics, allowed for the forensic DNA determination of the 'sex' and 'race' of an unknown originator of crime scene samples (Chapter 4 discusses the situation in the USA in this regard). The amendment belongs to the category of so-called 'window' legislation, as it leaves room for other externally visible traits, for example colour of hair or eyes, to be included through an Order in Council.

The most recent Dutch law was enacted in 2005: the DNA Convicted Persons Act (Government of the Netherlands 2004). Since this Act came into force, persons convicted of offences (i.e. sentenced to imprisonment, community service orders, hospital orders, placed in psychiatric hospitals or in penal institutions for systematic offenders or institutions for juvenile offenders) carrying statutory maximum prison sentences of at least four years will be obliged to provide DNA samples. Currently, new amendments are being considered, pertaining, for example, to familial searching and DNA dragnets (Ministry of Justice 2008a). The Netherlands is also a signatory to the Prüm Decision, which merged Dutch DNA profiling into a European data-sharing endeavour (see Chapter 2).

The Dutch DNA database

In 1997, the Dutch forensic DNA database was established. By September 2009, the database contained more than 132 000 DNA

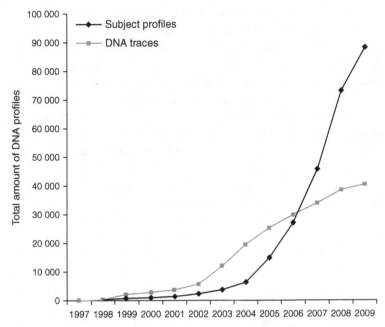

Figure 9.1. Dutch DNA database 1997 to September 2009. From DNA: sporen naar de toekomst (DNA: traces into the Future), the Dutch DNA database; updated data can be seen at http://www.dnasporen.nl/index.asp.

profiles, all from a population of some 16 million (Figure 9.1). Of DNA profiles currently in the database, 88 026 are reference profiles, also known as subject profiles. Reference profiles are derived from persons whose identity is known to the authorities. The remaining 40 192 DNA profiles are derived from crime scene traces.[2] The 2001 amendment and the 2005 law in particular, contributed importantly to this number of DNA profiles.

Most reference profiles (subject profiles) originate from convicted individuals, but they also include suspects and deceased victims. The DNA profile of a convict is stored for 20 years after a person is convicted for a crime laid down in the Penal Code that carries a sentence of less than six years of imprisonment. In contrast, DNA profiles are stored for 30 years when a person is convicted for a crime carrying a sentence of six years or more of imprisonment. Retention periods for

[2] Available from: www.dnasporen.nl (accessed 14 October 2009).

reference (subject) and evidentiary (traces) samples are the same as for DNA profiles. The DNA profile and the biological sample of a suspect are removed if the suspect is not convicted. The DNA profiles obtained from crime scene traces are removed when the originator of the sample has been convicted or when the public prosecutor decides not to use the match to prosecute the suspect. The DNA profiles and samples of deceased victims, as well as crime scene profiles and traces, are stored for 12 years if they are related to crimes carrying a sentence of less than six years of imprisonment. They are stored for 20 years if they are related to crimes carrying a sentence of six years or more of imprisonment, and 80 years in cases of crimes punishable with life imprisonment. About 90% of matches between crime scene profiles and subject profiles pertain to volume crimes such as burglary and (car) theft (Netherlands Forensic Institute 2008).

The trajectory of DNA profiling in the Netherlands is in some ways the same as in many other jurisdictions that have introduced forensic DNA technologies. The Dutch situation, however, sheds light on how DNA profiling practices were transformed in a legal system with an inquisitorial orientation; initially they were used on a case-to-case basis in severe, violent crimes, but now they are routinely deployed in volume crimes. This begs the question: Why and how did this happen? In addition, how did the function of DNA profiles change from a use as evidence in courts to informing processes of crime investigation? In exploring such transformations, three historical phases can be distinguished in the Netherlands. First, during 1989–1997, DNA profiling was introduced into the courts and became established as evidence for violent crimes. Second, during 1997–2001, new forensic DNA profiling technologies were introduced that led to the extension of DNA profiling beyond the scope of serious crimes to the realm of volume crimes. Third, the years post-2001 highlight the increasing room given to DNA profiling practices, with new criteria for 'body searches', a redistribution of responsibilities and competences, and DNA profiling becoming increasingly used for criminal investigation. These transformations are explored first before discussing some implied challenges for governance.

DUTCH DNA PROFILING PRACTICES: 1989–1997

When Alec Jeffreys and colleagues invented DNA profiling in the mid 1980s, the technique was dependent on biological materials containing large amounts of DNA, that is, DNA extracted from reference blood or DNA extracted from stains of blood and semen of about one square centimetre (Jeffreys *et al.* 1985). Usually, such amounts of (crime-related) DNA could be found at crime scenes or on the *corpus delicti* of serious violent crimes, but was absent in less-severe or non-violent crimes like burglary.

Soon, in the late 1980s, DNA profiling was introduced in the Netherlands. Initially, various suspects delivered blood samples voluntarily to prove their innocence. But it was not long before a compulsory body search was issued in a legal case brought against an individual suspected of rape. The suspect appealed successfully against this order. The case was finally brought to the Dutch Constitutional Court (Hoge Raad) and it led to the so-called 'saliva decision' (Dutch Constitutional Court 1990). This decision ruled that taking blood from the veins or saliva from the inner cheeks comprised a breach of the right to inviolability of the body, a basic right articulated in Article 11 of the Dutch Constitution and Article 8 of the *European Convention on Human Rights* (Council of Europe 1950).

To address this situation, and apply DNA profiling in serious crimes, Dutch law had to be changed (Toom 2006). In September 1994, the Forensic DNA Profiling Act came into force. For the first time in Dutch history, a compulsory body search could be issued to obtain blood, saliva and hair roots for DNA profiling. It was considered by the law to be proportionate if someone was suspected of a crime with a penalty of eight years of imprisonment or more. Crimes that incur eight or more years of imprisonment include severe violent crimes such as sexual assault, manslaughter and murder. Special measures were laid down to protect the rights of suspects. Only examining judges were allowed to order a bodily search for (both compulsory and voluntary) DNA analysis; DNA profiling had to be assessed as vital for finding 'the truth'; suspicion had to be backed up by facts and circumstances; a suspect needed to be asked to cooperate at least two times without success; only a licensed physician was allowed to take the sample; and samples were to be destroyed when the case at hand no longer demanded their availability. The law allowed digital storage of DNA profiles in a DNA database.

DUTCH DNA PROFILING PRACTICES: 1997–2001

Soon after the Forensic DNA Profiling Act came into force, new genetic technologies found forensic applications. First, polymerase chain reaction (PCR), a technology to copy specific strands of DNA millions of times, was introduced. Traces containing only small bloodstains, saliva or flakes of skin sufficed for reliable DNA profiling. Second, short tandem repeats (STRs) were shown to produce reproducible results between different laboratories using different DNA profiling methods, thereby remedying problems regarding production and interpretation of DNA profiling techniques as initially developed (see Chapter 12). Third, geneticists of the British Forensic Science Services combined PCR and STRs in standardised multiplex DNA profiling kits, which rendered DNA profiling cheaper and faster to use. A second-generation multiplex (SGM) DNA profiling system followed that was considered 'a highly discriminating and reliable individual identification tool suitable for both routine forensic applications and intelligence database construction' (Sparkes *et al.* 1996: 201). A later version of SGM (SGM+) is now applied in forensic laboratories internationally. In 1997, at the same time that the Dutch DNA database was activated, the NFI received accreditation for SGM from the Dutch Accreditation Council. This meant that SGM profiles could become legal DNA evidence.

A draft law proposal, the 'DNA & burglary' project and the 'Albin' case

Following these technological innovations, in 1997, Ministry of Justice policy makers drafted an amendment to the 1994 Forensic DNA Profiling Act (Ministry of Justice 1997). The drafted amendment provided that DNA profiles should be derived from saliva rather than blood samples. Taking a buccal swab was considered a less severe violation of the right of inviolability of the body. Therefore, it was proposed that an order to obtain a sample by an examining judge would be no longer required if a suspect provided a sample voluntarily, and that the objective of 'finding [the] truth', as laid down in the 1994 law, would no longer be required to obtain a DNA sample from a suspect. Instead, it was proposed that there was sufficient reason for taking a sample if it could be seen as being 'in the interest of the investigation'. These measures, it was argued, would lead to beneficial effects, such as a larger DNA database to combat crime. Nevertheless, despite the fact that taking saliva rather than blood samples was seen as a less severe violation of the right of inviolability of the body, it was not proposed at the time to extend the use of DNA profiling to other, less severe, crimes such as theft and burglary.

In the summer of 1997, the law proposal was submitted to key stakeholders such as police officials and legal experts.[3] The Netherlands Bar Association found the law proposal overly far-reaching on two counts. First, the violation of bodies was considered so severe that it was argued that high thresholds (i.e. 'vital for finding truth') should remain. Second, the Bar Association argued that only examining judges should be able to assess the necessity of a body search. Other consulted organisations found the proposal too conservative and advocated lowering the threshold for mandatory DNA profiling from crimes resulting in eight years of imprisonment to those resulting in four years, based on the argument that taking a saliva sample could hardly be recognised as a violation of the body and thus as an infringement of bodily integrity.[4]

The reason given in the draft law proposal for restricting DNA profiling to severe, violent crimes was that: 'DNA research usually seems to contribute little to the investigative process solving break-ins. In general, no blood or saliva left by the perpetrator is found at crime scenes of theft' (Ministry of Justice 1997, author's translation). This quote, however, surprised the former head of the Unit of Forensic and Technical Research of the Midden & West Brabant police district. When I interviewed this official in the summer of 2006, he disagreed with this line of reasoning.

> I had read the explanatory memorandum [of the draft law proposal]. It was said that the threshold would not be lowered to four years of imprisonment, because in practice it has been shown that few biological traces are found [at crime scenes of less severe crimes]. Then I thought: 'That is not true. We do not collect biological traces at volume crime scenes because they do not do so much with it at the NFI. You're not going to collect traces that won't be used.'

According to this police official, thieves, for example, *do* leave biological traces; those traces, however, are usually not collected because

---

[3] The draft law proposal was submitted to police organizations (Criminal Intelligence Service of the National Police Services Agency; Netherlands Police Institute; Board of Chief Superintendents), the NFI, the Board of Procurators General, Council for the Judiciary, the Netherlands Bar Association, the Dutch Organisation for Help to Victims and the Royal Dutch Medical Organisation.

[4] When the law proposal was submitted in November 1998, many Members of Parliament aired the opinion that taking a saliva sample could hardly be recognised as a violation of the body. Only the members of the Green Liberal Party (GroenLinks) and the Socialist Party agreed to maintenance of the high thresholds (House of Representatives 1999a).

crime scene investigators know that those traces will simply not be analysed. The reason for this is the limited resources available to the NFI, and the fact that typically most burglary suspects cannot be issued a compulsory body search to obtain blood for DNA profiling (to compare the crime scene stains with their DNA). After reading the draft law proposal, the official of the police district Midden & West Brabant organised a meeting with colleagues of the Utrecht police force and NFI officials. Together they decided to initiate a pilot project called 'DNA & burglary' (DNA bij inbraken) to support the case for lowering the threshold from eight to four years of imprisonment based on the argument that forensic DNA could well contribute to juridical 'truth-finding'. Before the project began in January 1998, the Offices of Public Prosecution (districts Breda and Utrecht) were enrolled into it. Forms were especially designed and printed in order to streamline the various administrative acts. It was decided to run the project for six months, but this was extended to a year after a donation of €45 300 from the Ministry of Justice. The project was carried out in accordance with the 1994 Forensic DNA Profiling Act. To limit the workload for the NFI, the submission of DNA traces (e.g. blood, cigarette butts, saliva and hairs) collected from volume crime scenes was restricted to a maximum of 60 traces monthly, meaning that both police districts (Midden & West Brabant, Utrecht) were allowed to submit 30 biological samples each month.

Three questions lay at the heart of the 'DNA & burglary' project. First, do burglars typically leave DNA traces? Second, would it be possible to produce usable DNA profiles from those traces? Third, do DNA profiles, after being stored in the Dutch DNA database, contribute effectively to solving volume crimes such as burglary? The project ran from January until December 1998. During this period, a total of 562 biological traces were collected and submitted to the NFI. Subsequently, 391 DNA profiles were produced and uploaded to the DNA database (Project DNA & Burglary 1999). The project demonstrated that biological traces could indeed be found at volume crime scenes and that most could be analysed, thus answering the first two questions.

The third question dealt with the DNA database, which had been operational since 1997. In its first calendar year, 49 DNA profiles, broken down into 28 reference DNA profiles and 21 DNA traces, were uploaded to the DNA database (Netherlands Forensic Institute 2008: 10). In 1998, during the second calendar year, and the year that the project 'DNA & burglary' ran, a total of 708 DNA traces were uploaded.

Table 9.1. *Log for cigarette butt T 123*

Sample number	PL1234/98–567890/001/002
Sample description	
Size	CIGARETTE BUTT
Collected	0 mm
Place of sample	SAMPLE CARRIER
Specificities	In front of stereo
DNA identification number	T 123

Source: Pro Justitia 1998.

The project 'DNA & burglary' contributed 391 DNA traces to the total amount and, with the use of the DNA database, 137 matches were established. In one case, it was found that a suspect's DNA profile matched DNA traces collected at 16 different crime scenes. This case involved a person referred to as 'Albin' (not his real name of course). Albin was suspected of having committed more than 100 burglaries in 1998.[5] His case illustrates how forensic DNA profiling became extended to the realm of volume crimes. The discussion here focuses on two identifying stickers that were placed onto a DNA trace (T 123) and a reference sample (P 9999), respectively.

Connecting crime scenes: T 123 and the DNA database

Let us revisit the following scene: one day in 1998, a burglary was reported in the police district Midden & West Brabant. Later that day a crime scene investigator from the Forensic and Technical Department secured from the crime scene, amongst other things, a cigarette butt. It was accounted for as shown in Table 9.1.

After collecting the crime scene trace and tagging it T 123, a Public Prosecutor of the Breda district submitted T 123 to the NFI for DNA analysis. It was one of the 562 traces collected as biological evidence within the framework of the project 'DNA & burglary', and

---

[5] The analysis is based on the criminal file compiled against 'Albin' (Pro Justitia 1998). The Dutch Board of Procurators General gave permission to use the file on condition that the privacy of individuals (suspects, victims, witnesses) not be jeopardised. Consequently, this chapter has altered registration numbers, uses an alias for the suspect and uses the author's DNA profile.

Table 9.2. *Numerical second-generation multiplex (SGM) DNA profile for T 123: basic profile hit of an individual*

Order nr.	Locus	Genotype
10	AMEL	X/Y
20	THO	6/9.3
30	D21	31/31.2
40	D18	12/15
50	D8	10/12
60	VWA	16/18
70	FGA	21/23

Source: Pro Justitia 1998.

one of the 391 DNA profiles produced and uploaded to the DNA database.

It should be mentioned here that the DNA database in 1998 was administered by the NFI and subject to legal regulation (Government of the Netherlands 1994), which, in the early 1990s, had oversight of three different systems in the DNA database (the current situation is discussed later in this chapter). The first was an analogue administrative system that stored information related to reference DNA profiles and DNA traces on datacards. Datacards included all relevant information regarding the sampling and securing of tissue and production of the DNA profile, administrative numbers, the DNA profile, the submitted testimony and the identifying stickers. The full name, date and place of birth, nationality, sex, and aliases were added to this information for reference DNA profiles. The second system was a digital system of reference DNA profiles. The third was a digital system that held DNA profiles derived from crime-related traces. DNA profiles that were uploaded to the two digital systems could be linked to the identity of the originator or sample by means of evidentiary stickers, such as T 123 or P 9999, and datacards. Information that could be traced back to individual identities was at that time not stored digitally (currently datacards are stored digitally). The format of the DNA profile that T 123 took when it was uploaded to the digital DNA database is a numerical representation that gives insight into the amount of repeats for each marker, as shown in Table 9.2.

Returning to the story; for analysis, the DNA profile T 123 was scanned for any previous entries. It soon appeared that it matched several other DNA traces, and as a result, different crime scenes were linked together. As it so happens, soon after T 123 was uploaded to the DNA database, Albin was arrested on suspicion of burglary by the Midden & West Brabant police force. During interrogation, Albin admitted that he, with accomplices, had committed over 100 burglaries. After being officially charged, reference fingerprints and shoe prints were obtained and compared with fingerprints and shoe prints collected at various crime scenes, which were found to match. The police suspicion of Albin's involvement in a large number of criminal offences was then backed up by evidence. Subsequently, DNA information was brought to the attention of the detectives by an NFI official, who told them about the DNA database matches. The detectives then summarised the legal case against Albin and the DNA matches in a written report (Pro Justitia 1998): 'When Albin was caught in the act of a break-in, a biological trace was collected at the scene of the crime which was subsequently DNA typed. It has been determined that that DNA profile matches DNA profiles typed from biological traces collected at ten other scenes of burglary.'

One of the functions of the forensic DNA database is that different crime scenes can be linked to each other, as occurred in Albin's case, by obtaining 'matches' between different crime scene samples. But how is it possible to link Albin's DNA profile to profiles derived from crime scene samples?

Connecting Albin to crime scenes: P 9999 and the DNA database

To be able to link Albin's DNA profile to DNA traces stored in the DNA database, police and prosecutor had to prepare Albin's case to bring it to the examining judge. The examining judge is the official who decides on issuing a body search to obtain blood for DNA profiling. The rules laid down in the 1994 Forensic DNA Profiling Act are thus applied.

At the time that the public prosecutor and police approached the examining judge, Albin was suspected of burglary, a crime with the liability of a maximum penalty of six years of imprisonment. However, if there is a suspicion that a burglary occurred *during the night* in combination with circumstances indicating that those break-ins were done *in cooperation with others* or by *forced entry*, a prison sentence of up to nine years can be imposed. Albin was suspected of breaking in during the night, and opening windows using tools in cooperation with others. Hence, the series of burglaries that Albin was suspected of was one of the only cases in that year of 1998 for which blood-taking was mandated in the context of a non-violent crime. Nevertheless, other issues had to be taken into account, too. Was mandatory blood-taking proportionate in this case? Was it vital for finding 'truth'? In addition, how would Albin respond when the examining judge requested him to volunteer a blood sample? In a nutshell, the examining judge regarded it as proportionate and vital for truth that a DNA profile of Albin be produced. As Albin volunteered a sample, it was not necessary to execute Article 195d of the Code of Criminal Procedure to obtain a blood sample by force.

After a blood sample was obtained by a licensed physician, as taking blood is considered a medical procedure, Albin's reference sample was labelled with sticker P 9999. It was submitted to the NFI and the DNA profile was determined and finally uploaded to the database. It appeared to match DNA profiles collected at 16 different crime scenes. During the court case against Albin, this DNA evidence, together with other evidence (footprints and fingerprints, confessions) led to a verdict of the judge that he found it both *legally* and *convincingly* proven that Albin had committed a total of 56 (attempted) burglaries and six car thefts. The judge sentenced him to five years of imprisonment, which was reduced to four years following appeal.

Albin's case was the first of its kind in the Netherlands, where one suspect was matched by means of the DNA database to so many different scenes of volume crimes. An NFI DNA expert, who was involved with Albin's case, wrote 'MEGAHIT' on a laboratory printout. When I interviewed this DNA expert, she told me:

OFFICIAL: [It was for the first time] during the project 'DNA & burglary' that we determined DNA profiles from traces collected at scenes of burglary [and uploaded them to the DNA database]. It did not take long before we started to observe matches, with this case with 16 matches as the highlight. It was the smash hit of the project.

INTERVIEWER: What did this 'megahit' mean for you?

OFFICIAL: This is the future. It works! It proves that our expectations regarding DNA profiling and databasing were right. And this is only the start; it will become huge. Now it is 10 years later and DNA profiling can be considered a business by itself, for instance regarding the Prüm Treaty.

DUTCH DNA PROFILING: 2001–PRESENT

When the project 'DNA & burglary' was terminated at the end of 1998, the participants evaluated the project and published the results in an official report in May 1999 (Project DNA & Burglary 1999). A month later, in June, with the results presented to the Ministry of Justice, the following response ensued (Ministry of Justice 1999):

> Minister of Justice A. H. Korthals is considering making it possible to order a body search for DNA typing regarding crimes with a penalty of four years or more imprisonment. The threshold is currently set on crimes with an eight-year or more sentence. Lowering the threshold means that DNA typing can be used more often [author's translation].

Subsequently, in November 1999, the Minister of Justice submitted a new law proposal to the House of Representatives (Tweede Kamer) of the Dutch Parliament, which a large majority supported (House of Representatives 1999b).[6] The new law, called the Forensic DNA Profiling in Criminal Proceedings Law, became effective in November 2001 (Government of the Netherlands 2001). Most importantly, it became legal for DNA profiling to be ordered when someone was suspected of having committed a crime with a penalty of four years or more, which included volume crimes such as (car) theft, breaking and entering, mistreatment of persons and animals, and making a public nuisance.[7]

With this expansion of DNA profiling, the NFI predicted that DNA profiling would be ordered up to 20 times more often when the 2001 amendment came into force (Netherlands Forensic Institute 2000: 10). This posed a challenge to Dutch DNA profiling practices in several ways. First, in an attempt to deal with all the expected requests, the NFI had to move to a new building with an extensive forensic DNA laboratory. It also had to hire new analysts and install new experts, buy

---

[6] In this document, the Minister gave three reasons for lowering the threshold: the success of the project 'DNA & burglary'; a decision made by the Dutch Constitutional Court, the so-called 'toothbrush decision' (Dutch Constitutional Court 2000); and the critique of several Members of Parliament who found that maintaining the high threshold of eight years was too conservative.

[7] Since November 2001, a compulsory body search for purposes of DNA profiling is linked to Article 67 of the Code of Criminal Procedure. This article describes the crimes for which someone can be taken in custody. This usually is the case when someone is suspected of having committed a crime with a penalty of four years or more.

new equipment and further streamline the production and administration of producing DNA profiles. Second, measures to protect basic rights of suspects, laid down in the 1994 law, were changed in two respects. First, it was decided to grant public prosecutors – the officials in charge of the process of criminal investigation – the same authority as the examining judge to issue an order to obtain a sample (saliva, blood or hair roots) from a suspect. This meant that responsibilities were shifted from an impartial examining judge to the leader of the process of crime investigation. Second, it was decided that medical doctors no longer were required to obtain a reference sample; police officers who had received special training were allowed to collect saliva or hair roots from cooperating suspects. Both measures can be understood as a move away from using DNA profiling as evidence in criminal proceedings towards an application in crime investigation.

Since 2001, more links have been established between DNA profiling and the police. In the introduction to this chapter, I described the Law on External Visible Personal Characteristics, which allows for the forensic DNA determination of 'sex' and 'race' of an unknown originator of crime scene samples (Government of the Netherlands 2003). Genetic information regarding 'sex' and 'race' is not so much used as evidence but is considered as an important lead that feeds into the process of criminal investigation. It enables detectives to focus their investigations on one group that shares a particular trait, such as 'race' and/or 'sex'. Cole and Lynch (2006: 53) use the term 'DNA photofits' for these traits and other DNA markers informative about external visible characteristics.

DNA photofits have informed the process of criminal investigation in the Netherlands several times. An example is the case of Milica van Doorn. She was found murdered in the city of Zaandam in 1992. The case was not solved. In December 2008, the Office of Public Prosecutor announced that a DNA photofit had been produced from biological material found near or on the body of the victim. The DNA photofit indicated that the originator likely came from Turkey or North Africa. It was then decided to organise a DNA dragnet. A total of 75 Turkish and North African men, aged between 16 and 30 who lived close to the crime scene at the time of the crime, were selected and requested to volunteer a DNA sample. At the time of this writing, 71 men were excluded as possible suspects. Yet, four persons still have to deliver a sample. The case remains unsolved to this day.

Organising DNA dragnets based on DNA photofits can raise important ethical and normative questions. First, there is the issue that the statistical chance that the originator of the sample is likely to come from Turkey or North Africa was translated in news reports to a firm claim that the perpetrator *was* Turkish or North African, which, of course, did not have to be true.[8] Second, and by association, DNA photofits have implications regarding privacy and discrimination (see also Chapter 4). Third, DNA photofits produce a population of 'interesting persons' or a 'suspect population' (Cole and Lynch 2006; M'charek 2008) that must be excluded as possible suspects. The DNA photofits link 'suspect populations' to crimes. Establishing such links is not a value-free exercise; it interferes with the relationship between civilians and the state, as the onus of proof, which traditionally rests with the Office of Public Prosecutor, is shifted to the 'genetic suspect' to prove his or her innocence by delivering a DNA sample. Currently, scientific research is being conducted in the Netherlands in an attempt to gain genetic knowledge about externally visible characteristics that may be used in crime investigation (see also phenotypic profiling discussion in Chapter 2). This leads the topic of the next section.

SCIENTIFIC RESEARCH AND FORENSIC APPLICATIONS

Population geneticists of a Dutch university recently published their research results regarding genetic markers for determining eye colour (Kayser *et al.* 2008 Liu *et al.* 2009). The genetic materials used in these studies are derived from two biomedical research projects. The first project, the Rotterdam Project, assesses the occurrence and determinants of chronic diseases; the second concerns the city of Rucpen and is part of a programme titled Genetic Research in Isolated Populations. Both groups who donated samples to these biomedical studies were populations of predominantly Dutch origin (Kayser *et al.* 2008: 412). The findings of the geneticists led them to the claim that they can predict brown and blue eyes with an accuracy of 93% and 91%, respectively, and other eye colours with an accuracy of 73% (Liu *et al.* 2009: R192). Subsequently, the results led to an announcement of the NFI, in March 2009, that the Minister of Justice would be asked for permission to apply this technique in forensic case work (Netherlands Forensic Institute 2009); this is expected to be authorised though an Order in Council in summer 2010.

---

[8] 'Turk vermoordde Milica van Doorn' [Turk killed Milica van Doorn] http://www. laatstenieuws.nl/ (accessed 31 March 2009).

This research is an example of how biomedical and genetic research can converge with forensic DNA profiling (see Chapter 5). But another route by which to gain genetic knowledge about externally visible characteristics has also been made available by means of the 2001 Forensic DNA Profiling in Criminal Proceedings Law, which involved two important decisions. First, reference samples (the DNA provided from subjects) no longer had to be destroyed when the case at hand was closed. Since 2001, more than 85 000 DNA profiles have been stored, meaning that a similar number of reference samples originating from criminals convicted in the Netherlands is available. Second, the Personal Data Protection Act (Government of the Netherlands 2000) is legally applicable to the DNA database, meaning that DNA profiles and biological (reference) samples both are understood as *information* (see Toom 2006; van der Ploeg 2007; M'charek 2008). This allows biological reference samples to be used for scientific research, for example to develop a genetic test that is informative regarding external visible characteristics. After results are proven (sufficiently) accurate, and the Minister of Justice allows use of the test through an Order in Council, the test can be used in the process of crime investigation to determine external visible characteristics from an unknown originator and consequently produce other 'suspect populations'.

This is not only a hypothetical possibility. A Dutch academic research institute currently uses – with the permission of the Minister of Justice (Ministry of Justice 2008b) – genetic material derived from reference samples from the DNA database. The samples are anonymised except for the country of birth. This is supposed to enable geneticists to develop new genetic tests informative about external visible personal characteristics from biological traces. This latter example and the above mentioned example of determining eye colour show how forensic and biomedical practices converge and both shape and are shaped by forensic DNA practices. The concluding section will reflect on the different aspects and implications of the introduction of forensic DNA technologies in the Netherlands that have been described above.

GOVERNING THE CHALLENGES OF FORENSIC GENETIC BODIES

This chapter has outlined the trajectory of Dutch DNA profiling practices. This technique has contributed importantly to, as Turner (2006: 228) called it, the common understanding that the 'code of the body' is becoming a major tool in criminal investigations. To become major tools in criminal investigations, bodies need to be discursively reconstructed (Williams and Johnson 2008: 97). Such discursive reconstruction has in the Netherlands been achieved, on the one hand, by legal measures and jurisprudence and, on the other hand, by genetic techniques such as PCR, STR and SGM. 'Bodies' in forensic DNA profiling practices have thus been 'enacted' (Mol 2002) by both a discourse of law and the practice of genetics as 'forensic genetic bodies' (see also Chapter 13). Using forensic genetic bodies as a metaphor allows for an association with 'growing'. This chapter describes how forensic genetic bodies in the Netherlands have been 'growing'. At first, the forensic genetic body only consisted of individuals suspected of severe, violent crimes. The forensic genetic body then expanded to volume crimes. Next, 'all individual bodies' of convicted criminals were ushered into the forensic genetic body. Finally, the forensic genetic body is expanding in the refrigerators of the NFI with multiple forensic reference bodies at the disposal of the authorities, which can be used to create knowledge about them. This raises the question of how large the forensic body should be allowed to 'grow', and where the limits are. For example, should the complete criminal population be included, or the population at large, or all males from 12 to 60 years of age, and/or every migrant or tourist entering the Netherlands? Until now, this issue has not been discussed in the Netherlands; however, successive policy makers seem to be driven increasingly by the possibilities generated through new DNA profiling technologies and changing ideas about a safe society and the 'war' against crime and terrorism.

The DNA database and markers for external visible characteristics thus gain in importance for crime investigation. Robust connections between DNA profiling and the police and the Office of Public Prosecution have been established and remain in place. Public prosecutors have gained authority to issue an order to obtain a saliva sample; police officers with special training are allowed to take saliva and hair root samples; DNA markers (or photofits) of external visible characteristics inform processes of crime investigation in 'suspect-less' forensic cases; and new DNA photofits have been proven scientifically accurate or are currently being developed. DNA photofits create new and complex relations between the process of crime investigation and forensic genetic bodies.

Turning to issues of DNA photofits, they not only reinforce concerns that DNA profiling and databasing could increase stigmatisation and discrimination but also, as outlined by M'charek (2008: 527), raise concerns that a result of this practice could be the 'lumping together' of 'groups of individuals ... into a racialized suspect population'. Informed by the DNA photofit, detectives can concentrate criminal investigations on suspect populations, as illustrated in the example of the Zaandam murder case. Individuals who are within a suspect population can be requested to voluntarily supply a sample for DNA profiling in DNA dragnets. This raises several further questions: What happens if an individual refuses? At the least, the individual – the genetic suspect – becomes 'interesting' for further investigation. If 'suspicious' clues are found (e.g. the genetic suspect has a criminal record or a 'facebook' linked to a victim or another suspect), the public prosecutor can designate this person as a legal suspect, thereby transforming his 'body' into 'a forensic genetic body'.

However, as we have seen, one-trait DNA photofits currently have a maximum accuracy of 93%. When traits are combined (e.g. sex, race, eye colour and geographical origin) in a DNA photofit, the statistical likelihood that the perpetrator and the combined DNA photofit actually converge decreases, thereby increasing the risk of plainly focusing a DNA dragnet on the wrong 'suspect population'. This leads to pressing questions. Is a combined DNA photofit with a low accuracy reliable enough to group individuals into a suspect population, and to ask all individuals in that population to volunteer a DNA sample? In addition, should DNA dragnets be made compulsory for 'suspect populations'? Then, what happens to innocent individuals who refuse to deliver? Will they automatically be treated as suspects?

In September 2008, two major Dutch political parties announced that they advocated compulsory participation for DNA dragnets.[9] So far, the Minister of Justice has rejected such a radical proposal. But that proposal is only one of the many implications for governance raised by the expansion of DNA databases and profiling, as indicated above. These are enough to reinforce the view that broad and informed public debate is required to better address and resolve these issues.

ACKNOWLEDGEMENTS

I would like to thank Richard Hindmarsh and Barbara Prainsack for their extensive reviewing and editing. I thank all informants for their cooperation, Kees van der Beek of the Netherlands Forensic Institute

---

[9] The parties are the Christian Democrats and the Liberal Party (www.cda.nl and www.vvd.nl, respectively; accessed 12 March 2009).

for his comments, and all my colleagues who inspired this article at the Amsterdam School for Social Science Research, in particular Amâde M'charek, John Grin and Mihai Varga. This chapter is based on my PhD research, which is funded by the Netherlands Organization for Scientific Research and is part of the Societal Component of the Genomics Program.

REFERENCES

Bal, R. (2005). How to kill with a ballpoint: credibility in Dutch forensic science. *Science, Technology and Human Values*, 30, 52–75.

Cole, S. A. and Lynch, M. (2006). The social and legal construction of suspects. *Annual Review of Law and Social Science*, 2, 39–60.

Council of Europe (1950). *Convention for the Protection of Human Rights and Fundamental Freedoms*. Strasbourg: Council of Europe http://conventions. coe.int/Treaty/Commun/QueVoulezVous.asp?NT=005&CL=ENG (accessed February 2010).

Dutch Constitutional Court (1990). *Wangslijmarrest, nr. 751.* [*Saliva decision, no. 751.*] The Hague: Hoge Raad der Nederlanden [Dutch Constitutional Court].

Dutch Constitutional Court (2000). *Tandenborstelarrest, nr. 10.* [*Tootbrush decision, no. 10.*] The Hague: Hoge Raad der Nederlanden [Dutch Constitutional Court].

Government of the Netherlands (1993). Wet van 8 november 1993 van de regeling van het DNA-onderzoek in strafzaken (Besluit DNA-onderzoeken). [Law of 8 November 1993 regarding DNA research for criminal proceedings (Forensic DNA Profiling Act).] *Staatsblad van het Koninkrijk der Nederlanden,* 522, 1–17.

Government of the Netherlands (1994). Reglement DNA-profielregistratie Gerechtelijk Laboratorium. [Regulation regarding the NFI DNA database.] *Staatscourtant, nr. 96.* The Hague: Ministry of Justice.

Government of the Netherlands (2000). Wet van 6 juli 2000, houdende regels inzake de bescherming van persoonsgegevens (Wet bescherming persoonsgegevens). [Law of 6 July 2000 regarding rules for protecting personal data (Personal Data Protection Act).] *Staatsblad van het Koninkrijk der Nederlanden,* 302, 1–25.

Government of the Netherlands (2001). Wet van 5 juli 2001 tot wijziging van de regeling van het DNA-onderzoek in strafzaken (Besluit DNA-onderzoek in strafzaken). [Law of 5 July 2001 regarding rules for DNA research in criminal proceedings (Forensic DNA Profiling in Criminal Proceedings Law).] *Staatsblad van het Koninkrijk der Nederlanden,* 335, 1–5.

Government of the Netherlands (2003). Wet van 8 mei 2003 tot wijziging van de regeling van het DNA-onderzoek in strafzaken in verband met het vaststellen van uiterlijk waarneembare persoonskenmerken uit celmateriaal. [Law of 8 May 2003 regarding changes of the law on DNA research in criminal proceedings regarding the determination of external visible personal characteristics from cell material.] *Staatsblad van het Koninkrijk der Nederlanden,* 201, 1–2.

Government of the Netherlands (2004). Wet van 16 september 2004, houdende regeling van DNA-onderzoek bij veroordeelden (Wet DNA-onderzoek bij veroordeelden). [Law of 16 September 2004 regarding DNA research from convicted criminals (DNA Convicted Persons Act).] *Staatsblad van het Koninkrijk der Nederlanden*, 465, 1–7.

House of Representatives (1999a). *Vergaderjaar 1998–1999, 26 271, nrs. 4 & 5.* [*Minutes of Dutch Parliament, 1998–1999, 26 271, no. 4 & 5.*] The Hague: Tweede Kamer der Staten Generaal [House of Representatives of the Dutch Parliament].

House of Representatives (1999b). *Vergaderjaar 1999–2000, 26 271, nrs. 6 & 7.* [*Minutes of Dutch Parliament, 1998–1999, 26 271, nrs. 6 &.*] The Hague: Tweede Kamer der Staten Generaal [House of Representatives of the Dutch Parliament].

Jeffreys, A., Wilson, V. and S. Thein (1985b). Individual-specific 'fingerprints' of human DNA. *Nature*, 316, 76–79.

Kayser, M., Liu, F. and Janssens, A. *et al.* (2008). Three genome-wide association studies and a linkage analysis identify *HERC2* as a human iris color gene. *American Journal of Human Genetics*, 82, 411–423.

Liu, F., van Duijn, K., Vingerling, J. R. *et al.* (2009). Eye color and the prediction of complex phenotypes from genotypes. *Current Biology*, 19, R192–R193.

M'charek, A. (2005). *The Human Genome Diversity Project. An Ethnography of Scientific Practice.* Cambridge, UK: Cambridge University Press.

M'charek, A. (2008). Silent witness, articulate collectives: DNA evidence and the inference of visible traits. *Bioethics*, 22, 519–528.

Ministry of Justice (1997). *Concept-wetsvoorstel tot wijziging van de regeling van het DNA-onderzoek in strafzaken.* [*Draft Law Proposal to Change the Law Regarding DNA Research for Criminal Proceedings.*] The Hague: Minister van Justitie [Ministry of Justice].

Ministry of Justice (1999). *Minister Korthals overweegt bredere toepassing DNA-onderzoek.* [*Minister Korthals is Considering Expanding the Scope for DNA Profiling.*] [Press release 21 June.] The Hague: Minister van Justitie [Ministry of Justice].

Ministry of Justice (2008a). *DNA-verwantschapsonderzoek bij aanpak criminaliteit.* [*DNA Familial Searching to Fight Crime.*] [Press release 16 October.] The Hague: Minister van Justitie [Ministry of Justice].

Ministry of Justice (2008b). *DNA-onderzoek uiterlijk waarneembare kenmerken.* [*DNA Research on External Visible Characteristics.*] [Letter of approval to the director of the NFI to conduct scientific research, ref. 5528833/08, 9 February 2008.] The Hague: Minister van Justitie [Ministry of Justice].

Mol, A. (2002). *The Body Multiple. Ontology in Medical Practice.* Durham, NC: Duke University Press.

Netherlands Forensic Institute (2000). *Jaarverslag 1999.* [*Annual Report 1999.*] Rijswijk: Nederlands Forensisch Instituut.

Netherlands Forensic Institute (2008). *Jaarverslag 2007: Nederlandse DNA-databank voor strafzaken.* [*Annual report 2007: Dutch DNA Database for Criminal Proceedings.*] The Hague: Nederlands Forensisch Instituut.

Netherlands Forensic Institute (2009). *Oogkleur voorspellen op basis van DNA* [*Predicting colour of the eyes based on DNA*]. Press release, 10 March. The Hague: Nederlands Forensisch Instituut.

Office of Public Prosecutor (2008). *Opnieuw grootschalig DNA-onderzoek in oude moordzaak.* [*Once More a DNA Dragnet in Old Murder*]. Amsterdam: Openbaar Ministerie http://www.om.nl/ (accessed 31 March 2009).

Pro Justitia (1998). *Nr. FA-123Ke. Inverzekeringstelling en voorlopige hechtenis/raad-kamer in het strafrechtelijk onderzoek naar 'Albin'* [*No. FA-123Ke. Arrest and Custody/Council Chamber on the Criminal Investigation regarding 'Albin'.*] The Hague: Openbaar Ministerie [Ministry of Justice].

Project DNA & Burglary (1999). *Eindrapportage DNA bij inbraken.* [*Final Report of the Project DNA & Burglary.*] Rijswijk: Gerechtelijk Laboratorium (NFI), politier-egio Utrecht, politie Midden en West Brabant, arrondissementparketten Breda, Utrecht, Nederlands Politie Instituut.

Sparkes, R., Kimpton, C. and Gilbard, S. *et al.* (1996). The validation of a 7-locus multiplex STR test for use in forensic casework. (II) Artefacts, casework studies and success rates. *International Journal of Legal Medicine*, 109, 195–204.

Toom, V. (2006). DNA fingerprinting and the right to inviolability of the body and bodily integrity in the Netherlands: convincing evidence and prolifer-ating body parts. *Journal of Genomics, Society and Policy*, 2, 64–74.

Turner, B. S. (2006). Body. *Theory, Culture and Society*, 23, 223–229.

van der Ploeg, I. (2007). Genetics, biometrics and the informatization of the body. *Annali dell' Istituto Superiore di Sanità*, 43, 44–50.

van Kampen, P. (1998). *Expert Evidence Compared. Rules and Practices in the Dutch and American Criminal Justice System.* Antwerp; Intersentia Rechtswetenschappen.

Williams, R. and Johnson, P. (2008). *Genetic Policing. The Use of DNA in Criminal Investigations.* Cullompton, UK: Willan.

JOHANNE YTTRI DAHL

10

DNA the Nor-way: black-boxing the evidence and monopolising the key

INTRODUCTION

In July 2004, the Norwegian Government appointed the so-called Strandbakken Committee to consider changes in the laws regarding the forensic DNA database,[1] which had become operational in 1999. In late 2005, exactly 10 years after it was first decided to establish a forensic DNA database in Norway, the committee submitted a White Paper to Parliament (Ministry of Justice and the Police 2005: 19). While the White Paper supported a substantial expansion of Norway's forensic DNA database, it also raised a number of issues about the use of DNA databases in criminal law administration. Approximately 35 stakeholders including government offices, other institutions and non-government organisation commented on the White Paper. These comments were collated in a proposition to the Odelsting (Odelstingsproposisjoner 19) (Ministry of Justice and the Police 2006),[2] which served as an advisory document to Parliament.

Two years after the publication of the White Paper, in December 2007, the Norwegian Parliament ruled to expand the forensic DNA database and appropriated 64 million kroner (approximately €7 million) to finance this 'DNA revolution'. According to Knut Storberget (2007), the Norwegian Minister of Justice, DNA analysis was an important

---

[1] The Strandbakken Committee comprises five members: a chief public prosecutor, a lawyer, a superintendent from the National Criminal Investigation Service in Norway, the director of the Norwegian Data Inspectorate and a law professor (who chairs the committee).

[2] When the parliament processes law issues, it is split into the Odelsting and the Lagting. Propositions to the Odelsting are used when the Government proposes new laws or cancellation of or amendments to existing laws (Chaffey and Walford 1997: 101).

Genetic Suspects: Global Governance of Forensic DNA Profiling and Databasing, ed. Richard Hindmarsh and Barbara Prainsack. Published by Cambridge University Press. Copyright © Cambridge University Press 2010.

tool in the battle against criminality, outperforming other forensic methods in efficiency and reliability by helping to free up police resources and raise detection rates for a variety of crimes, ranging from volume crime, serious crime and organised crime to national and international crime, which would lead to increased levels of security.

In accordance with the new law, the database has now expanded from including DNA profiles only of people convicted of serious crimes, such as robbery, sexual assaults, murder and grievous bodily harm, to including profiles of anyone convicted of a criminal offence leading to imprisonment.[3] This is an instance of 'function creep', that is, a widening of the scope of purposes for which DNA profiling and databasing is used (Dahl and Sætnan 2009; Chapter 2).

Since September 2008, when the new regulations became effective, the Norwegian forensic DNA database has consisted of three separate levels. First, there is the investigation database, containing profiles of people under reasonable suspicion of having committed a punishable criminal act. It is a temporary internal working database for the police. If suspicion is cleared the profile is eliminated. Second, there is the subject profile database, containing profiles of people convicted of an offence that could lead to imprisonment. These profiles are to be deleted five years after authorities become aware of the registered person's death. Third, there are crime scene profiles. In April 2009, the subject profile database contained 13 947 profiles, and the crime scene database contained 5025 profiles. This means approximately 0.3% of the Norwegian population of nearly 4.8 million have a profile stored on the forensic DNA database. In the course of 2008, a total of 791 hits were obtained with the help of the database; 601 of those were matches between subject profiles and traces from crime scenes, and 190 were matches between crime scene traces.

An interesting aspect of the Norwegian database is the concept of volunteers. In theory, anybody could volunteer to be on the register, without the need to provide any specific reason. In addition, profiles from intelligence-led mass screenings (dragnets), as well as profiles from victims and witnesses, are *not* uploaded to the register containing subject profiles unless the originators of the profiles have specifically requested for this to occur. While it is not specified in the statutory framework, it is assumed by the registrar that any such person who

---

[3] However, fear of a dramatic increase of cases has led to temporary regulations that state that only people sentenced to imprisonment for over 60 days will be registered (Director General of Public Prosecution 2008).

volunteers their profile to the database may later demand its deletion at any point in time. However, no such applications have of yet been submitted. To date, fewer than five people have volunteered their profiles to be uploaded.

While there was general political agreement that the DNA database should be expanded, related issues created some controversy. How far should the database be expanded? Who should be included on the database? Should DNA samples be stored after a DNA profile has been obtained? Last but not least, what institutions should be allowed to supply DNA analysis for the Norwegian DNA database and courts?

Whether samples from suspects should be stored after profiles are obtained was one of the most frequently asked questions in Proposition 19 (Ministry of Justice and the Police 2006). The question caused dissent in the Strandbakken Committee. Three of the Committee members wanted samples retained, while two wanted them destroyed.[4] However, the debate did not result in changes in the existing legislation; biological samples are still to be destroyed after profiles have been obtained. Another issue debated in relation to expanding the DNA database was the question of whether young offenders should have their profiles deleted from the database if they had not reoffended within a specific period of time. It was eventually decided not to include such a provision.

Such issues, and others referred to above, pose important challenges for responsible and transparent governance of Norway's expanding forensic DNA database. In this respect, they require further attention and critical discussion. This is the intention of this chapter. Special attention is paid to the fact that there is one DNA laboratory with a monopoly on DNA analysis in Norway, for two reasons: (a) the issue of monopoly was prominently debated in connection with the expansion of the DNA database, and (b) the prominence of the issue was reinforced by respondents in the in-depth interviews of key stakeholders (see below) that I conducted on the use of DNA evidence in courts and the expansion of the database. Discussion of these interviews forms the main substance of this chapter.

That said, the chapter first discusses the structure and main actors of the Norwegian system of forensic DNA analysis. It then introduces the interview methodology used and provides analysis of key

---

[4] The professor, the chief public prosecutor and the chief public prosecutor wanted samples retained, while the lawyer and the Director of the Norwegian Data Inspectorate wanted samples destroyed.

themes identified from these interviews. These themes include how DNA evidence and the Institute of Forensic Medicine and the Commission for Forensic Medicine (Den rettsmedisinske kommisjon) are perceived. Finally, the chapter addresses the implications for the governance of Norway's forensic DNA database arising from these themes or narratives, including the need for improved governance and a system more open to knowledge flows and mutual learning across actors in the criminal justice system. First, the structure of the system surrounding forensic DNA analysis and DNA evidence is presented.

DNA THE NOR-WAY

Before discussing how Norwegian stakeholders in the field of forensic DNA technologies perceive the structure of the Norwegian DNA system and how it works, some background information is provided.

Norwegian expert witnesses are appointed by the courts, a procedure that differs from that in several other countries, particularly in the Anglo-Saxon jurisdictions, where contesting parties bring in their own experts. Moreover, until 2005, there was only one laboratory performing forensic DNA analysis in Norway – the Institute of Forensic Medicine. This institute is state owned and functions as a unit within the University of Oslo. In 2005, a second laboratory, a private institute called Gena, opened and began to offer forensic DNA analysis. The owners hoped for changes regarding the parties that could conduct forensic DNA analysis. So far, however, Gena has only been used in a handful of cases and only as a provider of second opinion. Until 1 September 2008, it was specified in the Prosecution Instructions that the Institute of Forensic Medicine was to conduct forensic DNA analysis in Norway.[5] Thereafter, it was no longer specified. However, the government has ruled that conducting DNA analysis should still be a governmental matter. Hence, the situation is actually the same as it was prior to the changes in the Prosecution Instructions. The Institute of Forensic Medicine thus maintains a monopoly on conducting forensic DNA analyses. Nevertheless, the government, in accordance with a

---

[5] Prosecution Instructions are rules prescribed by the King pursuant to the Criminal Procedure Act of 1887 concerning the public prosecution system. These instructions date from 14 December 1934 with subsequent amendments. They are internal instructions over which the Criminal Procedure Act takes precedence in the case of conflict.

decision of the Norwegian Parliament, has recently granted funds for the organisation of another governmental DNA institute at the University of Tromsø.

Another important player in the Norwegian DNA system is the Commission for Forensic Medicine, which consists of four subgroups, one specialising in forensic genetics. The primary task of this subgroup is to conduct quality assurance tests on forensic DNA analysis reports before they are used in court. The Commission sends its assessment of the analysis to the courts and/or the prosecution and to the expert witness who has provided the expert testimony. Moreover, the Commission may request an additional report from the expert witness if it considers the original report unclear or incomplete. However, these requests are seldom responded to and there has been an ongoing debate between the Commission and the Institute of Forensic Medicine as to whether the latter is actually obliged to respond to such requests (Commission for Forensic Medicine 2009). Additionally, expert witnesses are required to submit a supplementary written report to the Commission if their statement in court exceeds what is written in the original report.

THE INTERVIEW METHODOLOGY

In exploring the debate on the expansion of Norway's forensic DNA database, an interpretive analytical approach has been employed, one that seeks to probe and understand stakeholder rationalities (values, beliefs and attitudes) in relation to the issue. Here, the investigator goes 'inside' the situation to gain a better understanding of meanings from the actors' own viewpoints. Attention is paid to language, narratives and storylines (Kvale 2006). A thematic and narrative analysis was conducted to explore how the interviewees (referred to as informants) perceived the Norwegian DNA system and the use of DNA evidence. Those interviewed represented a knowledgeable, diverse sample of key political and judicial stakeholders in the debate on forensic DNA in Norway, including lawyers, police, and parliamentary politicians, members of the Norwegian Commission for Forensic Medicine, expert witnesses and DNA laboratory workers. Members of the general public were not interviewed. Special attention was given to defence lawyers with experience in cases in which DNA evidence played a significant role. This was because they are important actors with regard to the use of forensic DNA evidence, and because they particularly highlighted the issue of the monopoly situation in forensic DNA in Norway.

All interviews were semi-structured to provide flexibility in order to encourage and attain the most open and informed responses. To protect the anonymity and confidentiality of those interviewed, they were not differentiated in any way in the analysis below. As the Norwegian community in this field is small, all those interviewed and the people they talked about are referred to as 'he'. All quotations have been translated from Norwegian.

ANALYSIS OF THE KEY THEMES

How DNA evidence is viewed: is DNA magic?

Several of those interviewed claimed that the use of forensic DNA technologies was presented in a rather one-dimensional manner both in the public debate and in Norwegian courtrooms. Most felt that 'that there is too little focus on the elements of uncertainty related to the use of DNA evidence'. Some claimed that the lack of focus on the uncertainties related to the use of DNA evidence is especially worrying as DNA is a relatively new type of evidence. While history has taught us that no forensic technique is foolproof or immune to error (Cole 2004: 80; Saks and Koehler 2005: 892), several informants responded that there was little focus on this in the public debate and in courts in connection with the use of DNA evidence. One of the main worries expressed by several defence lawyers interviewed pertained to prevalent misconceptions about what DNA evidence actually represents and what it is able to prove. They felt that there was a general image in society of DNA evidence as something 'magic', a 'silver bullet' that would solve all crime. One defence lawyer said that he had 'the impression that people think it is almost something magical, an answer to the problem of evidence and doubt'. Another defence lawyer argued that DNA evidence is overemphasised owing to the lack of focus on the uncertainties: 'My experience is that DNA – what should I say? It is given a bit too much weight. One doesn't see all the sources of errors.' As one of the lawyers interviewed said: 'DNA is so overwhelming that it is not possible to contest the evidence in courts'. The perception is that it is difficult to challenge DNA in the courtroom because most people think it is virtually infallible (Thompson 2006: 15). Another defence lawyer commented: 'I believe that, to a large degree, one overvalues what it is possible to derive with certainty from the [DNA] evidence, and a worst-case scenario then might be that we could get something that misguides us, rather than guides us about a fact. That is the main potential danger today.'

Several informants also stated that they needed to know more about DNA for guidance on what it might achieve.

Is there a need to know more about DNA evidence?

To a large extent, the excessive weight given to DNA evidence was explained by actors' lack of knowledge on the matter, as illustrated by the following.

> Hardly any questions are asked. They [judges, lawyers on both sides] accept what we [the expert witnesses] say ... I have very mixed feelings when I walk out of courtrooms. Of course I try to see whether people are following what I am saying. But it is one thing to understand the words. It is another thing is to understand the meaning and how to use it afterwards. We have seen examples [of where] they misunderstand. They do not have the knowledge to use the information they are given. So what they have grasped and used appears meaningless.

When defence lawyers talked about lack of knowledge they were vague, however, about what kind of knowledge they felt was missing. It appeared as if they themselves did not actually quite know what kind of knowledge they felt they did not have. This may in itself be a challenge. If they knew what kind of knowledge they lacked, then the obstacle of obtaining that knowledge would be smaller. More so, many of those interviewed seemed afraid of appearing inadequate on a subject perceived as important but difficult to grasp. The defence lawyers expressed a concern about the qualifications amongst all of the actors involved in the use of DNA evidence, including the police, judges and DNA analysts. As one lawyer said:

> Perhaps we ascribe greater evidentiary weight to it [DNA evidence] than there is factual foundation for. The involved participants understand to a very small extent what it really implies – judges, prosecutors, defence lawyers. They do not have enough knowledge to be able to ask the critical questions or check what the experts on forensic medicine have done.

However, defence lawyers expressed concerns not just regarding the qualifications of the other actors involved but also about those of themselves and their peers. As one lawyer admitted, 'I don't actually think that I have the skills to be able to plead that a piece of DNA evidence isn't valid'.

Which aspects of evidence are talked about in courts depends not only on the expert witnesses' presentation of the evidence in question but also on what questions lawyers, and sometimes judges, ask the expert witnesses. It is, of course, easier for a person to know which questions to ask, and what questions and answers may be relevant for a case, if the person questioning has a certain amount of knowledge of the topic. Consequently, a perception of having a lack of knowledge or understanding amongst defence lawyers may contribute to evidence not being adequately challenged, because they feel they are not sufficiently equipped to question and challenge it. Consequently, this may lead to aspects of the evidence not being presented adequately to judges and juries. An expert witness claimed that often no questions were asked in relation to DNA evidence, and explained it in the following way: 'When you don't know something, then you accept what is being said, and then you don't ask any questions; no control questions, no critical questions. You don't dig deeper into the underlying question. You don't see the value of everything that is being said.'

As several lawyers said, they accept – nearly without reservation – whatever the expert witness says, and typically they do not question the evidence as thoroughly as they feel they should have. Therefore, it appears as if DNA evidence is 'black-boxed' for many of those interviewed. The concept of *black-boxing* refers to the way scientific and technical practices are made invisible by their own success. When something runs efficiently, or when a question of fact is settled, the processes of interpretation and negotiation creating that 'fact' are often rendered invisible, leaving visible only inputs into and outputs from the processes but none of their internal complexity (Latour 1999: 304). Because of the complexity of certain statements, there may be several black-boxes in relation to one statement, and a person with limited knowledge of the matter may not have the qualifications to open all of them (Latour 1987: 80). When a person with limited knowledge of a technique, for example DNA evidence, wishes to question it, it may appear as if there is only one large black-box. When a person with more extensive knowledge starts to question or point out uncertainties in relation to DNA evidence, a black-box may be opened. However, opening one black-box may result in other black-boxes appearing. In this way, black-boxes may resemble Russian dolls: when one of them is opened, another appears. Nevertheless, the difference is that while there is only a limited number of Russian dolls, the number of black-boxes in relation to DNA evidence is potentially unlimited; it may depend on how much the person questioning the evidence knows about the matter. The less knowledge someone feels they have, the more elements in a story are likely to be black-boxed.

The following quote illustrates how a lawyer refrained from questioning the expert witness because he was afraid to reveal his own ignorance:

> There is too little knowledge. We don't have enough knowledge to ask good enough questions. There are only a few of us [defence lawyers] and of the prosecutors that know enough about DNA. This results in the questioning of things ... as a participant in court [an active participant, hence defence lawyer, prosecutor or judge] and not wanting to appear a fool because one is asking silly questions. Then one refrains from asking questions that might have made the expert witness say for example: Yes, we are very uncertain whether this is correct.[6]

Many of my informants expressed a need for more knowledge to enable them to defend their clients better. However, an adequate knowledge of forensics cannot be attained if adequate funding to educate the defence bar is not provided (Berger 2004: 121). Some also said that it was hard to learn about DNA because of busy schedules and other priorities; yet most of the defence lawyers I interviewed said they were self-educated (autodidacts) on the topic of DNA. Some of them had read books on the subject, while others had searched the Internet for information. A few had taken a course on the topic given by representatives of the Institute of Forensic Medicine. Mostly, knowledge about DNA evidence seemed to be obtained through a learning-by-doing process (Dahl 2009). Some said that it requires practice and experience to be able to understand and assess DNA evidence. Accordingly, what lawyers know about DNA evidence is dependent on the cases involving DNA evidence that they have had to deal with.

The reported lack of knowledge amongst the involved parties is a reason for concern, regardless of what kind of knowledge it is they are lacking. If the knowledge level is not adequate to challenge the presented evidence, or if the actors themselves feel they do not know enough about the evidence to question it, then a piece of evidence may not be given the right shape and place it should have had in a case. According to McCartney (2006: xx), this might have fatal consequences as '[m]iscarriages of justice will flourish in a culture which fails to properly scrutinise and question "scientific" evidence'.

---

[6] The informant continued: 'This is not a problem particular to DNA. It concerns all types of evidence'. This raises the question of whether the findings in this chapter are specific to DNA. It appears as though some of the findings may be transferable to other kinds of evidence; however, it also appears that the interviewed lawyers consider DNA evidence as exceptional and extraordinary evidence.

Johnsen (2007: 17) defines expert evidence as something that requires special knowledge to produce and that may require special knowledge on the part of the hearer to understand and use correctly. According to Giddens (1997), a consequence of modernity is an increasing use of experts. Because society has become so complex we cannot be fully knowledgeable in all areas, hence we are left to trust experts. This dependency on these specialists is very visible in courts, where professionals are used frequently, as here in relation to DNA evidence. While the court is not bound by the testimony of expert witnesses (Diesen and Björkman 2003), the involved actors may feel dependent on such testimony because the expert explains a topic about which they have limited or no knowledge and experience. As a result, experts, with their institutional authority and expertise on a topic, become the providers of a 'truth' in courts. Consequently, these witnesses obtain great power in the courtroom.[7] As one of the defence lawyers interviewed stated: 'I think we have a feeling that we can swallow everything the expert witness says'.

Perceptions around the Institute of Forensic Medicine: more magic?

According to some of the interviewed defence lawyers, it is not only DNA evidence that is considered something 'magical'; the government-owned Institute of Forensic Medicine was seen in a similar light. The lawyer felt that DNA and the Institute of Forensic Medicine had a 'magic' image because both appeared unquestionable. Another defence lawyer criticised the excessive authority bestowed on the Institute of Forensic Medicine by his peers and other actors in the criminal justice system: 'Clearly they make mistakes there, as everywhere else. All humans make mistakes, also at the Institute of Forensic Medicine, but it is hard to make people believe that.' A police officer, too, called for caution in this respect: 'It's important not to think that it is only the Institute of Forensic Medicine in Oslo that will be able to ensure it [quality and rule of law]'.

---

[7] Power is defined here as a capacity to have significant control of, or influence over, an important resource such as knowledge about DNA, be it economic, social, cultural or symbolic. Hence, power here consists of two central aspects: it implies first a good capacity to mobilise knowledge about DNA, and, second, the ability to communicate this form of knowledge. This definition of power refers to what Bourdieu (1990) called the 'energy of social physics'.

It could be argued that defence lawyers have a vested interest in tarnishing the reputation of the Institute of Forensic Medicine and, therefore, will talk of it negatively. However, a police officer has no apparent interest in undermining this reputation. Consequently, a police officer taking the same view of the institute as defence lawyers may be seen as objective and neutral. As discussed above, with a monopoly on the provision of forensic DNA analysis, and because it has so long been considered infallible, the Institute of Forensic Medicine has become the authority to define and manage what is perceived as the truth on forensic DNA in Norway. Being an 'obligatory passage point' (Latour 1987), it has become the translator of DNA evidence for everyone concerned – police, lawyers (defence and prosecution), jury members and judges. So, many of those interviewed also saw the Institute of Forensic Medicine as the key to the DNA blackboxes. One lawyer's commented: 'It is an obvious shortcoming; we have a system that is based on an ultimate truth … And because the perception is that there is only one answer, there is no point sending it [the DNA sample] to two places. As long as this is the perception, [the situation] will remain as it is.'

A supplier monopoly on DNA evidence also results in a monopoly on who may be an expert witness on DNA in courts. Several of those interviewed stated that the limited number of experts was disadvantageous, especially as defence lawyers reported feeling dependent on the expert witnesses. Similarly, another defence lawyer argued that, 'the problem is that in real life, it is mostly one person in Norway who attends all the court cases and talks about DNA. This doesn't lead to very much professional debate.' The fact that in almost all instances, the same person from the Institute of Forensic Medicine officiates as an expert witness on DNA in Norway illustrates the limited scope of the system.

That there is only one provider of DNA analysis and virtually one person providing expert testimony appears to have had the effect of raising the threshold for challenging that provider's expertise. As sole provider, the Institute of Forensic Medicine continually expands its experience with DNA analysis. It also continually accumulates recognition of its expertise by the actors involved in the DNA system. Accordingly, professional expertise is monopolised in one place and by very few people. Some of the lawyers, however, said that, for them, this situation gave no reason for worry, as expert testimony is always considered objective: 'Luckily, we live in the belief that the expert witness we are presented with is objective'.

The stance that there is only one truth and that expert testimonies are always objective is worrisome to those who consider DNA evidence to be less than absolutely certain and also to result from subjective interpretations. Cole (2004: 80), for example, calls for caution with regard to 'allowing law enforcement to monopolize expertise in the area of forensic DNA typing'. Recent history has demonstrated that DNA evidence has been subject to conscious and unconscious pro-prosecution bias (Thompson *et al.* 2003; Cole 2004: 80). One of the interviewed defence lawyers also called attention to this issue:

> They [the Institute of Forensic Medicine] work only for the public prosecutor and yet appear neutral in front of the court of justice, and this is a problem. There is nobody that you, as a defence lawyer, may turn to in order to get information or send something to [for analysis] ... we have to ask the police to do it, or send it abroad, and then also pay for it ourselves.

Gerlach (2004: 152) argues that monopolies in DNA analysis and interpretation in the criminal justice system create 'the impression that DNA is an exact and unambiguous science in which human interpretation plays no part. This tilts in favour of prosecutors.' This is also applicable to Norway's situation. DNA is considered by many to be an exact and unambiguous science. According to Norwegian law, court-appointed expert witnesses are to serve the court, but several lawyers felt that these witnesses typically support the prosecution. While some lawyers acknowledged the necessity of collaboration between the Institute of Forensic Medicine and the prosecution, they claimed that the consequences of the collaboration should be discussed in court. Because of their dependence on the expert witnesses, lawyers felt that they should be able to choose expert witnesses of their own who have not already been called upon by the prosecution in the same case. They articulated a desire to have the possibility of obtaining a second opinion, not only in court but also before the trial. As Norway is a small country, Norwegian experts have close working relationships with each other, which may complicate the situation (Johnsen 2007: 20).

Perceptions around the Commission for Forensic Medicine: quality control or second opinion?

Most of those interviewed expressed positive attitudes towards the Commission for Forensic Medicine. Nonetheless, some claimed that

its existence rendered obsolete the need for second opinions in forensic DNA analysis, as one public prosecutor expressed it:

> The Commission for Forensic Medicine *is* a second opinion [italic added].
> The routine is that the Institute of Forensic Medicine conducts the
> analyses and presents the DNA findings. They send all the reports to the
> Commission for Forensic Medicine, which then is to undertake a review
> and quality assurance of the DNA finding to ensure a steady course and
> that the results are based on professional practice. So it is a second
> opinion. And I assume nobody has anything against a second opinion as
> long as it is a professional and soundly argued second opinion.

Nevertheless, while the Commission for Forensic Medicine conducts
quality control only before a case reaches court, a person providing a
second opinion may also be present during a trial to hear what is said
by the other expert witnesses. As discussed above, any additional
information given on the DNA evidence in the expert witness testimony that is not in the written DNA report should be reported to
the Commission; however, according to a representative of the
Commission, this is hardly ever done:

> The assessments that are done today are done very smoothly, and are not
> being questioned, and they are not accounted for in the investigation.
> But it may be that they are accounted for in court, probably it is, pretty
> thoroughly, but if [the additional information provided in court] is not
> presented to the Commission of Forensic Medicine then we can't control
> it. Therefore we don't know anything about it either.

According to the Commission's annual report, several things may be
done to increase responsible governance of the use of DNA evidence in
Norway, for example, the Commission for Forensic Medicine should
know about the task requested of the Institute of Forensic Medicine by
the police. Further, assessments of the calculation of probability/like-
lihood ratios of evidence, according to international standards, should
be conducted and assessed more often. The police should also be sent
the tables illustrating the results from the DNA analyses and the
probability/likelihood ratios determined by the Institute of Forensic
Medicine. This may provide additional useful knowledge and information. The latter is not currently done because of political fear that local
police may construct their own databases. There has been an ongoing
debate between the Institute of Forensic Medicine and the Commission
for Forensic Medicine over whether the Institute is obliged to answer
requests for additional information and clarification. It is stressed that
such requests should always be responded to (Commission for Forensic

Medicine 2009: 20). Because the Commission conducts its quality control of a DNA analysis before the result reaches the courts, it has little chance of seeing how the evidence is contextualised. Therefore the quality assurance they conduct is considered limited, as a representative of the Commission explained:

> We are not a second opinion. We are not participants in the process in any way. We have the superior quality control and a supervising function ... It does not help to have the quality control if you only have a small window of the entire situation to look at ... For me what is interesting to see is the perception that there is no need for it [someone to conduct quality control] in relation to DNA, because it is so secure ... if this is the perception; what is there to question at all? That is actually a dangerous perception. There may well be miscarriages of justice in relation to DNA. That is really possible.

The Commission's representative thus supported the defence lawyers' perception of DNA evidence as something which Cole has observed in connection with fingerprint evidence, as 'a black-box whose outputs were scientific, unassailable, unproblematic and error free' (Cole 2004: 79). The Commission's representative also claimed there is little reflection of whether there are different interpretations of DNA evidence. This lack of reflection leads to the perception that there is no need for quality control, conducted either by the Commission itself or by second opinion. It is possible that this monopoly situation reinforces the conception of DNA as an unexceptionable truth. However, some informants thought there was indeed need for a second opinion, precisely to illustrate that DNA evidence is not an unquestionable truth. This is what a defence lawyer said about his own experience of realising how DNA evidence may be challenged:

> I think we all have a kind of conception that – when something comes from the Institute of Forensic Medicine, it is a nice name, it has authority within the system, and then what comes from there is correct. So when others say something that opposes what they say, it is a bit surprising, so it is a bit like: Yeah, yes, it is possible to think differently too. It surprises me to a certain extent, not intellectually, but ...

In this statement the informant is saying that while it does not surprise him intellectually that there may be more than one truth regarding DNA evidence, it somehow surprises him in practice as he is so accustomed to there only being one truth and having such large confidence in the Institute. As Gerlach (2004: 157) argues, the 'absence of a counter-expertise in forensic science – of DNA analysts readily available to the defence – further reinforces the sense that the science and technology are beyond challenge'. But a second opinion offers an opening for professional disagreement to take place in courts. As a defence lawyer put it: 'Clearly, as long as there are two parties disagreement may arise, and then you will have a better debate and a better analysis of the validity of the evidence'.

Lynch (2006: 167) writes about second opinions in forensic DNA analysis: 'When expert testimony is uncontested jurors may accept it as indisputable fact, and when it is contested jurors may become confused by contradictory expert claims'. The following quote from one of those intervieweds, an expert witness, describes the latter point as being beneficial, however:

> We have heard complaints that insignificancies and trivialities are emphasized [in court]. They [the expert witnesses] sit in front of a jury and a judge, hence a panel that doesn't have specialised knowledge, and if an expert witness manages to make a lot of fuss, even though it is based on stupid things it may be crucial. Consequently one tears down the confidence in the analyses that are conducted. I think the focus will be moved a bit, and that does not always serve criminal justice in all respects. It will require even more from the people who are judges that are given this material and that are to consider it and use it accurately.

Alternative laboratories

The debate about who should conduct DNA analysis in Norway has so far revolved around the question of whether DNA analysis should be carried out by a newly established government-owned laboratory as an integral part of the University of Tromsø, or whether private institutions should be potential contractors for DNA analysis as well. On this matter the Norwegian Minister of Justice Storberget stated, during a Parliamentary question period:

> I think it is important, when it comes to the development of evidence, that it be clear in relation to criminal cases, especially in relation to DNA, that it is the public authorities who should take care of it, especially to ensure that all considerations regarding the rule of law are maintained; moreover, to ensure that there are no suspicions that there are other motives in relation to the tests.[8]

---

[8] See http://www.stortinget.no/Global/pdf/Referater/Stortinget/2007–2008/s080409.pdf.

In general, for a 'counter-laboratory' to appear as trustworthy and to be accepted, it will need to prove itself better than the already existing laboratory (Latour 1987: 79), for example to prove forensic DNA competence through ISO accreditation.[9] The Institute of Forensic Medicine is not currently accredited, although the new, private institute, Gena, is ISO 17025-accredited for forensic DNA analysis and interpretation and complies with the Recommendation No. R (92) 1 of the Committee of Ministers on Forensic DNA. The Institute of Forensic Medicine instead relies on its standing of many years as a sole provider, its experience and reputation. A police officer whom I interviewed said the following about this crucial difference between the two institutes: 'I'm so incredibly critical of what's happening in relation to the monopoly of the Forensic Medicine Institute. Among other things it is due to what I mentioned: Lack of quality control, ISO approval and accreditation.'

Despite the governmental institute not being accredited and the private institute being accredited, most involved parties seem to have more faith in the state-owned institute. From political debates, public documents and interviews it appears that the high levels of trust in the Institute of Forensic Medicine stem partly from the fact that it is owned by the government. As opposed to privately owned institutes, which, in the Norwegian context, are seen as more prone to compromising objectivity and quality for profit interests (see also Chapter 13), government-owned institutions generally enjoy high levels of public trust. Undoubtedly, the high levels of trust in the Institute of Forensic Medicine are also a result of the good reputation of the professional experience of the Institute's members. An expert witness expressed scepticisms towards private institutes carrying out DNA analyses:

> I think this [DNA analysis] is a profession where no one should make money on other people's misery. It is the kind of asset a constitutional state must have. It should be considered a necessary evil, an enterprise such as the state must have to ensure the legal protection of the individual and the rule of law in general. So when this is privatised, someone with financial interests will be behind it. Nobody wants to conduct this kind of enterprise out of goodness of heart. Everyone wants to make money.

[9] Accreditation is an official statement of quality assurance that a laboratory is working according to a documented quality system and has satisfactory competence to conduct certain tasks. Laboratories are evaluated according to standards specified in NS-EN ISO/IEC 17025. Accreditation and annual quality controls of laboratory processes are conducted by Norwegian Accreditation.

In general, witnesses and evidence in a trial are questioned by the lawyers and sometimes by judges. Ironically, despite the fact that DNA evidence is a more reliable tool for exonerations than it is for incriminating, it is mostly used to incriminate (Cole 2007: 100). Thus: 'It will always be important for defence attorneys – and, preferably, independent defence experts – to scrutinise all aspects of DNA evidence carefully, including the recovery of evidence, laboratory procedures and proficiency, and statistical arguments' (Cole 2001: 301). Therefore the trustworthiness of the expert and the institution presenting the DNA evidence in court is important. Scrutinising such trustworthiness may be a way of scrutinising the evidence. This is one of the reasons why a new institute with no experience and not yet in the position of having achieved credibility among the involved participants carries less authority, in spite of having ISO accreditation. In addition, as one of the informants said, openly challenging the standing or the trustworthiness of the Institute of Forensic Medicine is not always easy: 'One doesn't really understand the background for DNA. It is very scientific, and if anybody starts questioning the established authority's statements on this, it is easy to lose track of the essential ideas and end up with what you think is correct and secure, hence it is not so easy to counter-argue the established academic and professional circle.'

When there is limited understanding regarding what is being said about evidence, trust in that evidence will depend not only on the evidence in itself but also on who presents it. Because of the very limited number of individuals dealing with forensic DNA in Norway, until recently the only option to get a second opinion has been to turn to experts abroad. When asked why they do not very often choose to go abroad with the DNA evidence, a defence lawyer mentioned cost considerations:

> It has to do with the traditions of the Norwegian judicial system. Most suspects can't afford to hire experts from abroad. And defence lawyers aren't paid to hire them ... the courts believe that we have someone in the country who can tell us this. We don't need another one. When the court has appointed one expert witness that should be enough. In that way it is hard to challenge the problem.

This implies that in cases where DNA evidence plays a crucial role, the legal protection of the individual may depend on whether the defendant has got the financial means to pay for an independent second opinion. That the financial problem is not limited to the Norwegian context is illustrated by Jasanoff (2006: 334): 'Indigent defendants, who cannot afford effective lawyering, may find their fates decided less by the strength of the scientific evidence as assessed by technical experts than by the vigour and ingenuity of the advocacy mobilized by their defence'. In Norway, there is the possibility of getting the state to pay for a second opinion, but then the defence lawyers have to convince a judge that this is required. However, the decision as to whether the state should pay for the second expert witness is often not made before the case reaches the court. This means the defendant has to take the risk of not having his expenses refunded, which may thus influence any decision about whether to get that second opinion; this, in turn, can influence the outcome of the case. With regard to having to go abroad for a second opinion, this may be not just a financial challenge but also a language problem. With regard to language, another challenge for judges and juries in understanding the complexities of DNA evidence arises because expert witnesses are often perceived as literally speaking another language, because their fields are so specific, scientific and also complex. In other words, black-boxes appear readily and often.

CONFIDENCE IN GOVERNMENT

The governance of the Norwegian forensic DNA system is closely related to Norwegian governmentality[10] – namely, extensive trust and confidence in the government (Listhaug 2005). One can see this reflected in the structure of the system: The police are to leave all DNA analysis to one institution, and in future possibly two, one government-owned institute and one university-based DNA institute. Retaining DNA analysis within the realm of government-owned institutions is perceived as helping to ensure the rule of law. This mode of governance may be seen as both trust based and trust preserving, which may be seen as a reflection of the Norwegian political system. However, according to several of those interviewed, the current mode of governance, with only one institute, leads to the perception that DNA analysis is conducted by a 'closed circuit' of only a few

---

[10] Foucault's (1991) concept of governmentality refers not only to the art of governance but also to the mentality of accepting being governed. It links the formation of the state with the formation of the subject. When people govern themselves through the knowledge they have gained from the institutions that produce the knowledge, governance may be seen as functional.

participants. This makes it difficult to learn or hear about DNA from anyone other than a few people employed by the state institute and even more difficult to get a second opinion. Hence, this mode of governance does not invite transparency, accountability or mutual learning across different actors in the criminal justice system. Consequently, some informants claimed this arrangement limits their perspectives and understandings of DNA and, again, results in black-boxing of DNA evidence, which again may explain the limited number of questions asked in court, as reported by my informants.

The fact that many lawyers claim they feel incapable of adequately challenging DNA evidence renders discussions about responsible governance of DNA profiling even more important. The Norwegian DNA system needs to be governed in a manner that ensures quality, openness and reflection. Improved governance of DNA profiling in Norway may be achieved by strengthening the position of the Forensic Commission and by ensuring that it works as it is supposed to. One possibility is to provide additional reports concerning the monitoring and reporting of any comments diverging from or exceeding written reports presented in court. Another possibility is to strengthen the Commission's position from merely conducting quality control to being a more visible knowledge provider to the involved actors. More specifically, to provide an education programme about DNA profiling and databasing for the involved actors that also aims to develop a knowledge society better informed about this contentious topic.

Turning again to the issue of the establishment of Gena, the accredited private forensic institute, this development has barely changed how DNA is perceived or questioned. Apart from the government ruling that conducting DNA analysis should still be a governmental matter, the use of Gena for a second opinion continues to be limited because defence lawyers (and their clients) risk having to pay for its services themselves. Additionally, paying for a second opinion may be risky as the private institute does not seem to elicit as much trust and have as much influence as the government-owned institute, partly because of its private ownership and partly because of its limited experience. Therefore, the prospect of opening another government-owned institute may be more beneficial for the Norwegian DNA system. It may be able to provide a second opinion, which could be perceived as more trustworthy and knowledgeable in relation to conducting DNA analysis, and it could also be useful for presenting, explaining and interpreting DNA analysis in court. This could lead to increased knowledge and mutual learning amongst the actors in the field. Such possibilities could easily occur within the Norwegian governmentality, that is, within a system with high levels of public confidence in the state. However, it will be crucial that any government-owned institutes are independent of each other and that if professional disputes and discord arise, they should be acknowledged in court. There are still a number of black-boxes and a monopolised key in relation to the use of forensic DNA in Norway. Increased transparency, use of second opinion, improved governance and increased knowledge is essential.

REFERENCES

Berger, M. (2004). Lessons from DNA: restricting the balance between finality and justice. In *DNA and the Criminal Justice System: The Technology of Justice*, ed. D. Lazer. Cambridge, MA: MIT Press, pp. 109–132.
Bourdieu, P. (1990). *The Logic of Practice*. Cambridge, UK: Polity Press.
Chaffey, P. and Walford, R. (1997). *Norsk-engelsk juridisk ordbok*. [*Norwegian–English Legal Dictionary*.] Oslo: Universitetsforlaget
Cole, S. A. (2001). *Suspect Identities*. Cambridge, MA: Harvard University Press.
Cole, S. A. (2004). Fingerprint identification and the criminal justice system. In *DNA and the Criminal Justice System: The Technology of Justice*, ed. D. Lazer. Cambridge, MA: MIT Press, pp. 63–91.
Cole, S. A. (2007). How much justice can technology afford? The impact of DNA technology on equal criminal justice. *Science and Public Policy*, 34, 95–107.
Commission for Forensic Medicine (2009).*Den Rettsmedisinske kommisjons årsmelding 2007–2008*. [*Forensic Commission Annual Report*] Oslo: Norwegian Civil Affairs Authority www.justissekretariatene.no/nb/Innhold/DRK/Arsmeldinger-og-vei ledere/ (accessed 29 May 2009).
Dahl, J. Y. (2009). Another side of the story: lawyers' views on DNA as evidence. In *Technologies of Insecurity. The Surveillance of Everyday Life*, eds. K. Aas, H. Gundhus and H. Lomell. London: Routledge-Cavendish, pp. 219–237.
Dahl, J. Y. and Sætnan, A. (2009). 'It all happened so slowly': on controlling function creep in DNA databases. *International Journal of Law, Crime and Justice* 37, 83–103.
Diesen, C. and Björkman, J. (2003). DNA-bevis är inte alltid starka. [DNA evidence is not always strong.] *Juridisk Tidsskrift*, 4, 890–904.
Director General of Public Prosecution (2008). *Ra 07–569 KHK/jaa 624.7* (Letter 15 August 2008): *Nye retningslinjer for registrering i DNA-registeret og innsamling av spor med DNA-analyse mv.* [Ra 07–569 KHK/jaa 624.7 (Letter of 15 August 2008): *New Regulations for Registration in the DNA Register and Collection of Traces for DNA Analysis etc.*] Oslo: Norwegian Prosecuting Authority www.riksadvokaten.no/ ra/ra.php?artikkelid=194 (accessed 16 October 2008).
Foucault, M. (1991). Governmentality. In *The Foucault Effect*, eds. G. Burchell, C. Gordon and P. Miller. Chicago, IL: University of Chicago Press, pp. 87–105.
Gerlach, N. (2004). *The Genetic Imaginary: DNA in the Canadian Criminal Justice System*. Toronto: University of Toronto Press.
Giddens, A. (1997). *The Consequences of Modernity*. Cambridge, UK: Polity Press.
Jasanoff, S. (2006). Just evidence: the limits of science in the legal process. *Journal of Law, Medicine and Ethics*, 34, 328–341.

Johnsen, J. (2007). Feilkilder ved ekspertbevis. Hvordan kan de påvirke utfallet av straffesaker? [Sources of error in expert evidence. How it may affect the outcome of criminal cases.] In *Rettsmedisinsk sakkyndighet i fortid, nåtid og fremtid* [*Forensic Medicine in the Past, Present and Future*] eds. P. Brandtzæg and S. Eskeland. Oslo: Cappelen, pp. 16-25.

Kvale, S. (2006). *Det kvalitative forskningsintervju.* [*The Qualitative Research Interview.*] Oslo: Gyldendal Akademiske.

Latour, B. (1987). *Science in Action.* Cambridge, MA: Harvard University Press.

Latour, B. (1999). *Pandora's Hope: Essays on the Reality of Science studies.* Cambridge, MA: Harvard University Press.

Listhaug, O. (2005). Oil wealth dissatisfaction and political trust in Norway: a resource curse? *West European Politics*, 28, 834-851.

Lynch, M. (2006). Circumscribing expertise: membership categories in courtroom testimony. In *States of Knowledge: The Co-production of Science and Social Order*, ed. S. Jasanoff. New York: Routledge, pp. 161-181.

McCartney, C. (2006). *Forensic Identification and Criminal Justice: Forensic Science, Justice and Risk.* Cullompton, UK: Willan.

Ministry of Justice and the Police (2005).: *Lov om DNA-register til bruk i strafferettspleien.* [*Law about DNA Database for Use in Criminal Law Administration.*] Oslo: Ministry of Justice and the Police.

Ministry of Justice and the Police (2006).*Odelstingsproposisjoner 19 (2006-2007): Om lov om endringer i straffeprosessloven (utvidelse av DNA-registeret).* [*Proposition to the Odelsting 19: About Changes of Laws in Criminal Procedure (Expansion of the DNA Database).*] Oslo: Ministry of Justice and the Police.

Saks, M. and Koehler, J. (2005). The coming paradigm shift in forensic identification science, *Science*, 309, 892-895.

Storberget, K. (2007). Vi skal oppklare mer. [We are going to detect more.] *Østlendingen* [Norwegian Newspaper].

Thompson, W. (2006). Tarnish on the 'gold standard': understanding recent problems in forensic DNA testing. *The Champion*, 10, 14-20.

Thompson, W., Taroni, F. and Aitken, C. (2003). How the probability of a false positive affects the value of DNA evidence. *Journal of Forensic Sciences*, 48, 47-54.

11

Portuguese forensic DNA database: political enthusiasm, public trust and probable issues in future practice

INTRODUCTION: ESTABLISHMENT OF THE FORENSIC DNA DATABASE

The Portuguese law on the forensic DNA database was established in February 2008 (Law 5/2008) (Government of Portugal 2008). The database is expected to be operational during 2010. The discussion in this chapter is informed by insights obtained from interviews carried out by Machado between March and May 2008 with law and forensic science experts who were part of the committee appointed by the Minister of Justice in January 2006 (Ministry of Justice 2006). The committee had the task of drafting a law regarding a DNA database for civil identification and criminal investigation. The draft law that resulted from the work of this committee was considered by the Portuguese Parliament in September 2007. All translations are by the authors if not indicated otherwise.

The Portuguese Government had already in 2005 announced its intention to create a DNA database for civil and forensic identification purposes. Initially, the intention had been to collect data from the entire Portuguese population. This idea was eventually abandoned in favour of a database restricted to the 'criminal' population, as in many other countries in Europe and elsewhere. According to the forensic experts who participated in the preparation of Portuguese law on the forensic DNA database, there were two main reasons why the initial plan to establish a universal database was ultimately abandoned. First, no other country had such a population-wide database. Second, in Portugal, the incidence of serious crime is relatively low and economic resources of the country are limited (Machado and Silva 2008a), which is a disincentive to expensive large-scale projects in this field.

Genetic Suspects: Global Governance of Forensic DNA Profiling and Databasing, ed. Richard Hindmarsh and Barbara Prainsack. Published by Cambridge University Press. Copyright © Cambridge University Press 2010.

In fact, the regulation of the Portuguese forensic DNA database is more restrictive in terms of data inclusion and information preservation than that of other European countries. These restrictions include profile entry criteria (DNA profiles are inserted only of those convicted of serious crimes with a sentence of three years or more), the obligation to remove profiles at the time of the definitive cancellation of criminal records and the requirement to destroy the actual DNA sample after the DNA profile has been obtained. If there is no record of a further crime committed by the same person, the DNA profile of the convicted offender is removed from the database five, seven or ten years after the sentence has been served if its duration was less than five years, between five and eight years, or more than eight years, respectively (article 15 of Law 57/98; Government of Portugal 1998). Thus there is no permanent retention of any criminal's DNA profile in the DNA database. In other words, rather than entering the field with a radically new approach to DNA databasing – one that comprised samples and profiles from the entire population – the Portuguese government instead chose a conservative solution. However, the political establishment's optimistic tone regarding the potential benefits of DNA technologies survived this shift. For example, when the Portuguese Minister of Justice announced the law proposal (for Law 5/2008) on the DNA database on 1 July 2007, he highlighted the database's power to prevent and solve crimes, to improve the efficiency and the speed of justice and to contribute to the common good and public trust. These characteristics, the Minister argued, would render the justice system better prepared to serve the public good (Ministry of Justice 2007); at the same time, it would be an effective tool to convict the 'real offenders' and to acquit the wrongly accused:

> Forensic science helps the law by making the courts more scientific and more rigorous … The credibility of the courts and criminal investigators is strengthened by the use of technical measures with high accuracy and reliability. The DNA database will contribute to the discovery of the true perpetrators of crimes and to the conduct of a successful prosecution. No less important, it will help to exonerate those who have been unjustly accused.

In the political discourse, the idea that science is an important element of justice and the most effective way to increase the effectiveness of courts and criminal investigation (Jasanoff 2006; Lynch et al. 2008) is blended with the appeal for the need of the 'development of Portugal'.

This was clearly implied by the Minister's emphasis on the need for the country to follow the example of more advanced countries in matters of DNA criminal investigation, as well as transnational database cooperation related to security policies and crime fighting.

Such pronouncements by the Minister of Justice illustrate the political atmosphere at the time of the introduction of the forensic DNA database in Portuguese society. The fact that the Portuguese Government presented the idea that the creation of a forensic DNA database would fulfil great expectations and would enhance public confidence is a good starting point for the examination of social, legal and ethical issues related to forensic DNA technologies in Portugal. In general, DNA database projects are more likely to obtain public support if the political discourses that present them to the wider society 'correspond with established narratives in a particular society' (Prainsack 2007: 86). This is of particular relevance considering that the government promised a solution to some salient problems of contemporary Portuguese society, which include a high level of fear of crime that is to a large extent created and fed by the media (Fox *et al.* 2007), associated growing feelings of insecurity and decreasing trust in the criminal justice system (Contini and Mohr 2007). As mentioned above, although Portugal is one of the poorest countries in Europe and it has a relatively low level of serious crimes compared with other European countries, the number of crimes is currently rising. The positive image of forensic DNA technologies, as portrayed to the public by the political elite, therefore, played an important role in justifying the technology's adoption (Williams and Johnson 2005; Williams *et al.* 2008) as a tool of fighting and preventing crime.

Despite the political enthusiasm accompanying the establishment of the Portuguese forensic DNA database, many citizens might feel uneasy about these developments. According to some national opinion polls (Santos *et al.* 1996; Cabral *et al.* 2003), Portuguese citizens tend to associate their justice system with themes such as vulnerability to pressure from powerful people, exposure to corruption and little guarantee of the confidentiality and security of the information found by criminal investigation (Costa 2003). In our opinion, there is also the question of proportionality between the possible benefits of a DNA database and its economic costs and whether the expenditure is warranted for serious, yet rare, crimes. At the same time, strategies for crime prevention and social rehabilitation of delinquents are limited, which raises the question whether part of the investment in DNA technologies would be better utilised in those areas (Machado and Silva 2008b).

The next section describes the legislative basis for the Portuguese forensic DNA database, which is presented by the Portuguese authorities as being the result of great concern about citizens' rights. The chapter then addresses the social, ethical and practical implications of the legal framework by focusing on important concerns and issues that were prominent in Portugal's public debate.

The aim of this chapter is to provide a contribution to the debate on emerging practices related to DNA databases for crime investigation and prevention through discussion of some of the important challenges in the foreseeable future regarding the governance and regulation of DNA technologies in criminal investigation in the context of the strong probability that the Portuguese forensic DNA database will grow larger in this period. That proposition is not only fuelled by the enthusiastic rhetoric of government but also is in accordance with expectations that the increase in number and scope of forensic DNA databases will form a general trend both in Europe and worldwide (Nuffield Council on Bioethics 2007). Of particular interest in the Portuguese case is to examine how a developing country in Europe with a DNA database law created only in 2008 was convinced that the imminent expansion of the database was a socially useful goal. The social, ethical and political implications of the legal framework of the Portuguese forensic DNA database are critically examined by exploring the loopholes in existing legislation and issues around the likely public acceptance of future governmental efforts to expand the inclusion and storage of DNA profiles. This last issue is particularly acute in a society with low levels of trust in the criminal justice system, commonly regarded as corrupt, discriminating, slow, inefficient and protective of the more powerful social groups (Cabral *et al.* 2003).

LEGAL AND REGULATORY FRAMEWORKS: PROTECTING CITIZENS'RIGHTS?

Law 5/2008 establishes the legal framework for the Portuguese forensic DNA database (Government of Portugal 2008). The law makes clear that the analysis of DNA can only be used for purposes of civil identification and criminal investigation (no. 1 of article 4 of Law 5/2008), and prohibits the use of any information stored in the DNA database for any other purposes other than forensic. The purposes of civil identification are pursued by comparing the DNA profiles of the samples of biological material collected in person, from the corpse, from part of corpse or in a place where collections with those purposes are carried out. In the context of civil identification, a comparison can be made between those profiles and the profiles in the DNA database (no. 2 of article 4 of Law 5/2008). Individuals can be identified by comparing the DNA profiles of the samples of biological material collected at crime scenes with those of individuals who, directly or indirectly, may be associated with them, as well as by comparing these profiles with the existing profiles in the DNA database (no. 3 of article 4 of Law 5/2008). The law also allows the use of the data in the database for purposes of scientific research or statistics only after irreversible anonymisation, which means that the contractor receiving the data must not have any information about the identity of the individuals recorded (article 23 of Law 5/2008).

The use of personal information for forensic identification raises different issues of a social and ethical nature depending on whether the focus is on civil identification or on identification in the context of criminal investigation. A forensic DNA database for criminal investigation purposes raises controversies around the proper balance between preventing, detecting and prosecuting crime and individual rights to liberty, autonomy and privacy (Nuffield Council on Bioethics 2007). A DNA database for civil identification that may contain genetic information on all or part of the population is valuable for the identification of missing persons or for identifying a deceased person or a body part. However, the practices related to forensic identification, either for civil or for criminal investigation aims, can be seen as a way of strengthening control or inspection of citizens by means of a physical, moral and social surveillance of their bodies and behaviours (Monahan 2006).

This chapter focuses on the issues of criminal investigation identification, arguing that the Portuguese forensic DNA database presents features that are restrictive by comparison with similar European databases. This argument is based on the criteria for a comparative analysis of legislation in Europe provided by the Report on Member Countries' DNA Database Legislation Survey (European Network of Forensic Science Institutes 2006). The Portuguese legal framework is examined with respect to profile entry criteria, collection of samples, profile removal criteria and the practice of elimination of reference samples.

The National Institute of Forensic Medicine processes the DNA samples (article 17 of Law 5/2008) and conveys the results to the competent judicial authorities (article 19 of Law 5/2008). The Institute is responsible to the Ministry of Justice (which is also the custodian of the DNA database) for coordinating medico-legal services at the national level. It has three main laboratories that provide a wide range of forensic services, including toxicology, genetic analysis and forensic psychiatry, when ordered by Portuguese courts of law. All the activities developed by the Institute that are related to the forensic DNA database are supervised and controlled by an independent administrative body (Supervisory Council for the DNA Database (*Conselho de Fiscalização da Base de Dados de Perfis de ADN*)), with powers of authority conferred by the Portuguese Parliament. The law stipulates that this supervisory body will be constituted by a group of three individuals 'of recognized reputation and holding full capacity of their civil and political rights' (no. 3 of article 29 of Law 5/2008).

With regard to profile entry criteria, the Portuguese forensic DNA database can include profiles for purposes of civil identification from volunteers and relatives of missing persons; from professionals who collect and analyse DNA samples; from a corpse or part of a corpse; and, in criminal investigation, from convicted individuals with a sentence of three years or more (no. 1, 2 and 3, respectively, of article 18 of Law 5/2008). In the case of volunteers, relatives and professionals, Portuguese law establishes that DNA profiles obtained from the analysis of samples, as well as corresponding personal data, can be integrated into the database only on the free, written and informed consent of these individuals (no. 1a and 1b of article 18 of Law 5/2008).

With regard to collection of samples for criminal investigation, this is performed at the request of the defendant or is ordered by a judge (no. 1 of article 8 of Law 5/2008). Samples can also be collected from convicted individuals with a sentence of not less than three years (no. 2 of article 8 of Law 5/2008). The DNA profiles from suspects are never included in the forensic DNA database for criminal purposes; they are included only upon conviction and if ordered by a judge (article 18 of Law 5/2008).

Samples from accused defendants and convicted offenders can be collected by force (without consent). Nevertheless, the law states that the individual has the right to be informed, specifying the content and possible uses of his or her genetic information. However, no guidelines exist on how to proceed if a person refuses to sign the informed consent (see Chapter 3). In addition, there is no reference to, or special safeguard concerning, the rights of minors or individuals incapable of giving informed consent. This is contrary to the recommendations of the Portuguese National Council of Ethics for the Life Sciences (Henriques and Sequeiros 2007), and of other ethics committees, who advised that special attention be given to the treatment of children by legal systems (Nuffield Council on Bioethics 2007).

Portuguese law also stipulates that all the samples coming from volunteers and convicted offenders are to be destroyed immediately after the DNA profile has been obtained (no. 1 of article 34 of Law 5/ 2008). In addition, the law says that the preservation of samples covers only the analysis and counter-analysis needed for purposes of civil identification and criminal investigation (article 32 of Law 5/2008). The samples collected from corpses, missing persons and relatives of missing persons are to be destroyed if the missing person has been identified. If identification is not made, samples are to be destroyed 20 years after their collection. The samples collected from forensic professionals must be destroyed 20 years after the cessation of their official function (e.g. at retirement; no. 1 of article 26 of Law 5/2008).

Profiles are eliminated from the DNA database in the following situations: profiles derived from crime scene samples that do not match with the profile of the accused are eliminated 20 years after collection of the respective samples (no. 1e of article 26 of Law 5/2008), and profiles from convicted offenders are eliminated at the time of the definitive cancellation of criminal records, maximally 10 years after the sentence has been served. The profiles of volunteers and relatives of missing persons are kept for an unlimited time, unless they revoke their previous consent; the profiles from corpses are eliminated after identification (no. 1e of article 26 of Law 5/2008).

As mentioned previously, the law also states that samples should be destroyed once genetic profiles have been obtained from them (no. 1e of article 15 and no. 1 of article 34 of Law 5/2008). Nonetheless, the law also calls for the creation of a biobank (article 31 of Law 5/2008) with the National Institute of Forensic Medicine as its custodian. The law stipulates that samples should be kept in a safe place without the possibility of immediate identification of the person (no. 1 of article 31 of Law 5/2008). In addition, agreements with other entities for the use of samples can be made, but the law only indicates that these contracts should 'guarantee conditions of security and confidentiality' (no. 2 of article 31 of Law 5/2008).

Finally, another important feature of the legislation on the Portuguese forensic DNA database is the concept of volunteers (see Chapters 3 and 4). Law 5/2008 states the possibility of a gradual, phased construction of the DNA database from collecting samples from volunteers, either the relatives of missing persons or unidentified victims, and from anyone who wishes to donate a sample (no. 1 of article 6 of Law 5/2008). The DNA profiles collected from volunteers are to be preserved for an unlimited time and will be uploaded to the general database, to be removed in the case of explicit revocation of the previously given consent or in the case of victim identification. The collection of samples from volunteers is to be made with free, informed and revocable consent and following a sample collection request in writing, which must be addressed by the volunteer to the competent entities for laboratory DNA analysis – the Laboratory of the Scientific Police and the National Institute of Forensic Medicine (no. 2 of article 6 of Law 5/2008).

The regulation of the Portuguese forensic DNA database appears more restrictive in terms of data inclusion and information preservation than those in other European countries, which might mean at the least a more protective orientation towards citizens' rights (Dias 2007). Nevertheless, according to law experts and forensic scientists who were members of the commission charged with producing a DNA database law proposal, the uses of this technology, as well as the scope of the forensic DNA database, will 'inevitably' increase in the next few years (Machado and Silva 2008a).

UNCERTAINTIES IN THE LAW AND PROBABLE
ISSUES IN FUTURE PRACTICE

Following the description above of the legislative provisions, this chapter now moves on to examine the weaknesses and ambiguities found in the legislation as gleaned by analysing the law itself and from interviews with experts directly involved in the preparation of the draft law. Examining the different narratives emerging from those interviews provides valuable insights to forms of governance consisting of a mix of legal forms and informal regulation by cultural customs and unwritten codes and practices. Since the forensic DNA database was not operational at the time of writing, although it was expected to start its activities during 2010, these narratives project the social perspectives or perceptions of the experts who prepared the first version of the legislation but not the actual enactment of DNA profiling and databasing in practice. Consequently, the narratives offer a potentially strong basis to assess probable future practice.

It has been our argument that the legislative framework of the Portuguese forensic DNA database is relatively restrictive in its uses for a number of reasons: samples are not retained, the criteria of profile entry only allows the inclusion of offenders convicted for serious crimes with a sentence of three years or more and subject profiles will be removed at the time of the maximum expiration deadline for the relevant criminal proceedings. Although this restrictiveness in the inclusion and storage of data will likely limit the effectiveness and efficiency of the database once operational, high expectations and ideas of a 'super efficient mechanism for identifying offenders' (Williams *et al.* (2004: 12) with regard to the UK database) prevail in the narratives of the Portuguese experts interviewed. According to these informants, the particular provisions in the law establishing the Portuguese forensic DNA database responded to the need to obtain consensus and support among political parties. It was considered necessary to produce a law that would not create too much apprehension about violations of the rights of individual citizens. One of the means by which this objective would be met was the deliberate decision to let the database grow 'slowly'. A law expert commented: 'It was already a miracle to have this DNA database law. At the beginning the database will not be efficient or effective, but I am in favour of starting slowly.' A forensic science expert said: 'All countries start with more restrictive laws and then go into the expansion. That is the tendency and this is logical.'

These perceptions, that the Portuguese forensic DNA database would 'inevitably expand', can be associated with the concept of 'function creep', a term frequently used when discussing surveillance technologies and which refers to additional uses beyond those envisaged or mentioned at the time of installation (Innes 2001; also Chapter 2). Several questions arise. What then, in the foreseeable future, will be the acceptable uses, forms and levels of surveillance and social control proceeding from the forensic DNA database? What safeguards will be needed?

One way of dealing with the many legal, ethical, social and political issues that have been raised with regard to the future expansion of the Portuguese forensic DNA database could be through a committee to monitor and assess the DNA database practices in terms of accountability, security, quality assurance and ethical standards (as discussed below). These are needed especially with regard to accountability, the transparency and effective functioning of the administrative body that will control and supervise the activities of the national forensic DNA database; the concept of 'volunteers' (donating their DNA to the database by choice); quality assurance monitoring of crime scene forensic examinations; the interpretation of bioinformation and assessment of scientific evidence in criminal courts; and the regulation of the circulation of genetic information in international cooperation treaties and agencies. Each of these issues will be examined in turn.

Accountability

Law 5/2008 establishes an independent administrative board (Supervisory Council for the DNA Database) to supervise and control the future activities of the Portuguese forensic DNA database. The composition of this council was established in February 2009 (Resolution 14/2009) by nomination by The Portuguese Parliament (2009). At present, this board is composed of three law experts and it has powers of authority conferred by the Portuguese Parliament and the responsibility for reporting publicly its assessment of the activities of the DNA database through an annual report presented to the Parliament. The law stipulates that this supervisory body will comprise three individuals 'of recognized suitability and reputation and holding full capacity of their civil and political rights' (no. 3 of article 29 of Law 5/2008), but no further details are provided. However, what is meant by 'suitability and reputation' in this context? What are the criteria that identify such individuals?

Compromise is already suggested in that this allegedly independent body will develop its activities by using the human and technical resources and the facilities of the entity it monitors: the National Institute of Forensic Medicine (no. 3 of article 30 of Law 5/2008). Our concerns regarding the competences ascribed to this board are twofold: first, we think there should be clear guidelines as to the safeguarding of the board's independence and impartiality, as well as to its scope of powers and the objectives of monitoring; second, the functioning of this board should foster public confidence in the database in order that the public is confident that data held in it are not misused (Nuffield Council on Bioethics 2007: 92). Although the composition of the administrative board was not known at the time of the interviews, the opinion of our informants is that ordinary citizens should not be participants. Instead, they unanimously felt that only law experts, forensic scientists or bioethicists should be appointed as board members, as the following comment from a law expert exemplifies:

> In my opinion this body must be constituted by specialists. No, no, there is no need to integrate lay people. You see, the committee can only have three [members]. One would choose a judge or a public prosecutor, and a forensic expert. Or maybe a lawyer, one who is a specialist in human rights.

We do not agree with the suggestion to limit board membership to professional experts. Given that transparency and clear accountability are required, the inclusion of non-expert citizens in the board would benefit transparency and improve responsiveness to the wider civil society (Williams *et al.* 2004: 63). Following on from this, it seems a strong need exists for a nationwide public awareness campaign about the forensic DNA database. This conclusion could be applied to a broader European context, as recommended by a report prepared for the UK Human Genetics Commission carried out in collaboration with the ESRC Genomics Policy and Research Forum in Edinburgh, and the Policy, Ethics and Life Sciences Research Centre in Durham and Newcastle (Murtuja *et al.* 2008).

Volunteers

Portuguese law stipulates that one type of DNA profile recorded in the DNA database could stem from volunteers (see also Chapter 3). But there is no clarification of why people would volunteer, or under which conditions this could be done. When asked about the possible intentions of people willing to donate a sample, one of the law experts who participated in the production of the draft law to be submitted to the Portuguese Parliament thought that this was a 'nice idea that will not be put into practice'. Instead, the expert suggested, the possibility that some high-risk professionals might want to see their DNA profile stored in the forensic DNA database:

> It became very obvious that the previous draft of a universal database was unfeasible, because of financial considerations and ethical concerns. But we did not want to abandon the idea of universality and participation in the collective wellbeing [implicit in the first project of creating a universal database]. Maybe professionals at high risk, such as firemen, fishermen, pilots, military want be recorded into the database, because the risk of disappearing or being involved in a disaster.

All interviewed experts expressed some doubts regarding the expectation that a considerable number of citizens would volunteer to provide a sample for such purposes. In the words of a forensic scientist, the creation of the volunteer for donation of a sample for the DNA database had a hidden intention on the part of the legislators who produced the law on the forensic DNA database, which was to demonstrate the idea that there are people who believe that a forensic DNA database is a 'good' thing that nobody should fear: 'Will someone want to be a volunteer? Who knows [smile]? I'm not very confident about that ... I think that the idea of the volunteers was mainly to silence certain politicians who are always afraid of risks to civil rights, of discrimination, and that sort of thing.' This comment indicates that the idea of the volunteer within the planning of the Portuguese forensic DNA databasing served as a device with the aim of projecting an image of legislation that is protective of the rights of citizens and that will contribute to the common good. The comment also seems to convey a conviction by some public officials that all Portuguese citizens can decide to donate their DNA to the state so that the DNA may be analysed and integrated in a database that aims to combat criminality and guarantee public tranquillity (Silva and Machado 2009). But arguably, nobody really expects there will be many volunteers. This seeming paradox can perhaps be understood if we shift from the level of pragmatism and policy to the level of values.

The request of the volunteer, which must be in writing, for DNA profile inclusion may thus have a symbolic role indicating the maximisation of choice and a sense of individual responsibility in the maintenance of social order by those citizens who decide to donate their DNA sample to the state. The sample is received by the state as a voluntary gift, a contribution for the expansion of a database that aims to combat criminality and guarantee public security and tranquillity. Does this concept of volunteers then provide evidence as to the emergence of a new morality (Marx 2006) that 'obliges the "good" citizen to provide a sample of his or her body as a gift towards common good' (Rose and Novas 2005: 440)?

If that is so, then it is our contention that the concept of volunteer as someone who wishes to give a sample feeds the notion that participation of the ordinary citizen in the construction of the forensic DNA databases is needed and should be valued (Evans and Plows 2007). In addition, this sort of public participation fosters the increase of mandatory volunteerism as good citizenship (Marx 2006), which discloses and reproduces modes of categorising citizens and creating hierarchies, from criminals to law-abiding citizens. This reinforces the notion that individuals and institutions both shape and are shaped by a range of moral and control imperatives (Haimes and Williams 2007); which seems the case with the Portuguese DNA database law regarding the disposition concerning informed consent for DNA sample collection.

Quality control: taking DNA samples

Portuguese legislation states that the method for collecting DNA samples is by mouth swab, which is defined as a non-intimate part of the body, with the informed consent of the individual (article 10 of Law 5/ 2008). But, to reiterate, the law does not indicate which procedures are to be followed in the absence of consent. This loophole in the law may indicate that the refusal of consent is not seen as a real option for the individual. In other words, as something unlikely to happen, an assumption seemingly confirmed by one forensic science expert: 'Usually no one refuses consent, if the individual is properly informed. If consent is refused I would not take the DNA sample, unless there is a court order. If a judge orders the DNA taking, I will have to do it.'

The possibility that courts will draw from a refusal any inferences they deem appropriate, and to treat refusal as supporting other prosecution evidence, was also addressed by one law expert: 'The judge might well think, "You refused consent, you are the criminal". But this is a very long discussion that has been going on for several years. What to do in cases of refusal of consent for forensic examination is an issue that is independent of the forensic DNA database law.'

Law enforcement authorities contend that the DNA database law is protective of citizens' rights, but they are silent about social inequalities, which might be amplified by the practices of forensic DNA databasing. The absence of clear guidelines as to how to operate in the case of refusal of consent allows the potential for some abuses to arise, including additional suspicion placed on those who refuse to cooperate. This issue is even more relevant in contexts where there is evidence that judges tend to rely too heavily on forensic bioinformation (Machado 2008); have little or no training regarding the interpretation of statistical evidence relating to match probabilities, and who perceive that courts are incompetent to decide on the weight of the evidence; or who question the competence of the laboratories (Costa *et al.* 2003).

Interpretation of scientific evidence in an inquisitorial criminal justice system

In inquisitorial judicial systems, the judge plays a dominant role in the examination process and in imposing rules of evidence and court procedures. Often, the judge will perceive genetic expert reports as a type of evidence close to an absolute truth (Chapter 13), or at least as constituting all that is worth knowing (Jasanoff 2006). Unlike adversarial legal systems, which allow for the presentation of opposing viewpoints before a relatively passive judge who then adjudicates, inquisitorial trials actively ask parties for factual truths and expert reports are perceived as one way of going about this (Cooper 2004).

Law 5/2008, which regulates the establishment and functioning of the Portuguese forensic DNA database (Government of Portugal 2008), refers to the need for assurance of the quality of the forensic laboratories performing DNA analysis in criminal investigation. The law states that 'all laboratories should meet the scientific, technical and organizational procedures that are internationally established' (no. 3 of article 5 of Law 5/2008).[1] In the harmonisation and standardisation of such technical procedures, genetic markers and benchmarks of quality of performance of laboratories are valued in emphasising the importance of regulating the transnational circulation of genetic information. The importance of standardisation of local procedures, therefore, cannot be overestimated.

A study carried out in 1997 in a Portuguese forensic laboratory brought to light acute problems, ranging from a lack of standardisation and harmonisation of procedures between different laboratories and varying daily working circumstances depending on the availability of financial resources to situations of sheer negligence, equipment failures and lack of sterilisation and maintenance (Costa 2003: 153–156). Although that study is over a decade old, our recent interview evidence suggests that these problems remain relevant to the Portuguese forensic science scene. In addition, a recent high-profile criminal investigation case, of Madeleine McCann,[2] is illustrative of existing difficulties in criminal investigation in Portugal regarding evidence collection and protection of crime scenes. With regard to the latter, it is common that the first police officers arriving at the crime scene are not trained sufficiently and tend to overlook potential evidence (Costa 2003). Furthermore, if insufficiently trained or non-forensic agents collect, transport and store evidence, this can often result in defective labelling, degradation or contamination.

---

[1] The Portuguese laboratories that can make forensic analysis by using data from the national forensic DNA database follow the regulation established in the *Standardization of DNA Profiling Techniques in the European Union* and the *European Network of Forensic Science Institutes*.

[2] The Madeleine McCann case concerns a three-year-old British child that was reported missing in Portugal. In May 2007, a British couple (Kate and Gerry McCann) were spending holidays in the Algarve in a resort in Praia da Luz called *Ocean Club* with their three children (Madeleine three, and Sean and Amelie, two-year-old twins). It was thought that the little girl had been abducted from the room where she was sleeping with her siblings while her parents were having dinner in a restaurant in the holiday complex. Along with the largest police operation ever undertaken in Portugal to find a missing person, massive media interest was triggered in the weeks that followed Madeleine's disappearance. The fact that the first Portuguese police officers that arrived at the crime scene did not secure it and thus did not keep the pertinent evidence uncontaminated was a basis for the accusation of incompetence by some Portuguese and British media.

While the Portuguese legislation on DNA databasing explicitly invokes the need to monitor the qualities of the laboratories through standardisation and harmonisation of procedures based on international requirements, the law does not mention the need to monitor the quality of crime scene isolation and preservation, or the collection, transport and storage of the biological samples. The law only provides that '[it] is required that all the data entered in the database have followed the chain of custody steps' (no. 4 of article 18 of Law 5/2008 (Government of Portugal 2008)), without providing any concrete guidelines for quality assurance and for how scientific evidence should be admitted in court. The legislation fails to acknowledge the possibility of errors in crime scene examination and the potential limitations of DNA profiles and biological sample analysis resulting from contamination or misinterpretation of mixed samples.

This blindness towards potential risks in the uses and interpretation of biological information appears to stem from the myth of the infallibility of genetic identification. It also becomes associated with poor or no training for legal professionals to provide them with a minimum understanding of statistics regarding DNA evidence (Nuffield Council on Bioethics 2007: 74). Consequently, although DNA evidence has great authority in many countries, and we believe that this can also be the case in Portugal, additional measures need to be taken to ensure that criminal investigation is not limited to DNA evidence (Cole and Lynch 2006). On 15 July 2008, the Medical-Legal Council of the National Institute of Forensic Medicine approved the regulation of operation of the Portuguese forensic DNA database (Deliberation no. 3191/2008) (National Institute of Forensic Medicine 2008). That document established that 'the DNA profile is evidence to be considered in conjunction with other evidence in the case' (article 2 of Deliberation no. 3191/2008). Overall, more attention needs to be paid to the situation that such evidence requires careful interpretation in the context of a court case as a whole, and that guilt cannot be assumed on the basis of DNA evidence alone (Jasanoff 2006; Nuffield Council on Bioethics 2007: 65).

Circulation of genetic information and international cooperation

Earlier in this chapter, it was noted that the establishment of national forensic DNA databases can result from the international pressure for the expansion of these databases and the relevance of this to Portugal was discussed. There are several implications from the modes and levels of intercountry sharing and exchange of genetic information as set forth in international protocols and treaties. International cooperation and exchange of genetic information among countries are reinforced by the increasing interest in expanding to a Europe-wide DNA profile database (Chapter 2). Technological standardisation of laboratory techniques and procedures in provision of DNA services amongst diverse national jurisdictions does not eliminate local issues, such as resources and operating conditions of the laboratories, legislative differences, the diversity of institutions and practices of judicial systems and the level of criminal investigation.

However, concerns regarding terrorism and cross-border criminal activities have pressured national jurisdictions into serious consideration of the need to create international cooperation mechanisms in these areas. One instrument is the Prüm Decision, signed in May 2005 by seven Member States of the European Union: Germany, Austria, Belgium, Spain, France, the Netherlands and Luxembourg. By 2007, the contracting parties had risen to 11, with the addition of Italy, Finland, Portugal and Slovenia. The Prüm Decision's main objective is to develop greater cooperation in terms of combating terrorism, international crime, organised crime and illegal immigration through international networking for the sharing and exchange of genetic information (Council of the European Union 1997, 2001). The principal elements that characterise the Prüm Decision are its distinctive 'pan-European' dispositions regarding police and judicial cooperation and the definition of access of other Member States to a country's databases that support criminal investigation, from information on DNA profiles, fingerprint data and vehicle registration data.

The provisions of the Prüm Decision have raised some concerns, including the absence of a standardisation and monitoring policy for procedures connected with cooperation activities, and the collection, retention, treatment, interpretation and legal application of DNA profile information. In this context, pressure could easily be brought to expand a database such as the Portuguese one, which is relatively restrictive in terms of whose profiles are included and retained. The Portuguese Government's eagerness to 'catch up' with developments in other countries makes this scenario likely. In addition, Portuguese legislation on the forensic DNA database only states that the country will have to comply with its obligations regarding international cooperation in the domains of civil and criminal investigation using DNA profiles. The legislation only makes such procedures absolute by stating, '[in] any case it isn't allowed to transfer biological material' (no. 2 of article 21 of Law 5/2008).

CONCLUSIONS

The establishment, in February 2008, of a national forensic DNA database in Portugal was accompanied by much political enthusiasm, especially from the Portuguese Government, about applying such technologies in crime prevention and prediction and in criminal investigation, as well as about furthering the idea of 'public welfare'. Three types of narrative have been employed to justify its introduction. First, there is the view that science is an important element in the provision of justice and represents the most credible way to improve the efficiency of courts and criminal investigation. Second, there is an emphasis on the need to follow more advanced countries in matters of DNA criminal investigation and transnational database cooperation with regard to security policies and crime fighting. Third, there is a high expectation of a contribution to the common good provided by DNA profiling technology, through crime fighting and prevention as well as through the identification of dead bodies and missing persons. These arguments support the creation and the expansion of DNA databases in countries at the forefront of such development, especially the UK.

Despite such political enthusiasm about the establishment of the Portuguese forensic DNA database, public confidence in the efficiency of the forensic DNA databasing appears to be weak. A crucial reason for this is the low confidence that citizens have been found to have in public institutions in general and in the justice system in particular, which is generally seen as corrupt and vulnerable to certain social groups (Santos *et al.* 1996; Cabral *et al.* 2003; Contini and Mohr 2007). In addition, there is the question of the proportionality between the possible benefits of the forensic DNA database and its economic costs in a developing country with a relatively low level of serious crime (in the European context).

Perhaps on the positive side, the creation of a forensic DNA database in Portugal appears more restrictive than in most European countries regarding profile entry criteria, profile removal criteria and the practice of eliminating samples. For instance, as the criteria of profile entry allows only the inclusion of offenders convicted for a serious crime with a sentence of three years or more, profiles of criminal offenders will be removed at the time of the definitive cancellation of a criminal record. This practice means that Portuguese legislation has stipulated the removal of the profiles after a certain period of being loaded in the database, in order to insure the opportunity for ex-prisoners to be 're-integrated' into society with a 'clean' trajectory, 'cleared' of any official record that could directly identify him or her as a perpetrator of past crime. In this context, it is also needed to evaluate the potential for social rehabilitation and reintegration that the Portuguese forensic DNA database can actually offer.

Nevertheless, and according to some statements from law and science experts directly involved in the preparation of the draft law for the establishment of the Portuguese forensic DNA database, it seems that the Portuguese forensic DNA database will 'inevitably' increase in breadth and scope in the next few years (Machado and Silva 2008a). The analysis of these narratives provides valuable insights for the further mapping and understanding of some cultural assumptions, values, unwritten codes and practices occurring in the production of a soft or informal type of governance of a forensic DNA database (Kaye and Gibbons 2008), as the concept of volunteers and the views on the civic accountability and the uncertainties exemplified.

Turning to the legislation for providing guidelines for the certification of quality and safety of procedures and standardisation of the techniques according to international standards, there are uncertainties and potential problems associated with the constitution and management of the Portuguese database. Such ambiguities promise to raise important ethical and governance questions. For example, what control and supervision will be developed to guarantee the transparency and effective functioning of the Supervisory Council of the Portuguese forensic DNA database? What moral and control imperatives will shape the concept of 'volunteers'? How will the monitoring of any potential errors be undertaken during the collection and management of biological material? What difficulties might arise in interpreting bioinformation by judicial actors and what might be the probable weight given to this type of evidence in criminal courts? And, how will the potential problems associated with transnational circulation and sharing of genetic information be resolved? A further issue may be that the complexities related to 'post-genomic' knowledge are currently being omitted, or at least clearly downplayed, in the legislation, in public debates and in official representations of science and DNA technologies (Wynne 2005).

Yet another concern is that a moral imperative is constructed by emphasising the idea of the involvement of volunteers in the creation of a forensic DNA database; this imperative requires lay people to participate in the 'fight' against crime and the defence and consolidation of public security in the position of supporters of a collective project. However, this is beyond the control and accountability of lay people, in particular the volunteer in the name of desired confidence in the state. The ideology of the neutrality and truth of science and of the power of justice to defend equality and individual rights is used by the Portuguese Government to garner the confidence of its citizens, in particular in the capacity of the criminal justice system to fight crime and to consolidate public security. However, mechanisms of civic accountability and participation in the modes of organisation and maintenance of genetic data are missing.

Overall, it is our assessment that the given quality, efficacy and safety of certain techniques and/or procedures will result in contradictory and ambivalent processes. We argue for the need for a new policy, one that provides a firm awareness of ethical complexities and that fosters participatory democracy to best address these complexities. Appropriate participatory avenues for decision making should be built for effective public engagement that can account for the heterogeneity of knowledge and expectations, certainties and uncertainties raised by the forensic DNA database. It is imperative to create regulatory responses that can involve both the citizens and the experts in partnerships for agenda building. There is a need to establish ethical and legal guidelines that can help different audiences to present their opinions in a way that makes sense to a diverse range of people and which enhances agreements and social cohesion on the purposes to be achieved. Giving voice to public perspectives and their representation is vital for public confidence in, and good governance of, the Portuguese forensic DNA database.

ACKNOWLEDGEMENTS

We would like to thank the Foundation for Science and Technology (Portuguese Ministry of Science, Technology and Higher Education) for financing this research through postdoctoral fellowship SFRH/BPD/34143/2006. We gratefully acknowledge the work of Filomena Louro from the Scientific Editing Programme of Universidade do Minho in revising earlier versions of this article.

REFERENCES

Cabral, M., Vala, J. and Freire, A. (2003). *Desigualdades sociais e percepções da justiça.* [*Social inequalities and perceptions of justice.*] Lisbon: Instituto de Ciências Sociais [Institute of Social Sciences].
Cole, S. A. and Lynch, M. (2006). The social creation of suspects. *Annual Review of Law and Social Science*, 2, 39–60.

Contini, F. and Mohr, R. (2007). Reconciling independence and accountability in judicial systems. *Utrecht Law Review*, 3, 26–43.

Cooper, S. (2004). Truth and justice, inquiry and advocacy, science and law. *Ratio Juris*, 17, 15–26.

Costa, S. (2003). *A Justiça em Laboratório. A Identificação por Perfis Genéticos de ADN entre a Harmonização Transnacional e a Apropriação Local. [The Justice in the Lab Identification by DNA Genetic Profiles between Transnational Harmonization and Local Appropriation.]* Coimbra: Almedina.

Costa, S., Machado, H. and Nunes, J. (2003). O ADN e a Justiça: A Biologia Forense e o Direito como Mediadores entre a Ciência e os Cidadãos. [DNA and justice: forensic biology and the law as mediators between science and citizens.] In *Os Portugueses e a Ciência*, ed. M. E. Gonçalves. Lisbon: Publicações Dom Quixote, pp. 200–227.

Council of the European Union (1997). Council Resolution of 9 June 1997 on the exchange of DNA analysis results (97/C 193/02). *Official Journal C 193*, 24/06/1997 P. 0002–0003.

Council of the European Union (2001). Council Resolution of 25 June 2001 on the exchange of DNA analysis results (2001/C 187/01). *Official Journal C 187*, 03/07/2001 P. 0001–0004.

Dias, F. (2007). *Direito Penal. Parte Geral. Tomo I. Questões fundamentais: A Doutrina Geral do Crime. [Penal Law.]* Coimbra: Coimbra Editora.

European Network of Forensic Science Institutes (2006). *Report on Member Countries' DNA Database Legislation Survey.* The Hague: European Network of Forensic Science Institutes www.enfsi.eu/get_doc.php?uid=241 (accessed 12 June 2007).

Evans, R. and Plows, A. (2007). Listening without prejudice? Re-discovering the value of the disinterested citizen. *Social Studies of Science*, 37, 827–853.

Fox, R., van Sickel, R. and Steiger, T. (2007). *Tabloid Justice: Criminal Justice in an Age of Media Frenzy.* Boulder, CO: Lynne Rienner.

Government of Portugal (1998). *Lei 57/98 de 18 de Agosto de 1998. [Law 57/98, of 18 August 1998].* Lisbon: Government of Portugal www.apav.pt/portal/pdf/ident_criminal.pdf (accessed 9 March 2010).

Government of Portugal (2008). *Lei 5/2008 de 12 de Fevereiro de 2008. [Law 5/2008 of 12 February 2008].* Lisbon: Government of Portugal http://www.mj.gov.pt/sections/pessoas-e-bens/base-de-dados-geneticos8948/proposta-de-lei-que/ (accessed 28 February 2010).

Haimes, E. and Williams, R. (2007). Sociology, ethics, and the priority of the particular: learning from a case study of genetic deliberations. *British Journal of Sociology*, 58, 457–476.

Henriques, F. and Sequeiros, J. (2007). *Parecer sobre o Regime Jurídico da Base de Dados de Perfis de ADN. [Report: Opinion on the Legal System for DNA Profiles Database.]* Lisbon: Conselho Nacional de Ética para as Ciências da Vida [National Council of Ethics for the Life Sciences] www.cnecv.gov.pt/cnecv/en/opinions/ (accessed 16 April 2010).

Innes, M. (2001). Control creep. *Sociological Research Online (Special Issue)*, 6(3) http://www.socresonline.org.uk/6/3/innes.html (accessed 16 April 2010).

Jasanoff, S. (2006). Just evidence: the limits of science in the legal process. *Journal of Law, Medicine and Ethics*, 34, 328–341.

Kaye, J. and Gibbons, S. (2008). Mapping the regulatory space for genetic databases and biobanks in England and Wales. *Medical Law International*, 9, 113–130..

Lynch, M., Cole, S. A. and McNally, R. (2008). *Truth Machine. The Contentious History of DNA Fingerprinting.* Chicago, IL: University of Chicago Press.

Machado, H. (2008). Biologising paternity, moralising maternity: the construction of parenthood in the determination of paternity through the courts in Portugal. *Feminist Legal Studies*, 16, 215-236.

Machado, H. and Silva, S. (2008a). Confiança, voluntariedade e supressão dos riscos: Expectativas, incertezas e governação das aplicações forenses de informação genética. [Trust, voluntariness and deletion of risks: expectations, uncertainties and governance of the forensic uses of genetic information]. In *A sociedade vigilante: Ensaios sobre vigilância, privacidade e anonimato* [*The Surveillance Society: Essays Around Surveillance, Privacy and Anonymity*], ed. C. Frois. Lisbon: Institute of Social Sciences Press, pp. 152-174.

Machado, H. and Silva, S. (2008b). A Portuguese perspective. Commentary on the Nuffield Council on Bioethics Report *The Forensic Use of Bioinformation: Ethical issues. Biosocieties*, 3, 99-101.

Marx, G. (2006). Soft surveillance: The growth of mandatory volunteerism in collecting personal information: 'Hey buddy can you spare a DNA?' In *Surveillance and Security: Technological Politics and Power in Everyday Life*, ed. T. Monahan. New York: Routledge, pp. 37-56.

Ministry of Justice (2006). Despacho n. 2584/2006, de 19 de Janeiro de 2006. [Dispatch no. 2584/2006 of 19 January 2006.] *Diário da República* (II série), 24, 1518-1519.

Ministry of Justice (2007). *Public pronouncement on 1 July 2007*. Lisbon: Ministry of Justice www.portugal.gov.pt/6C17/Governo/Ministerios/MJ/Intervencoes/Pages/2007 0601_MJ_Int_Bases_Dados_Perfis_ADN.aspx (accessed 16 April 2010).

Monahan, T. (ed.) (2006). *Surveillance and Security: Technological Politics and Power in Everyday Life*. New York: Routledge.

Murtuja, B., Adris, K. and Ahmed, J. (2008). *A Citizens' Inquiry into the Forensic Use of DNA and the National DNA database. Citizens' Report*. Blackburn, UK: Vis-à-Vis Research Consultancy www.hgc.gov.uk/UploadDocs/DocPub/Document/Citizens%20Inquiry%20-%20Citizens%20Report.pdf (accessed 10 May 2009).

National Institute of Forensic Medicine (2008). Deliberaçãon. 3191/2008, de 3 de Dezembro de 2008. [Deliberation no. 3191/2008.] *Diário da República* (II série), 234, 48881-48886.

Nuffield Council on Bioethics (2007). *The Forensic Use of Bioinformation: Ethical Issues*. London: Nuffield Council on Bioethics www.nuffieldbioethics.org/fileLibrary/pdf/The_forensic_use_of_bioinformation_-_ethical_issues.pdf (accessed 10 March 2008).

Portuguese Parliament (2009). Resolução da Assembleia da República n. 14/2009 de 13 de Março de 2009. [Resolution no. 14/2009 of 13 March 2009.] *Diário da República* (Isérie), 51, 1678.

Prainsack, B. (2007). Research populations: biobanks in Israel. *New Genetics and Society*, 26, 85-103.

Rose, N. and Novas, C. (2005). Biological citizenship. In *Global Assemblages: Technology, Politics, and Ethics as Anthropological Problems*, eds. A. Ong and S. Collier. Oxford: Blackwell, 439-463.

Santos, B., Marques, M., Pedroso, J. *et al.* (1996). *Os tribunais nas sociedades contemporâneas. O caso português*. [*The Courts in Contemporary Societies. The Case of Portugal*.] Porto: Afrontamento.

Silva, S. and Machado, H. (2009). Trust, morality and altruism in the donation of biological material; the case of Portugal. *New Genetics and Society*, 28, 103-118.

Williams, B., Entwistle, V., Haddow, G. *et al.* (2008). Promoting research participation: why not advertise altruism? *Social Science and Medicine*, 66,
1451-1456.

Williams, R. and Johnson, P. (2005). *Forensic DNA Databasing: A European
Perspective.* [Project Interim Report.] Durham, UK: University of Durham.
www.dur.ac.uk/p.j.johnson/EU_Interim_Report_2005.pdf (accessed 14 June
2007).

Williams, R., Johnson, P. and Martin, P. (2004). *Genetic Information and Crime
Investigation. Social, Ethical and Public Policy Aspects of the Establishment,
Expansion and Police Use of the National DNA Database.* London: The Stationery
Office.

Wynne, B. (2005). Reflexing complexity: post-genomic knowledge and reduc-
tionist returns in public science. *Theory, Culture and Society*, 22, 67-94.

JAY D. ARONSON

12

On trial! Governing forensic DNA technologies in the USA

INTRODUCTION

When DNA profiling was first introduced into the American legal system in late 1987, judges and prosecutors heralded it as the 'greatest advance in crime fighting technology since fingerprints' (*People of New York* v. *George Wesley and Cameron Bailey* 1988). Press accounts proclaimed that the technique would 'revolutionize' law enforcement (Lewis 1988; Marx 1988; Moss 1988). In many ways, it has. In the past two decades, DNA evidence has been used in the USA to solve countless violent crimes that might otherwise have been relegated to cold case files, putting thousands upon thousands of rapists and murderers behind bars and inducing guilty pleas from thousands more. Arrest warrants have been issued based solely on crime scene DNA profiles (Bieber 2002; Denver District Attorney 2008). Millions of profiles are now stored in a national DNA database, ready to be used in the aid of criminal investigations around the country. DNA evidence has even been used to free more than 250 'wrongfully convicted' individuals from prison – until recently a uniquely American phenomenon that signals both the problems inherent in the country's criminal justice system and its faith in the power of science to bring the truth to light (Innocence Project 2010).

Yet, the introduction and development of DNA profiling in the USA has been far from perfect. Although closure, if not complete resolution, has been achieved in the majority of debates over molecular biology and population genetics that once affected the legal admissibility of the technique (Thompson 1993; Derksen 2003; Aronson 2007; Lynch *et al.* 2008), several issues remain inadequately addressed by the scientific and legal communities. From the scientific and technical perspective, there is still concern over the interpretation of mixed

Genetic Suspects: Global Governance of Forensic DNA Profiling and Databasing, ed. Richard Hindmarsh and Barbara Prainsack. Published by Cambridge University Press. Copyright © Cambridge University Press 2010.

samples and the possibility of undetected error in forensic DNA profiling (Tobin and Thompson 2006). From the governance perspective, the forensics community (under the leadership of the Federal Bureau of Investigation (FBI)) is the de facto regulator of the technique and is responsible for issuing the guidelines that all laboratories follow to ensure valid and reliable results. While such self-regulation is itself problematic to many observers, it becomes even more so when one realises that there is no wide-scale, regular proficiency testing in American DNA laboratories, despite a major study funded by the National Institute of Justice showing that it is possible to do so (Peterson and Gaensslen 2001).

From an ethico-legal perspective, there is still much disagreement on how to handle the collection and retention of biological samples for use by law enforcement. There is also a great deal of dissention about who should be included in DNA databases, when profiles should be removed and to what uses these databases can be put. This matter is of grave importance now that law enforcement agencies are engaging in so-called familial searching (genetic proximity testing) – that is, tracking down criminal suspects through relatives (who match at most but not all genetic loci) in the database (see also Chapter 2). This practice also brings to mind a very American dilemma that predates modern genetics by more than a hundred years: that of race and its troubled relationship with biomedical science.[1]

This chapter charts the history of DNA profiling and the national DNA database within the context of the American legal system and American society. Particular attention is paid to the initial introduction of the technique from the late 1980s through to the mid 1990s, because it was in this period that most major patterns regarding governance and oversight were established. Towards the end of the chapter, the discussion turns to recent issues that have emerged in the context of the expansion of the DNA database and the creation of ever more powerful technologies for recovering and analysing DNA from potential criminal suspects.

In examining DNA profiling and the national DNA database in the USA, it is possible to gain a sense both of what is unique about the American context and of what lessons are offered for other countries attempting to develop good governance of forensic DNA technologies. What immediately stands out is that the introduction of the technique

---

[1] For more see the October 2008 special issue of *Social Studies of Science* on Race, genetics and disease'.

into the courtroom was much more contentious in the USA than in most other countries, while the expansion of the national DNA database and other associated ethical and legal issues have received less attention in the USA than elsewhere in the world. Further, the involvement of private companies in the US DNA profiling market has created significant challenges for the legal and scientific communities.

COURTROOM CONTROVERSIES

Use of DNA profiling was introduced into the American legal system from 1987 onwards by two competing biotechnology companies – Lifecodes Corporation and Cellmark Diagnostics (Aronson 2007). These two companies were initially very successful in getting DNA evidence admitted in court. Beginning with the serial rape trial of Tommie Lee Andrews in mid 1987 (*Tommie Lee Andrews* v. *State of Florida* 1988), the technique was admitted with no significant defence challenge in more than 200 rape and murder cases over an 18-month period. In those early trials, the defence community was unequipped to challenge the prosecution's assertions about the reliability and validity of the technique (Aronson 2007). Defence lawyers lacked both scientific knowledge and contacts with academic scientists who could help them to understand the technique better and testify in court about its shortcomings. Prosecution lawyers, by comparison, had the advice and scientific connections of the private biotechnology companies who were seeking to have their product admitted into jurisdictions around the country. This disparity meant that judges charged with determining admissibility of DNA typing were solely exposed to positive commentaries about the technique.

By mid 1989, however, this situation began changing dramatically. A small but effective group of defence attorneys succeeded in casting doubt on the admissibility of DNA evidence on both technical and social grounds (Derksen 2003; Aronson 2007; Lynch *et al.* 2008). These attorneys and their high-powered expert witnesses argued that both Lifecodes and Cellmark had rushed to court prematurely in order to gain market share as quickly as possible.

In the landmark trial of *People* v. *Castro* (1989), New York-based defence lawyers Barry Scheck and Peter Neufeld worked closely with biologist and mathematician Eric Lander, of the Massachusetts Institute of Technology, to attack what they saw as the 'sloppy' laboratory work of Lifecodes, which had conducted DNA testing for the prosecution (Parloff 1989; Aronson 2007; Lynch *et al.* 2008). Scheck, Neufeld and Lander forcefully argued that the company's results were marred by lack of adequate quality controls, scientifically untenable procedures for declaring a match between two profiles, and lack of appropriate, generally accepted standards by which to ensure reliable results in individual laboratories (Lander 1989; Parloff 1989). As the trial progressed, prosecution experts, including soon-to-be Nobel Prize winner Richard Roberts, were shocked by problems uncovered during the defence's challenge to DNA profiling. By the end of the trial, the prosecution experts said they felt Lifecodes officials had been dishonest with them in explaining how the company carried out DNA testing (Lewin 1989). Ultimately, scientists from both sides decided to meet out of court to summarise the shortcomings of Lifecodes' DNA evidence. They issued a joint statement outlining their findings (Lewin 1989). This statement also called for an outside body such as the National Research Council (NRC) to evaluate the technology and lay out recommendations for improving its scientific methodology and statistical techniques as well as its quality control and quality assurance practices (Lewin 1989).

In the context of this unusual joint statement, the judge in the case had little choice but to side with the defence on the issue of admissibility (Aronson 2007). The judge accepted the defence's claim that the nature of forensic samples allowed challenges at each stage of the DNA profiling process, which opened doors for other academic scientists to testify about potential problems caused by damaged and degraded DNA samples. The judge also chided Lifecodes for its inadequate quality control and record-keeping practices and demanded that the company make changes to these aspects of its work as a precondition for future admissibility of evidence submitted by them. Ultimately, the judge determined that, while particular DNA results in the case were inadmissible, there was nothing fundamentally wrong with DNA profiling as a technology. At the conclusion of the trial, Lander published an article chronicling his concerns in the journal *Nature*, which would ignite a five year controversy about the validity and reliability of DNA evidence (Lander 1989). Echoing the joint statement in the case, Lander argued that private forensic companies should not be allowed to regulate themselves, and he advocated a much stronger role for the academic scientific community in the oversight of the technique and its use in the legal system (Lander 1989: 501).

In a series of cases in 1989 in which William C. Thompson, a psychologist and lawyer from the University of California at Irvine, was involved as a defence lawyer or consultant, especially *State* v. *Schwartz* (1989) and *State* v. *Cauthron* (1993), the defence raised a fundamental objection to the fact that the two private companies were generally unwilling to release information about their products, even though legally obliged to do so during the discovery process.[2] The companies argued that the data was proprietary and would represent a risk to their business if it became public (New York State Forensic DNA Analysis Panel 1989). Thompson and others argued that the inability of the defence to gain access to these data made it impossible to independently review work by the companies (Aronson 2007).

The defence community also claimed that work by Lifecodes and Cellmark was fundamentally unscientific because no common technical, procedural or interpretative standards existed that could be used to evaluate specific DNA typing results in court. Without well-defined standards agreed upon by the scientific community, defence attorneys and experts argued in court cases, in congressional testimony and in interviews with the mass media that it was impossible to tell whether a private laboratory had done its analysis properly and with an appropriate level of scientific rigor (Lander 1989). This situation, they noted, was in marked contrast to genetic testing in the biomedical context, which was governed by strict monitoring and quality control regimes mandated by the federal government in the Clinical Laboratory Improvement Amendments of 1988 (e.g. Lander 1989: 505; US Senate Subcommittee on Constitution 1992). Although the private companies noted in courtroom testimony that they had undertaken significant efforts to validate and standardise their systems, the defence community argued that proper validation and standardisation could only be carried out by scientists outside of the laboratories actually doing the DNA testing (*People of New York* v. *George Wesley and Cameron Bailey* 1988).

Courtroom challenges by the defence and Lander's *Nature* article (1989) created a crisis of confidence within the American criminal justice system. Many prosecutors and law enforcement agents believed it was only a matter of time before the judicial community and the public began to accept the defence's arguments and to question the validity and reliability of DNA profiling (Derksen 2003; Aronson 2007; Lynch *et al.* 2008).

---

[2] In the US legal system, both parties in a dispute are allowed to request disclosure of relevant information and documents from the other side before the trial begins. This process is called discovery.

THE FEDERAL BUREAU OF INVESTIGATION TAKES OVER

By late 1989 and into the early 1990s, two issues had become increasingly important in the debates over forensic DNA evidence: the lack of standards and who had the authority and expertise to set them. Because the US Federal Bureau of Investigation (FBI) was almost synonymous with forensic science in the USA, people within the criminal justice system increasingly looked to the Bureau to develop procedural standards and standardised materials that could be used in forensic DNA laboratories around the country.

The FBI began its investigations of DNA profiling almost immediately after the discovery of the technique in 1984. Hoping to avoid the kind of controversy that had emerged over the recent introduction of a new blood analysis technique,[3] the FBI (unlike the two private companies) decided to take a very cautious approach to developing their DNA profiling system. It initially planned to publish peer-reviewed studies demonstrating the technical validity of its technique, and to cultivate widespread support for its technique in the scientific community so that the technique would be judged admissible in the courtroom under the *Frye* 'general acceptance' standard (Aronson 2008).[4] This approach was abandoned, however, when it became clear that the shortcomings of the systems of the private companies were having a noticeably negative impact on judicial and public opinion regarding the validity and reliability of the technique. As FBI official John Hicks would comment several times in the late 1980s, 'developments in the private sector caused us to accelerate our efforts [to develop a DNA profiling regime]' (Hicks 1989: 209). Thus, with no publication record and very little awareness of their methodology within the scientific community, the FBI began offering DNA evidence to law enforcement agencies in December 1988 (Aronson 2007)

---

[3] The multisystem blood protein analysis test (see Aronson 2006 for more details).
[4] At the time, the leading case on the admissibility of scientific evidence in almost all jurisdictions was *Frye* v. *United States of America* (1923). In this case, the defendant had appealed the trial court's decision to exclude results from a lie-detector test that was favourable to him. The Court of Appeals for the Washington, DC Circuit upheld the trial court's decision, stating that the hallmark of admissible evidence was its 'general acceptance' within a relevant scientific community. In practice, the *Frye* rule meant that a judge was charged with the responsibility of determining whether or not a particular idea or technique was accepted by enough relevant scientists to be considered valid and reliable in a court of law.

In the course of developing its DNA profiling system and in creating standards and technology for the nascent national DNA database (see below), the FBI team chose a genetic marker system incompatible with the existing DNA typing regimes used by Cellmark and Lifecode.[5] As the FBI's system almost immediately became the new accepted standard in the USA (through the FBI's leadership position in the American forensic community and because they were the dominant provider of forensic testing to law enforcement agencies around the country), the private companies were forced to adopt it in order to assure continued business (Aronson 2008).

The FBI thus 'took over' the DNA profiling market, the monopoly of which was enhanced in other ways as well, perhaps most importantly through the creation of the FBI's Technical Working Group on DNA Analysis Methods (TWGDAM), which met for the first time in November 1988 (Derksen 2003). The group was composed primarily of representatives of state and local public crime laboratories and law enforcement agencies (18 of 31 participants) and FBI employees (11 of 31), with the final two slots filled by academics who had worked with the FBI in the past. No representatives of the two private companies or the defence community were invited to join the working group.

In an interesting display of a general trend against strict regulation and for self-policing in the forensic sciences in the USA, TWGDAM was not given any regulatory authority. Rather, the results of TWGDAM deliberations would merely be considered 'suggestions or guidelines to assist individual crime laboratories in the establishment of their DNA programs' (Derksen 2003). Therefore, voluntary adoption of the guidelines, rather than explicit or mandatory adoption of regulation, was the way chosen to proceed, which reflected other developments in the regulation of the life sciences going back at least to the Asilomar conferences on recombinant DNA technology of the early 1970s (Wright 1994; Krimsky 2005).[6]

---

[5] Specifically, the FBI adopted a different restriction enzyme to cut up the DNA into pieces (HaeIII) to those used by either Cellmark (HinfI) or Lifecodes (PstI). While the choice was framed in purely practical and technical terms (its low cost and the fact that it created smaller fragments that were better visualized with the current restriction fragment length polymorphism technologies), it meant that neither company's results could be loaded into the new database or be compared with existing profile.

[6] In February 1975, a group consisting primarily of molecular biologists met at the Asilomar Conference Center in California to discuss the potential environmental

While the FBI certainly never advocated an explicit regulatory function for itself or TWGDAM, once the Bureau had made the decision to establish and run a nationwide database, it was clear that any jurisdiction wishing either to load profiles to the database or to check the database for matches to an unknown profile recovered from a crime scene would have to conform to the Bureau's chosen standard (Aronson 2008). In subsidising TWGDAM, in controlling about a third of its membership and in setting the agenda, the FBI positioned itself as the unofficial regulator of forensic DNA analysis.

WHO SHOULD GOVERN THE TECHNOLOGY?

The TWGDAM was not the kind of oversight desired by defence attorneys, defence experts or the private biotechnology companies involved in DNA testing. Subsequently, a significant debate emerged in Congress over who possessed the necessary authority and expertise to guide or regulate the production of forensic DNA evidence (US House Committee on the Judiciary, Subcommittee on Civil and Constitutional Rights 1990; US Senate Subcommittee on the Constitution 1992).

Underlying the question of who had the authority to set stand-ards was broader disagreement about what constituted 'peer review' in the context of forensic DNA analysis. The FBI claimed that peer review of DNA identification regimes should be done by parties with an inter-est in the success of the technique. In contrast, the defence community and many academic scientists believed that the hallmark of peer review was organised scrutiny by scientists and other individuals disinterested in the success of the technique (e.g. a federal agency such as the Food and Drug Administration or the National Institute of Standards and Technology) (Aronson 2008).

Of course, the FBI-controlled quasi-regulatory system was not the only possible way of governing DNA profiling. Just before the FBI became the dominant force in the DNA profiling market in late 1988, state and local crime laboratory directors in California and New York undertook separate, proactive efforts to evaluate the strengths and weaknesses of the technique and suggested ways of improving the technological system. In late 1987 and early 1988, the California Association of Crime Laboratory Directors created a blind proficiency

and health hazards of recombinant DNA technology. At the end of the meeting, they crafted strict, but voluntary, guidelines to control research in this domain, rather than waiting for potentially more restrictive governmental regulations to be put into place.

test for Lifecodes and Cellmark to 'provide sound advice to [local law enforcement agencies in California] regarding the value and limitations of a service available from private vendors before they have acquired the skills to provide it in their own laboratories' (California Association of Crime Laboratory Directors 1987: 2). In the test, Cellmark made several clerical errors (including mixing up two samples). Lifecodes, although it made no overt mistakes, was unable to type several samples because of limitations in their technology. Curiously, this test, in which the results were far less than ideal, was the first and last major effort within the forensic science community to undertake true blind proficiency testing of DNA laboratories.

In 1987, the New York State Crime Laboratory Advisory Committee went even further, recommending that laboratories wishing to conduct DNA testing in the state be closely overseen and regulated. The Advisory Committee put together a panel including a diverse array of stakeholders from forensic science, law, law enforcement, academia and government. The panel crafted recommendations and sent them to the Commissioner of Criminal Justice Services for New York State. The recommendations included the establishment of a state board to set minimum procedural, quality control and quality assurance standards; minimum qualifications for laboratory workers; and a system of accreditation for all laboratories proffering DNA evidence in New York courts. Perhaps the most striking feature of the panel's final report was its cautionary conclusion that DNA profiling was still 'science in the making' (New York State Forensic DNA Panel 1989: iv). Indeed, the panel explicitly called for active participation from all stakeholders in the criminal justice system, including defence attorneys and legal academics, in shepherding its continued development (New York State Forensic DNA Analysis Panel 1989). In general, the panel advanced the need for open peer review and regulation of DNA profiling by people outside the forensic science community. In their view, having scientists who were familiar with forensic science, but not a part of it, was a valuable way to ensure the credibility and reliability of DNA evidence (New York State Forensic DNA Analysis Panel 1989). It also seemed to be a way to ensure the credibility of the proposed New York State Board.

Remarkably, many of the most far-reaching recommendations of the panel were eventually put into place in New York State. This occurred in July 1994 after the passage of Executive Law Article 49-B entitled *Commission on Forensic Science and Establishment of DNA Identification Index* (Barber and Gur-Arie 1994). That said, for reasons that will become very clear below, the FBI was incensed that it was going to be placed under the authority of a state regulatory agency, especially one populated so heavily by non-forensic scientists. After intense lobbying by the FBI of New York Governor Mario Cuomo, which stalled passage of the law for some time (Bureau of National Affairs 1990), a compromise was reached whereby the state's regulations did not apply to 'any lab operated by any agency of the federal government, or to any forensic DNA test performed by any such federal lab' (Commission on Forensic Science and Establishment of DNA Identification Index, 1994).

THE POPULATION GENETICS CONTROVERSY

However, it was not until 1994 that a national solution to the question of oversight emerged, as other concerns took the spotlight in early 1990. Specifically, in a series of trials culminating in the Toledo, Ohio, federal court case of *United States of America* v. *Stephen Wayne Yee et al.* (1991; hereafter referred to as the Yee case),[7] the defence community argued that the method used by the FBI to calculate the probability of a random match between a suspect's DNA profile and the DNA profile found at the crime scene was dangerously flawed. The central defence claim was that the FBI did not adequately take into account a phenomenon called 'population substructure'[8] when estimating the rarity of a given genetic marker within major racial groups (the reference population used in forensic casework to determine the rarity of a particular gene) (Lewontin and Hartl 1991). This failure, they argued, could lead the FBI to grossly overestimate the rarity of a DNA profile in a particular community, and thereby overstate its evidentiary value in the courtroom. Indeed, the FBI assumed that all major ethnicities within a given 'race' (such as Swedes, Norwegians, Irish, Italians, etc. in the Caucasian group) tended to intermarry and, subsequently, had gene frequencies that did not diverge significantly from the racial group average. While these subpopulations may have slight differences from the overall racial profile, for the FBI and their supporters, such deviations were only of academic interest and did not affect the kinds of probability calculation made in forensic casework (Chakraborty and Kidd 1991).

---

[7] Yee involved the murder of a video store clerk by a motorcycle gang who mistakenly identified the victim as a member of a rival organization.

[8] Population substructure arises when gene frequencies are not identical in subpopulations of a given species or group as a result of barriers to random mating.

In presenting the defense case in the Yee case, defence lawyers Barry Scheck and Peter Neufeld argued that the FBI's DNA typing regime had not been subject to adequate peer review and scrutiny within the academic (that is, non-forensic) scientific community. While being mindful of the notion that the courtroom is not a research laboratory, Neufeld and Scheck argued that it was nonetheless necessary 'to use the legal process as a surrogate for scientific procedure so that critical predicate data could be obtained, and for the first time independently assessed' (Neufeld and Scheck 1990: 3). To do so, the pair obtained the services of two of the most important population geneticists in the field, Daniel Hartl and Richard Lewontin. In addition to testifying on behalf of the defence in a pre-trial hearing (in which the DNA evidence was ruled admissible), Hartl and Lewontin also published their criticisms of the FBI's statistical approach in the journal *Science* (Lewontin and Hartl 1991). This article led to one of the fiercest controversies in the recent history of science, with numerous accusations of impropriety and unethical behaviour being posited by both defence and prosecution advocates. Hartl even accused James Wooley, the lead prosecutor in the Yee case, of attempting to intimidate him in an effort to persuade him not to publish the *Science* article (Kolata 1991; Roberts 1991a, 1991b). Further, so much pressure was put on *Science's* editor-in-chief to neutralise the effects of the critique that he ultimately delayed publication of Lewontin and Hartl's article until a more pro-DNA typing 'perspective' piece could be commissioned to rebut or blunt many of its claims (Aronson 2007). The outcome was that the expert witness in the Yee case, Ken Kidd, and long-time FBI collaborator Ranajit Chakraborty wrote an article declaring the FBI's methods fundamentally sound, which preceded Lewontin and Hartl's in the 20 December 1991 issue of *Science* (Chakraborty and Kidd 1991). In the same issue, *Science* ran a news article that described the contentious events surrounding this unusual dual publication. This article noted that debate over DNA typing had become 'decidedly nasty', and that the stakes were much higher than in a typical scientific disagreement (Roberts 1991a). Accounts of this episode then appeared in countless media outlets around the country.

DEBATE OVER GOVERNANCE AND REGULATION OF DNA TECHNOLOGY IN WASHINGTON

Such disputes served as the backdrop for hearings in the US Congress throughout 1991 on two bills introduced to regulate and fund DNA profiling in the USA. In these hearings, the FBI and supporters continued to advance the argument that forensic science was a unique branch of science (because of the conditions under which it operated – that is, outside of the traditional academic laboratory and with samples less than pristine), and could, therefore, only be regulated from 'within' the forensic science community itself. Defence-oriented witnesses argued that forensic science should be treated like any other science and be open to external scrutiny. No clear consensus emerged from these hearings.

Towards the end of 1991, in response to the failure of both the House and Senate bills to win unqualified approval of the forensic science community and the FBI, Representative Don Edwards introduced a late-session replacement, HR. 3371, the DNA Identification Act of 1991. This legislation gave the FBI no official regulatory role but instead made it the institution that issued guidelines and standards for the forensic community. The FBI was thus given political legitimacy and authority to constitute the expert panels that would set rules for the field. To ensure that these rules and standards would be followed nationally (after all they were only suggestions), Congress tied funding of state laboratories to compliance with FBI-issued standards. The Edwards Bill thus served as the basis for the DNA Identification Act of 1994, which passed both houses and became the first major federal law regulating forensic DNA analysis in the USA. In short, this Act legitimised the FBI's de facto ability to define the parameters of the regulation of forensic DNA analysis work in the USA without making it an official regulatory agency.

But while DNA profiling was being debated in Congress, the importance of deciding how to regulate the technique had taken a back seat to the debate over population genetics stimulated by the Yee trial and the NRC's first report, issued on 14 April 1992. The long-awaited report, *DNA Technology in Forensic Science*, had come out more than three and a half years after it was initially proposed, and was seven months overdue (National Research Council 1992). The report's scope was broad, covering everything from technical and statistical issues to legal admissibility to the social and ethical issues associated with the development of DNA databanks.

Although the majority of the NRC committee's recommendations met with little resistance in the forensic, academic and legal communities, some became the subject of severe criticism from all directions. The bulk of this disapprobation was leveled at the NRC committee's idea of creating a 'ceiling principle', which represented a statistically conservative method for determining the probability of a random match between two DNA profiles. This principle, which rather arbitrarily limited the individualising power of each genetic marker in a DNA profile, was meant to be a temporary fix until the FBI could complete a large-scale empirical investigation of the extent of population substructure in US populations. Although critics charged that this solution was nonsensical, courts around the country began to reject DNA evidence that was not analysed using the ceiling principle. This greatly angered the FBI and the law enforcement agencies, and they put a tremendous amount of political pressure on the NRC to empanel a second committee to resolve the issue more favourably.

RETURN TO THE COURTROOM AND CLOSURE

As the second NRC committee began its work, Scheck, Neufeld and Thompson again became involved in a high-profile courtroom battle over DNA testing; the double murder trial of ex-football star and actor Orenthal James (OJ) Simpson (*People v. Orenthal James Simpson* 1995). This time, though, they decided that it was in their best interest to forgo the kind of admissibility hearing orchestrated in *People v. Castro* and the Yee case (Aronson 2007). A strong trend was emerging in appellate decisions that DNA profiling was fundamentally a sound technique fully admissible in court. Further, a *Nature* article co-authored on the eve of the Simpson trial by the unlikely pair of Eric Lander and FBI scientist Bruce Budowle (Lander and Budowle 1994), proclaiming that the 'DNA wars' were over (because the scientific problems that had been identified by Lander in the past were resolved), also reduced the likelihood that the defence would succeed in their challenge to the validity of DNA evidence in general.

Instead, the defence sought to question the human and institutional dimensions of DNA profiling, leaving the technical core of the system relatively unchallenged (Lynch, 1998). Their argument was that DNA profiling was, in principle, capable of producing valid and reliable results, but that a potential for human error existed every step of the way. Ironically, while the defence had previously argued that the potential for human error was an inherent part of the DNA forensic technological system, they now sought to characterise DNA profiling as a mature, well-tested technology that produced errors only when the people using it (in this case the corrupt investigators from the Los Angeles Police Department) made mistakes or committed outright fraud. As the final verdict suggests, the defence was highly successful in casting doubt on the DNA evidence. Even the non-genetic evidence, including the bloody footprints that matched an unusual pair of shoes owned by Simpson, was neutralised in the case. The jury deliberated for less than four hours before proclaiming Simpson not guilty on 2 October 1995.

After the Simpson trial, Scheck and Neufeld began to make a strong distinction between the technology itself and its use in specific situations. While a few notable defence lawyers and experts continued to question the overall validity and reliability of DNA evidence (most notably William C. Thompson and collaborators), Scheck and Neufeld decided that challenging DNA evidence was no longer in the best interests of promoting justice for defendants. In their public pronouncements, they began to suggest that the important issues surrounding the validity and reliability of DNA profiling had been resolved. In interviews, they pointed out proudly that the Simpson trial had raised the consciousness of crime laboratories and fundamentally altered the way in which biological evidence from the crime scene was collected and processed (Scheck 2003). Such a shift made sense given the increasing success of the Innocence Project, a non-profit legal clinic they had set up at Cardozo School of Law in New York City in 1992, which used DNA evidence to exonerate the wrongfully convicted, leading to their release from prison.

The second NRC report was finally published in 1996. Its most important conclusion was that recent empirical studies conducted by the FBI of allele frequencies in subpopulations had rendered the use of the ceiling principle to limit allele frequencies unnecessary (National Research Council 1996). Because of the composition of the committee and its very strict mandate, concerns about proficiency testing, laboratory accreditation and governance of DNA profiling were either downplayed or completely ignored in the second NRC report. Therefore, the committee did nothing to question (or undermine) the FBI's control of DNA profiling, which was solidified by the DNA Identification Act of 1994.

A NATIONWIDE DNA DATABASE

Just as debates about DNA profiling using restriction fragment length polymorphisms (RFLP) were coming to a close, the next generation of DNA profiling technology was being developed: short-tandem repeat (STR) analysis (for an excellent overview, see Butler 2005). Although Thomas Caskey first developed STRs in the USA in the early 1990s, their implementation was much faster in Europe. Today, however, almost all US DNA profiling laboratories conduct STR testing.

Primarily in response to the problems faced by RFLP testing, the STR markers commonly used in DNA profiling today have been very well characterised, and all population data are freely available or clearly referenced on the Internet (National Institute of Standards and Technology 2008). Further, the FBI and its allies went to great lengths to ensure that the STR loci chosen were not subject to unusually high levels of substructure in particular populations and were inherited relatively independently of one another. Because of careful attention paid to population genetics issues during the development of STR technology, there are now few concerns about population genetics in DNA profiling.

From a technical perspective, defence challenges to STR testing have primarily focused on the proprietary nature of the kits used to produce DNA profiles and the computer algorithms used to interpret test results, as well as the difficulty in analysing samples containing more than one source. Several academics argue that the lack of well-developed proficiency testing schemes for DNA profiling and other forensic techniques leaves DNA evidence vulnerable to potentially undetectable errors (e.g. Koehler 1997; Peterson and Gaennselen 2001; Tobin and Thompson 2006). These arguments have met with little success and DNA evidence currently enjoys its status as the gold standard of forensic identification in the US legal system. It is unlikely that such issues will be taken up in a serious way by the forensic science and law enforcement communities unless judges begin to make their resolution a requirement for admissibility.

More recently, significant debates between civil libertarians and law enforcement agencies have emerged over various ethical and legal issues surrounding the ever-expanding national DNA database, known as the Combined DNA Index System (CODIS). The formation of this system was officially announced in October 1998, although the FBI began building its nationwide database infrastructure in 1990 as a pilot project involving 14 state and local laboratories (Butler 2005: 438). The FBI's nationwide index is now used in more than 170 laboratories around the country and contains more than 7 940 321 offender profiles and 306 000 crime scene profiles, with all 50 states and the District of Columbia participating. According to FBI statistics, as of February 2010, CODIS had 'produced over 107 600 hits assisting in more than 109 900 investigations' (Federal Bureau of Investigation 2009).

The inclusion criteria for CODIS are a major area of concern. Under the US federal system, individual states can determine who is to be included in state-level databases. Some states limit inclusion to serious felony convictions; some include misdemeanor convictions, and others load a person's profile onto the database as soon as s/he is arrested under suspicion of committing a felony. The DNA Fingerprinting Act of 2005 allows CODIS to include/receive all samples collected under the authority of state law, and also permits inclusion of DNA from federal arrestees and from non-US detainees. The US Department of Justice is currently seeking to expand the inclusion criteria to encompass individuals arrested under federal law.[9] Because of the inconsistency of state laws and the potential for bias in selecting certain kinds of crime for inclusion and not others, some commentators have advocated a universal database in which everybody's profile is included whether they have committed a crime or not (e.g. Watson 2003: 290; Kaye and Smith 2004).

Another topic of much debate is the potential violation of privacy rights (Steinhardt 2004; Genetics and Public Policy Center 2008). Serious disagreement exists among relevant stakeholders about whether the taking of biological materials from private (that is, the body, home, car or other personal property of the source) or public (e.g. saliva on discarded cigarette butts, empty soda cans or spit on the sidewalk) spaces for the purposes of DNA profiling constitutes a 'search'. This issue is relevant because the Fourth Amendment of the US Constitution guarantees 'the right of people to be secure in their persons, houses, papers, and effects, against unreasonable searches and seizures'. Courts have generally found that the taking of biological materials from the body via blood tests or cheek swabs does in fact constitute a search (and, therefore, requires probable cause to be carried out). However, Fourth Amendment challenges to the taking of DNA profiles have been generally unsuccessful because courts have found a compelling 'special need' (i.e. it was in the interest of public safety) for law enforcement to take the sample, that is, to carry out a 'search', thereby bypassing traditional Fourth Amendment protections (Winickoff 2004). The surreptitious taking of biological materials from public spaces has also not been deemed a search because these materials are considered as 'abandoned' by the individual leaving them behind (Genetics and Public Policy Center 2008).

While most state legislatures have not addressed the storage issue, a few require long-term retention of biological materials used to generate DNA profiles. Only Wisconsin mandates destruction of the sample once DNA analysis has been completed (Genetics and Public

---

[9] See Justia Regulation Tracker (http://regulations.justia.com/view/108529/).

Policy Center 2008). Civil libertarians, on the one hand, are concerned about the potential uses of these samples after DNA profiles are extracted, while, on the other hand, law enforcement officials often argue for the need to retain biological materials in the event that technologies change and new profiles need to be generated to repopulate the database (Steinhardt 2004).

So far, little serious discussion has occurred about the use of non-forensic biomedical databases for forensic purposes in the USA. However, the increasing prevalence of samples collected in the forensic context being used by academic and corporate scientists to investigate potential genetic predispositions to violent or antisocial behaviour are sure to increase awareness of the link between forensic science and biomedical research in the coming years (Kaye 2006). Only eight states expressly prohibit this practice, while one explicitly authorises it and the remaining 41 do not give clear guidance on the issue (Axelrad 2005).

Perhaps the most heated discussion taking place in the USA regarding DNA databases revolves around the issue of familial searching (Bieber et al. 2006; Greely et al. 2006) – that is, looking for very similar but non-matching DNA profiles when comparing crime scene samples and individuals included in forensic databases. Familial searching is based on the assumption that there is a high likelihood that people who share a large majority of genetic markers will be closely related. This practice has raised several ethical concerns that are as yet unresolved. Most generally, the practice shifts the locus of genetic surveillance from the individual to the family (Bieber et al. 2006; Greely et al. 2006).

More specifically, there is a high degree of correlation in the USA between one's likelihood of being incarcerated and the previous or current incarceration of a parent or sibling (Duster 2004). The use of familial searching could exacerbate this situation by dramatically increasing the odds that a person who has close relatives in CODIS will be implicated in a particular crime compared with a person who has no relatives in the database (Duster 2008).

It is virtually impossible to ignore the issue of race in the context of any debate on crime and justice taking place in the USA. There are significant racial disparities at all levels of the criminal justice system, from the likelihood of being arrested and convicted of a crime to sentencing and to representation in the prison population (Duster 2004). In all of these areas, African Americans and Hispanics fare poorly (arrest rates are more than six times higher for blacks and Hispanics than for whites), suggesting that these minority populations will soon be overrepresented in CODIS as well, leading to the same kinds of concern raised in the context of familial searching (Austin *et al.* 2007). Levine *et al.* (2008) liken this development to the period 1876–1965, when racial segregation of blacks and whites was the law of the land in the US South. In these authors' words, CODIS will soon become 'Jim Crow's database' (Levine *et al.* 2008: 11). In this regard, it is perhaps unsurprising that although there have been very few (there are no reliable statistics on how many) 'DNA dragnets' in the USA almost all have targeted specific racial groups (Duster 2008). A final development with potential to fundamentally alter the investigation of crimes in the USA is the use of so-called ancestry informative markers to attempt to establish the geographic origins of an individual's genome and to reconstruct his or her phenotype from this information (Ossario 2006).

CONCLUSION: WHAT SHOULD WE (NOT) LEARN FROM THE US CONTEXT?

Although DNA profiling and databanking in the USA share many similarities with their counterparts in other parts of the world (as many chapters in this volume show), their technical and regulatory development offers a window into the peculiarities of the US legal system and culture more broadly. It is clear that CODIS has generally expanded more slowly than European databases until recently, because of concerns over liberty and privacy in the USA, and that US law enforcement agencies have been reluctant to use DNA dragnets and mass screening techniques. In terms of the development of DNA profiling, there was significantly more public controversy in the USA than in the other countries covered in this volume. In England and Wales, for instance, the majority of technical concerns were identified and addressed through collegial interactions among forensic and academic scientists outside the courtroom. There are, of course, many explanations for this situation, ranging from the extremely adversarial nature of the US legal system to the differing professional norms within national scientific communities.

Yet, whether efforts to ensure the validity and reliability of DNA profiling are collegial or adversarial (or both), what matters most is that there is a serious review of methods, practices and protocols by individuals who do not start from the assumption that the technique is without fault. Problems of bias can arise when peer review and standards setting is conducted by individuals or organisations with a vested interest in the success of the technique, whether as a producer of DNA evidence or as a closely affiliated scientific partner.

This concern is doubly warranted in the context of procedural, ethical and racial issues surrounding the expansion of the database and familial searching in the USA, because US public discourse and policy decisions are so thoroughly dominated by law enforcement. Quite simply, there is little room for civil libertarians or DNA critics at the policy-making table. Even powerful and well-financed groups such as the American Civil Liberties Union are left to voice their concerns at academic meetings, in low-circulation publications and in quotations in media reports written by sympathetic journalists. This, I think, is not a lesson that should be learnt by other countries attempting to develop good governance of forensic DNA technologies, but instead is a situation that ought to be avoided at all costs in a situation where DNA analysis is playing an increasingly important role in law enforcement around the world.

REFERENCES

Aronson, J. (2006). The 'starch wars' and the early history of DNA profiling. *Forensic Science Review*, 18, 59–72.

Aronson, J. (2007). *Genetic Witness: Science, Law, and Controversy in the Making of DNA Profiling*. New Brunswick, NJ: Rutgers University Press.

Aronson, J. (2008). Creating the network and the actors: the FBI's role in the development of DNA profiling. *Biosocieties*, 3, 195–215.

Austin, J., Clear, T., Duster, D. *et al.* (2007). *Unlocking America: Where and How to Reduce America's Prison Population*. Washington, DC: The JFA Institute.

Axelrad, S. (2005). *Survey of State DNA Database Statues*. Boston, MA: American Society for Law, Medicine, and Ethics www.aslme.org/dna_04/grid/guide.pdf (accessed 19 December 2008).

Barber, G. and Gur-Arie, M. (1994). *New York's DNA Data Bank and Commission on Forensic Science*. New York: Mathew Bender.

Bieber, M. (2002). Meeting the statute or beating it: using 'John Doe' indictments based on DNA to meet the statute of limitations. *University of Pennsylvania Law Review*, 150, 1079–1098.

Bieber, F., Brenner, C. and Lazer, D. (2006). Finding criminals through DNA of their relatives. *Science*, 312, 1315–1316.

Bureau of National Affairs (1990). Landmark DNA law stalled. *BNA Criminal Practice Manual*, 4, 491–492.

Butler, J. (2005). *Forensic DNA Typing: Biology, Technology, and Genetics of STR Markers*, 2nd edn. Amsterdam: Elsevier.

California Association of Crime Laboratory Directors (1987). *Position on DNA Typing of Forensic Samples*. Personal collection of William C. Thompson, Irvine, CA.

Charkraborty, R. and Kidd, K. (1991). The utility of DNA typing in forensic work. *Science*, 254, 1735–1739.

Commission on Forensic Science and Establishment of DNA Identification Index (1994). Executive Law Article 49-B. NY CLES Exec 995 [1994: ch. 737], 995-e.

Denver District Attorney (2008). *John Doe DNA Case Filings/Warrants.* www.denverda.org/DNA/John_Doe_DNA_Warrants.htm (accessed 8 January 2009).

Derksen, L. (2003). Agency and structure in the history of DNA profiling: The stabilization and standardization of a new technology. PhD Thesis, University of California, San Diego.

Duster, T. (2004). Selective arrests, an ever-expanding DNA forensic database, and the specter of an early -twenty-first-century equivalent of phrenology. In *DNA and the Criminal Justice System: The Technology of Justice*, ed. D. Lazer. Cambridge, MA: MIT Press, pp. 315–334.

Duster, T. (2008). DNA dragnets and race: larger social context, history and future. *GeneWatch*, 21, 3–5.

Federal Bureau of Investigation (2010). *CODIS–NDIS Statistics.* Washington, DC: Department of Justice www.fbi.gov/hq/lab/codis/clickmap.htm (accessed 16 April 2010).

Genetics & Public Policy Center (2008). *DNA Forensics and the Law: Issue Brief.* Washington, DC: Genetics and Public Policy Center www.dnapolicy.org/images/issuebriefpdfs/DNA,%20Forensics,%20and%20the%20Law%20Issue%20Brief.pdf (accessed 14 October 2008).

Greely, H., Riordan, D., Garrison, N. *et al.* (2006). Family ties: the use of DNA offender databases to catch offenders' kin. *Journal of Law, Medicine and Ethics*, 34, 248–262.

Hicks, J. (1989). FBI program for the forensic application of DNA technology. In *DNA Technology and Forensic Science*, eds. J. Ballantyne, G. Sensabaugh and J. Witkowski. Cold Spring Harbor, NY: Cold Spring Harbor Laboratory Press.

Innocence Project (2010). *Website.* www.innocenceproject.org/ (accessed 16 April 2010).

Kaye, D. (2006). Behavioral genetics research and criminal DNA databases. *Law and Contemporary Problems*, 69, 259–299.

Kaye, D. and Smith, M. (2004). DNA databases for law enforcement: the coverage question and the case for a population-wide database. In *DNA and the Criminal Justice System: The Technology of Justice*, ed. D. Lazer, Cambridge, MA: MIT Press, pp. 247–284.

Koehler, J. (1997). Why DNA likelihood ratios should account for error, *Jurimetrics Journal*, 37, 425–437.

Kolata, G. (1991). Critic of 'genetic fingerprinting' tests tells of pressure to withdraw paper. *New York Times*, 20 December, A20.

Krimsky, S. (2005). From Asilomar to industrial biotechnology: risks, reductionism and regulation. *Science as Culture*, 14, 309–323.

Lander, E. (1989). DNA fingerprinting on trial. *Nature*, 339, 501–505.

Lander, E. and Budowle, B. (1994). DNA fingerprinting dispute laid to rest. *Nature* 371, 735–738.

Levine, H., Gettman, J., Reinarman, C. *et al.* (2008). Drug arrests and DNA: building Jim Crow's database. *GeneWatch*, 21, 9–11.

Lewin, R. (1989). DNA typing on the witness stand. *Science*, 244, 1033–1035.

Lewis, R. (1988). DNA fingerprints: witness for the prosecution. *Discover*, June, 44–52.

Lewontin, R. and Hartl, D. (1991). Population genetics in forensic DNA typing. *Science*, 254, 1745–1750.

Lynch, M. (1998). The discursive production of uncertainty: the OJ Simpson 'dream team' and the sociology of knowledge machine. *Social Studies of Science*, 28, 829–868.

Lynch, M., McNally, R., Cole, S. A. et al. (2008). *Truth Machine: The Contentious History of DNA Fingerprinting*. Chicago, IL: University of Chicago Press.

Marx, J. (1988). DNA fingerprinting takes the witness stand. *Science*, 240, 1616–1618.

Moss, D. (1988). DNA – the new fingerprints. *ABA Journal*, May, 66–70.

National Institute of Standards and Technology (2008). *Short Tandem Repeat DNA Internet Database*. Gaithersburg, MD: National Institute of Standards and Technology www.cstl.nist.gov/div831/strbase/ (accessed: 13 October 2008).

National Research Council (1992). *DNA Technology in Forensic Science*. Washington, DC: National Academy Press.

National Research Council (1996). *The Evaluation of Forensic DNA Evidence*. Washington, DC: National Academy Press.

Neufeld, P. and Scheck, B. (1990). *Defendants Post-Hearing Memorandum on the Inadmissibility of Forensic DNA Evidence*. United States v. Yee, 3. Personal collection of Richard C. Lewontin.

New York State Forensic DNA Panel (1989). *Final Report*. Albany, NY: Office of the Governor.

Ossario, P. (2006). About face: forensic genetic testing for race and visible traits. *Journal of Law, Medicine and Ethics*, 34, 277–292.

Parloff, R. (1989). How Barry Scheck and Peter Neufeld tripped up the DNA experts. *American Lawyer*, December, 50–56.

Peterson, J. and Gaensslen, R. (2001). *Developing Criteria for Model External DNA Proficiency Testing: Final Report*. Chicago, IL: University of Illinois Press.

Roberts, L. (1991a). Fight erupts over DNA fingerprinting. *Science*, 254, 1721–1723.

Roberts, L. (1991b). Was science fair to its authors? *Science*, 254, 1722.

Scheck, B. (2003). Conversations with History Series at UC-Berkeley: *DNA and the Criminal Justice System* [Interview with Harry Kriesler.] Berkeley, CA: Institute of International Studies http://globetrotter.berkeley.edu/people3/Scheck/scheck-con3.html (accessed 17 August 2006).

Steinhardt, B. (2004). Privacy and forensic DNA databases. In *DNA and the Criminal Justice System: The Technology of Justice*, ed. D. Lazer. Cambridge, MA: MIT Press, pp. 173–196.

Thompson, W. (1993). Evaluating the admissibility of new genetic Identification tests: lessons from the 'DNA war'. *Journal of Criminal Law and Criminology*, 84, 22–104.

Tobin, W. and Thompson, W. (2006). Evaluating and challenging forensic DNA evidence. *The Champion*, July, 12–21.

US House Committee on Judiciary, Subcommittee on Civil and Constitutional Rights (1990). *FBI Oversight and Authorization Request for Fiscal Year 1990 (DNA Identification)*. 101st Cong., 1st sess. [Hearing took place in 1989.] Washington, DC: Government Printing Office.

US Senate Subcommittee on Constitution (1992). *DNA Identification*. 101st Cong., 1st sess., House Serial 30/Senate Serial J-101-47. [Hearing took place in 1989] Washington, DC: Government Printing Office.

Watson, J. (2003). *DNA: The Secret of Life*. New York: Knopf.

Winickoff, D. (2004). *The Constitutionality of Forensic DNA Databanks*: 4th Amendment Issues, updated 2005. Boston, MA: American Society of Law and Ethics www.aslme.org/dna_04/reports/winickoff_update.pdf (accessed 14 October 2008).

Wright, S. (1994). *Molecular Politics: Developing American and British Regulatory Policy for Genetic Engineering, 1972–1982*. Chicago, IL: University of Chicago Press.

CASES

Frye. v. *United States of America* (1923). 293 F. 1013 (D.C. Circ. Court).
People v. *Joseph Castro* (1989). 545 N.Y.S.2d 985 (Bronx County Sup. Court).
People of New York v. *George Wesley and Cameron Bailey* (1988). 533 N.Y.S.2d 643 (Albany County Court, 1988).
People v. *Orenthal James Simpson* (1995). Los Angeles County Superior Court, BA 097211, WL 672670.
State v. *Richard C. Cauthorn* (1993). 846 P.2d 502 (Supreme Court of Washington, 1993).
State v. *Thomas Robert Schwartz* (1989). 447 N.W.2d 422 (Supreme Court of Minnesota).
Tommie Lee Andrews v. *State of Florida* (1988). 533 So.2d 841 (Fl. Court Appeals, 1988). [Original trial was unpublished: *State of Florida* v. *Tommie Lee Andrews* (Orange County Circuit Court, 1987).]
United States of America v. *Stephen Wayne Yee et al.* (1991). 134 F.R.D. 161 (US District Court for Northern District of Ohio, 1991 (adopting Magistrate's Report)).

13

Biosurveillance and biocivic concerns, from 'truth' to 'trust': the Australian forensic DNA terrain

INTRODUCTION

Bloodstains were used to establish a crime or provide corroborating evidence as early as 384 AD (New South Wales Ombudsman 2001: 5). In the modern era, that procedure was significantly advanced in 1901 by the Austrian medical researcher and later American Nobel Prize winner Karl Landsteiner. He discovered antigens in the blood, which led to the classification of what we know as the ABO blood group system. While the first and obvious benefit of this system was to avoid death from transfusion and thus to make surgery safer, it also occurred to Landsteiner that this classification could be used for forensic purposes. But it was not until 1985 that (UK) police first used blood – along with semen, saliva, other body fluids and hair – for forensic DNA profiling, after it was discovered that individuals could be identified from DNA by restriction fragment length polymorphism (RFLP) (Jeffreys *et al.* 1985a, 1985b).

In 1988, English baker Colin Pitchfork was the first person convicted of murder through the use of DNA evidence. In the same case, suspect Richard Buckland became the first person to have innocence established by DNA evidence (Sanders 2000). However, this method of profiling did not involve DNA amplification and, therefore, required a relatively large amount of DNA – 25 or more hairs or a cent-sized bloodstain – the fresher the better. This could be a drawback in criminal cases, where DNA is often taken from human tissues degraded or contaminated by exposure. Nevertheless, other successful convictions of criminals soon occurred in the USA and the UK through DNA evidence, which subsequently became portrayed as a new 'language of truth' for criminal investigation (Walsh 2005: 54). Unlike fingerprints,

Genetic Suspects: Global Governance of Forensic DNA Profiling and Databasing, ed. Richard Hindmarsh and Barbara Prainsack. Published by Cambridge University Press. Copyright © Cambridge University Press 2010.

which are relatively easy to avoid leaving, for example by wearing gloves or wiping traces off, DNA traces are harder to avoid leaving as DNA derives from multiple sources almost impossible to account for. The contribution of biology to criminology through taxonomic and classificatory thinking is often attributed to the empirical studies by Italian army physician Cesare Lombroso (1913), who posited that serious criminals displayed various 'primitive' facial and bodily features that distinguished them from others (Hil and Hindmarsh 2006). Later, Lyon (2001), in reflecting on forensic profiling for investigating possible clues to criminal actions and for classifying them, conceptualised the body as a 'forensic resource', or as a forensic site for (bio) surveillance and differentiation, where the body, 'is treated like a text. It becomes a *password*, providing a document for decoding' (Lyon 2001: 77). Such conceptualisation aligns with the Foucauldian perspective that biosurveillance would represent a political technology to reinforce the subservience of citizens to the state, achieved through state institutions in collaboration with allied sectors of society (Foucault 1990: 95), here, for example, law enforcement networks.

Following the taxonomic approach, Williams and Johnson introduced a taxonomy of two categories of surveillance techniques that reflected 'a powerful new development in the forms of surveillance available to those tasked with the government of the conduct of contemporary subjects' (Williams and Johnson 2004: 4). The first category, comprising a *preconstructive* mode of surveillance, gathered information about members of a population whose bodily actions and appearances were observed and recorded – traditionally referred to as 'panoptic'. The most obvious instance of such a technique is the use of closed circuit television recording equipment, but the increasing use of low-energy microwave scanning techniques at airports is a more recent innovation. The second category, comprising a *reconstructive* mode of surveillance, sought to identify individuals whose bodily presence and actions were *invisible* to observational techniques and was applied retrospectively. DNA profiling, like latent fingerprint collection, fits here in that it does not directly *watch* present individuals, but instead is used to infer their past presence or actions, especially where no witnesses are available to identify potential suspects. Williams and Johnson (2004: 6) recognised that the power of reconstructive surveillance rests on the ability of a DNA database to be an 'automatic' bio-identification archive: speedy, efficient, automatic and accurate. This was unmatched in the history of policing, these authors claimed, as it enabled 'seemingly indefinite retrospective identifications' where subjects became identifiable and detectable through DNA as an 'omnipresent witness' both 'of and as the body', which intensified the 'gaze of surveillance' (Williams and Johnson 2004: 10).

However, challenging the 'worth' of forensic science and the 'applicability and reliability of using DNA in the legal context' are 'ideological concerns and concerns regarding ethics and integrity' (Walsh 2005: 52). Walsh (2005) based the challenge of ideological concerns on the finding of Corns (1990) that DNA evidence might be interpreted by at least three models: first, as a 'legal model' centred 'on the need to convict the guilty' (which might indicate intent, perhaps bias, in that endeavour; e.g. with undue weight given to DNA evidence (Goodman-Delahunty and Tait 2006: 100)); second, as a 'libertarian model', which questions whether the utilitarian benefits of DNA outweigh a person's 'right to silence, burden of proof resting upon the Crown and the right against self-incrimination'; and third, as a 'scientific model', focused 'on traits such as sampling error, probability theories and so on' (Corns 1990; see also Thompson 2007; Lynch et al. 2008; Chapter 6). Corns highlighted that 'what is missing from these [three] models are links between the deployment of DNA technology and broader socio-political trends in the context of criminal justice'. In other words, Corns 'raised the notion that science (with DNA as central example) [was] "appropriating" the criminal justice process, systematically and ideologically', which led to 'more repressive criminal laws and practices' (Corns 1992, cited in Walsh 2005: 53). Corns saw that shift occurring through a 'future "ratchet effect" where police powers are continually extended resulting in the loss of traditional civil rights such as the right to silence and the privilege against self-incrimination' (Walsh 2005: 53). In the contemporary era, increasing concerns about 'function creep' through expanding DNA databases reflects such concerns, as the investigation in this chapter demonstrates.

Turning to ethics and integrity, Thompson (1997) argued that empirical assessment of forensic DNA evidence highlighted three major credibility weaknesses. The first two pertained to commercial promotion and gain, with DNA services companies and their 'commercially embedded' scientists operating more as a trade guild than participants in science. However, whether commercial or public operators (with the latter more the case in Australia), scientific credibility was compromised by forensic scientists in DNA service providers often adopting the goals of their clients, such as police and prosecutors, and in seeing themselves as part of law enforcement, and thus arguably focused more on intent than objectivity. That enmeshment in Australia is signalled in the organisation of Australia's DNA databases (see below), and in the similarity of narratives by police and forensic scientists (media) about DNA forensic technologies (also discussed below). Despite such concerns (and associated ones, as Chapter 2 outlines) – which are found mainly in specialist circles – the DNA forensic approach remains successful and has been found to contribute to public trust in the criminal justice system (e.g. Kellie 2001, cited in Walsh 2005: 53).

This chapter addresses these concerns, especially through the notion of biocivic concerns, or 'civic concerns/issues with respect to the life sciences' (Hindmarsh 2008a: 273), here specifically forensic profiling and databasing. These biocivic concerns include what many have posed as biosurveillance increasing through extended DNA forensic profiling and databasing, and the associated socio-political implications of that development for good governance in a democratic society characterised by the integrity of and public trust in governance (see Chapter 1), or what Levitt and Weldon (2005: 311) refer to as 'trustworthy governmental arrangements'.[1]

In the contemporary era, such biocivic concerns are increasingly seen as requiring transparency, accountability and inclusive participatory approaches for policy formation (e.g. Peters 2000; Fischer 2006). These features especially refer to a particular conjunction of policy areas, recently referred to as the *politics of life,* including the life sciences, health, agrifood, energy and the environment, where fundamentally destabilising taken-for-granted assumptions about life itself often arise. All refer to dimensions of life and innovation difficult to manage (as they are under human control only to a limited extent), or where the public has good reason to suspect that serious limitations exist for socio-political control and policy steering.[2] As Hindmarsh and Du Plessis (2008: 175–176) highlighted, '[t]hese are fields of innovation in

---

[1] One Australian example (albeit from the private sector) of the need for good governance of the life sciences can be seen in the case of blood samples from newborn screening programmes in relation to regulating human tissue collections (or biobanks) for human health research. Lawson (2008: 524) noted: 'The need for public trust (and legitimacy) to maintain these public health programs was starkly illustrated by the decreased participation in [the State of] Western Australia after the police obtained blood samples from newborn screening cards without the parents' consent in conducting their investigations'.

[2] The Participatory Governance and Institutional Innovation (PAGANINI) project is an EU-supported research programme investigating problem solving in 'the politics

which established strategies for socio-political control appear to be inadequate, where traditional mechanisms of governance can be seen to hamper policymaking ...'.

The chapter introduces DNA profiling and databasing in Australia before exploring the key media narratives of DNA profiling and databasing that represent the Australian debate. This representation is also examined through looking at key policy responses. The concluding section provides suggestions as to how the tensions arising between forensic DNA-driven law enforcement and biocivic concerns might be better addressed for good governance, which includes a plea for a transition from 'DNA as a language of truth' to 'DNA as a language of trust'.

THE AUSTRALIAN TERRAIN OF DNA PROFILING AND DATABASING

In 1989, the first Australian court case involving DNA evidence occurred in the Australian Capital Territory. Desmond Applebee was convicted of three counts of sexual assault after a DNA sample matched him to blood and semen on the victim's clothes. Later that year, a rapist was convicted through DNA evidence in the State of Victoria. The Victorian Government promptly enacted the first provisions enabling the collection of DNA samples from crime scenes and suspects (Victorian Parliament Law Reform Committee 2004: 60).

To consolidate such promising law enforcement developments across Australia meant storing DNA samples on state and territory databases, and this led to moves for the drafting of uniform legislation to address a range of issues including national standardisation of quality control techniques for sampling and the reliability of laboratory procedures and admissibility of DNA evidence; interjurisdictional sample security; privacy; and the powers given to law enforcement officers for the acquisition of body samples (Easteal and Easteal 1990: 8; Victorian Parliament Law Reform Committee 2004: 60–61). In 1990, the Australian Police Minister's Council (APMC)[3] received a report (the

of life' policy areas (http://www.univie.ac.at/LSG/paganini/; accessed 12 March 2010). Rose (2007: 3) discusses the emergence of 'a politics of life itself', which is concerned with 'growing capacities to control, manage, engineer, reshape, and modulate the very vital capacities of human beings as living creatures'.

[3] In 2006, the Australasian Police Ministers' Council (APMC) changed its name to the Ministerial Council for Police and Emergency Management – Police (MCPEMP), following a direction from the Council of Australian Governments. Its role is to promote a coordinated national response to law enforcement issues that includes a

Easteal Report) from criminologist Patricia Easteal entitled *Forensic DNA Profiling: Need for Australian Databases* (Easteal 1990; see also Stringer 2002: 26). Subsequently, several developments occurred either prompted or seemingly prompted by that report and other materials emerging at that time on the issues.

In 1991, under the auspices of the Standing Committee of the Attorney General, the Model Criminal Code Officers Committee (MCCOC), with representatives from most Australian jurisdictions with expertise in criminal law and justice matters, began working on a national model criminal code, including a Model Forensics Procedures Bill (which included reference to DNA profiling and data-basing), for adoption by all Australian jurisdictions (New South Wales Ombudsman 2001: 6).

In 1992, the newly formed National Institute of Forensic Science began to develop national standards of quality control and accreditation of forensic laboratories. In 1993, the APMC received a report (the *Ross Report*) from the Director of the National Institute of Forensic Science entitled *Considerations of the Easteal Report*. The APMC subsequently commissioned a working party to consider the standardisation of state and territory legislation on obtaining body samples, and to make recommendations on legislation relating to privacy issues and database integrity (Stringer 2002: 26). In 1995, a supportive report, which also invited submissions from forensic laboratories and privacy and human rights commissioners, was sent by the working party to the APMC. In 1995, the MCCOC presented the (second version of the) Model Forensic Procedures Bill to the Standing Committee of the Attorney General, which endorsed the bill by a majority and forwarded the draft Bill to the APMC; this then recommended that legislation be established for a national DNA profile database.

This proposal was timely as it followed a decade of development of DNA profiling that culminated in the discovery of short tandem repeats (STRs) of non-coding DNA sections. This new technique, together with automated sequencing technology involving the polymerase chain reaction to amplify the STRs before their fluorescent detection, made it possible to obtain DNA profiles from very small samples and from old samples where the DNA might have degraded (Martin *et al.* 2001; Williamson and Duncan 2002).

wide range of national law enforcement policy development and implementation activities including DNA legislation; see http://www.ag.gov.au/www/agd/agd.nsf/Page/Committeesandcouncils_Ministerialcouncils_AustralasianPoliceMinisters Council(APMC) (accessed 3 June 2009).

Reduction of costs resulting from partial, and later full, automation of DNA analysis, with the capacity to process hundreds of samples in a day, paved the way for national STR DNA databases (Jobling and Gill 2004).

Worldwide, the first national DNA database was implemented in 1995 in the UK, soon followed by New Zealand, several European countries and the USA and Canada (Walsh et al. 2004: 36). Globally, DNA evidence as a new 'language of truth' for criminal investigation had arrived. For Australia, the retrospective capacity of the new technology was soon emphasised. In 1996, the State of South Australia convicted Rodney Winters of the rape and murder of a woman 14 years earlier. Such developments prompted the State of Victoria to extend the range of DNA sampling to serious offenders as a class, in 'the belief that, having committed one serious offence, they may re-offend or they may have already committed other undetected offences' (Victorian Parliament Law Reform Committee 2004: 165), with no time limit on retroactivity. These devleopments also the prompted the Office of the Federal Privacy Commissioner (OFPC) in 1996 to recommend a coherent, consultative policy approach on privacy questions raised by genetic testing (see Weisbrot 2003: 91).

In August 1997, a meeting of senior police and forensic scientists discussed issues relating to a national convicted offender DNA database. Keynote speaker Chief Constable Ben Gunn (of Cambridgeshire Constabulary, UK) represented the (British) Association of Chief Police Officers, which developed and maintains the UK convicted offender DNA database. The two key outcomes of the meeting were to have the database issue placed on the October 1997 Crime Commissioner's Conference agenda and that a nationally common system for DNA profiling should be adopted (Stringer 2002: 26–27). Subsequently, in November 1997, a working party of two crime commissioners and three senior scientists was formed to establish a national database. Concomitantly, in alignment with international moves to standardise DNA methodology to enable cross-comparison of DNA profiles and thus cross-jurisdictional investigation of crime (Walsh et al. 2004: 36), all state and territory government forensic laboratories agreed to adopt a commercially available multiplex polymerase chain reaction system involving the analysis of nine STRs and the amelogenin sex test (Stringer 2002: 27; Walsh et al. 2004).

Developments in the USA were also influential. In 1998, the US Government's Federal Bureau of Investigation set up a national DNA database system, the Combined DNA Index System (CODIS), enabling city, county, state and federal law enforcement agencies to compare DNA profiles electronically (Walsh 2005; see Chapter 12). Subsequently, the Australian Government committed A$50 million to establish the CrimTrac Agency (CrimTrac) to develop a new national criminal investigation system with an online national DNA database as a central element. CrimTrac, to be located in the Attorney General's portfolio, was to be developed in cooperation with state and territory governments (through an intergovernmental agreement with all police ministers) to facilitate the exchange of forensic information despite jurisdictional variations in matching rules. The DNA database would be subject to the Commonwealth's Privacy Act 1988. Concomitantly, strengthened state and territory privacy practices were being introduced or prepared.

Such moves were timely for proponents of a national database, as in 1999 the first Australian cold hit was made though Victoria's DNA database (Moor 1999). In February 2000, the MCCOC released its final report, entitled the *Model Forensic Procedures Bill and the Proposed National DNA Database* (Model Criminal Code Officers Committee 2000; Gans 2002). This Bill proposed comprehensive legislation to protect the rights and interests of individuals from whom DNA samples could be collected, by specifying uniform procedures for the authorisation, collection and matching of DNA profiles, which would facilitate the establishment of CrimTrac. Soon thereafter, on 1 July 2000, CrimTrac was established; between 2000 and 2002, CrimTrac initiated the National Criminal Investigation DNA Database (NCIDD), a new National Automated Fingerprint Identification System and a National Child Sex Offender System. But as CrimTrac became operational, the Commonwealth's Crimes Amendment (Forensic Procedures) Bill 2000 (involving amendments to part 1D of the Commonwealth's Crimes Act 1914 based on the MCCOC's *Model Forensic Procedures Bill and the Proposed National DNA Database*[4]) was subjected to an Australian Senate inquiry.

SENATE INQUIRY: BIOCIVIC CONCERNS EMERGE

The Senate inquiry found that although Commonwealth legislation had a limited capacity to influence state and territory legislation, the Bill should be passed to encourage jurisdictions to standardise forensic

---

[4] Crimes (Forensics Procedures) Bill 2000: Explanatory Memorandum (Cth). Available from: http://www.austlii.org/au/legis/cth/bill_em/capb2000367/memo1. html (accessed 5 June 2009).

practices. An influential submission (Office of the Federal Privacy Commissioner 2000: 2–3), had argued:

> Unless the Model Bill is adopted uniformly, the arrangements for the DNA system as a whole would allow an agency in one State to obtain information collected in another jurisdiction in circumstances that would not be allowed in its own State. This would be a diminution of the rights of the citizens of that State as established under that State's laws.

The OFPC, in strongly supporting the view that the Model Bill provided 'an adequate minimum standard for the authorisation and collection of DNA samples if it is enacted by all participating jurisdictions', was wary of compromise as some jurisdictions had introduced divergent legislation since the release of the final draft of the Model Bill. In some cases, those divergences had reduced privacy protections incorporated into the Model Bill, for example with regard to definitions and procedural requirements for collection and analysis of 'intimate' and 'non-intimate' samples. For example, in the Northern Territory, mouth (buccal) swabs were deemed 'non-intimate', unless the suspect was younger than 14 years of age, and could be taken without consent and in the absence of a court order. Likewise, in Queensland, the Police Powers and Responsibilities Act 2000 (Qld) enabled police to carry out 'intimate' forensic procedures on suspects who refused to consent, without first requiring police to obtain a magistrate's order. Another key concern of the OFPC related to the efficacy of privacy and accountability safeguards in the national DNA system, where the existing system of checks and balances was disparate owing to jurisdictional boundaries and divergences in relevant legislation (Office of the Federal Privacy Commissioner 2000: 4). The OFPC, therefore, recommended an assessment of the adequacy of third party oversight of the national DNA database system.

A contrasting submission to that of the OFPC was from public interest group Justice Action. At the time of the inquiry, Justice Action (2000: 1) was developing advice for prisoners and non-prisoners that it claimed addressed 'much of the misleading information which has been propagated about forensic DNA testing and the inadequacies of "informed" consent as defined under the NSW Crimes (Forensic Procedures) Bill 2000 and proposed under the Commonwealth Crimes Amendment (Forensic Procedures) Bill 2000'.

In its submission, Justice Action also highlighted disparities in legislation in different jurisdictions, particularly with regard to destruction of forensic material and data obtained from it, which Justice Action

(2000: 2) asserted, '[i]n combination with the new [DNA] technology has ... created the potential for breaches of privacy on a scale unimaginable only a few decades ago'. Other issues the group raised included the reliability of forensic evidence, cross-contamination of DNA samples as a result of inadequate laboratory hygiene, informed consent with respect to DNA testing that might 'implicate someone in a crime with which they have no involvement', and privacy issues with regard to family members in relation to a subject being tested (Justice Action (2000: 5). The issue of police misuse of DNA samples was also raised, as was the need for independent oversight of the proposed new procedures and for forensic collection to be done by non-police specialists independently of investigating police.

Following acceptance (on 20 June 2001) by the federal legislature of the amendments to the Commonwealth's Crimes Act 1914, efforts to coordinate state and territory efforts, however, faced ongoing inconsistent implementation of their databases. Concomitantly, global civic concerns were enhanced by shifts to, and increasing police calls for, entire population DNA databases. More broadly, in Australia, a federal agency survey had found that 80% of general public respondents expressed some level of concern about gene technology, especially about the inadequacy of regulatory structures to deal with social and ethical questions, such as discrimination and privacy (Biotechnology Australia 2001). In 2000, a study by Barlow-Stewart and Keays (2001) attracted significant media attention in identifying 48 anonymously reported cases of genetic discrimination, in areas that included life insurance and employment ...' (Barlow-Stewart 2009: 15).

The first author of this study later claimed that '[t]his was the impetus for the Australian Federal Government' (Barlow-Stewart 2009: 15), more specifically the Attorney General and the Minister for Health and Aged Care, in February 2001, to call for a governmental inquiry on how best to protect genetic privacy, guard against discrimination, and ensure the 'highest' ethical standards in research and practice. The Australian Law Reform Commission (ALRC) and the Australian Health Ethics Committee (AHEC) of the National Health and Medical Research Council conducted the inquiry (Australian Law Reform Commission 2001), which became the largest public inquiry worldwide of its kind at that time. The final report, *Essentially Yours: The Protection of Human Genetic Information in Australia*, was presented to the Australian Parliament in May 2003 (Australian Law Reform Commission–Australian Health Ethics Committee 2003).

Not surprisingly, such developments continued to attract intense Australian media scrutiny, following a steady growth of media articles on DNA profiling and databasing from 1995, the year that marked the advent of national DNA databases (Hindmarsh 2008a). This media activity is discussed in the following section.

KEY MEDIA NARRATIVES

A study conducted between 1995 and 2006 examined media voices (or narratives) on this topic (Hindmarsh 2008a); this is a useful way to explore the debate as the media is found consistently to be an important site in the construction of public discourse and especially about new genetics-related topics (e.g. Conrad and Gabe 1999; Hornig Priest 2001; Marks *et al.* 2007; Salleh 2008). The study found that approximately 50% of all Australian media articles (391) on DNA profiling and databasing occurred during 2000–2002. Media voices were dominated by the narratives of police, science and government agencies. Other voices, those of civil libertarians, prisoners' rights and public interest groups, and legal academics and ethicists, were less represented. Overall, three key narratives were evident: DNA database implementation, biocivic concerns and persuasion. All were directed at consolidating support for a particular point of view. These key narratives were identified by using a coding framework that categorised the narratives of the articles according to whether they offered narratives of 'ways of doing' and/or 'ways of knowing' and by which social actors they were associated with. Often an article was found to contain several narratives.

The first key narrative of *DNA database implementation*, found in 57% of the media articles, comprised subnarratives of articles on the organisation of DNA testing, sample collection, legislation and the storage of samples. The subnarrative of sample collection especially reflected police pressure for expansion of DNA databases of suspects and serious offenders to include minor but high-volume crimes. Rising statistics on repeat offenders were quoted as a reason to test all prisoners in order to clear up unsolved crimes. Another method advocated was mass voluntary testing. In Australia, this occurred in 2000 in the country town of Wee Waa, New South Wales (NSW) after the brutal rape of an elderly woman, and again in Claremont, a suburb of Perth, in the pursuit of a serial killer. Divergences in legislation across Australia were also highlighted with reference to implementing the CrimTrac national DNA database, as were concerns about increasing police powers.

The second key narrative of *biocivic concerns*, found in 44% of the media articles, comprised subnarratives of civil liberties, integrity of process, privacy and human rights. In subnarratives of civil liberties and integrity of process, issues centred on the need for safeguards to protect civil liberties with respect to sample collection and analysis and their use in court. Concerns were expressed about potential police misuse of DNA databases, laboratory errors and contamination, the possibility of DNA samples being planted at crime scenes, overall sample security, the scientific reliability of DNA testing and false DNA matches. Finally, there was concern about potential encroachments upon civil liberties through expanded police power to take and retain DNA, 'function creep' and expanding DNA databases, coerced acquisition of DNA samples and DNA dragnets being uneconomical and unethical. Concerning the last, '[t]he Chairperson of NSW Law Society's Human Rights Committee argued the mass testing was a "frightening glimpse of a future police state in NSW"' (cited by Saul 2001: 78).

Privacy subnarratives reflected concern about genetic privacy, discrimination and bodily privacy, with many voices opining that people had the right to keep information contained in their DNA private, and that DNA tests might also reveal information about relatives.[5] Bodily privacy was raised, particularly in relation to mouth swabs. Overall, concerns related to adequate safeguards to protect privacy. Such concerns were moderately reduced after 2002 with the increasing normalisation of DNA databases, legislative implementation and with inquiries occurring. Other subnarratives posited that a DNA database could violate people's rights as an excessive form of government surveillance, and that collecting DNA samples from suspects of a crime diluted the presumption of their innocence.

The third key narrative, of *persuasion*, found in 47% of the media articles, comprised subnarratives that sought to deliberately persuade about the positive or negative aspects of forensic DNA profiling and databasing. Following Gerbner (1972), they reflected deliberately

---

[5] As Thompson (2007: 12) points out: 'The key questions raised by familial searches, from a civil liberties perspective, are how often they lead to testing of innocent people – i.e., people who do not have the matching profile – and how often they might falsely incriminate innocent people through coincidental matches' (also see Chapter 2).

Table 13.1. *The main media narratives of persuasion and their claimants, in order of frequency*

Narratives (in descending order)	Claimants
DNA databases enhance the solving of crimes	Police, government agencies, scientists,
DNA databases are a deterrent to crime	Police, scientists
DNA profiling proves innocence as well as guilt	Media, scientists
DNA profiling is just like fingerprints ('a non-controversial technique')	Media, government agencies, police
DNA profiling saves time and money: it gets cases solved quickly	Police, government agencies, scientists
DNA databases are justified by the changing nature of policing ('going high-tech keeps up with criminals')	Government agencies, police, scientists

one-sided statements that aim to block or downplay alternatives. In the media articles examined, these narratives were clearly dominated by those voicing a positive spin, as Table 13.1 shows. Police narratives (which were dominant in the articles examined) were on how DNA databases would solve crimes, while government agencies appealed to both law and order and citizen protection, although they were clearly more in alignment with police narratives.

Returning to the analysis, as Figure 13.1 shows, it seemed in 2003 that public distrust was receding, as represented by a sharp dip in media narratives of biocivic concerns, but then in 2004 and 2006 biocivic concerns about integrity of process increased again, while privacy issues also started to increase again in 2005. Both do not seem to have abated to date. Why might this be so? This is examined below by way of key policy developments from 2000 onwards as well as a closer reading of media articles post-2004.

KEY POLICY DEVELOPMENTS

The post-2002 dip in biocivic media narratives coincided with the commencement of the 2001 ALREC–AHEC inquiry into protection of human genetic information, the Crimes Amendment (Forensic Procedures) Act 2001 (which had issued from the Model Forensic Procedures Bill), the 2003 Victorian Law Reform Committee Inquiry on forensic sampling and DNA database in criminal investigations

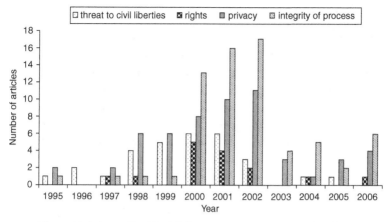

Figure 13.1. Australian biocivic issues, 1995–2006. These data first
appeared as a table in Hindmarsh (2008a).

(Victorian Parliament Law Reform Committee 2004) and the
ALRC–AHEC inquiry's report *Essentially Yours* (Australian Law Reform
Commission–Australian Health Ethics Committee 2003). Also bolstering public confidence was that the new legislation introduced in a
number of jurisdictions during this time began to yield positive results.
For example, by 2003, of the 30 608 DNA samples collected since new
Queensland laws were introduced, 461 were linked to previously
unsolved crimes (Lappeman 2003); in Western Australia, a man was
arrested for an alleged sexual assault committed 13 years previously.
Subsequently, DNA profiling became increasingly promoted as
'fighting crime regardless of time spans' (Pearce 2003).

Less positively, authorities noted potential police misuses of DNA
samples, which needed regulation (Giles 2003), and media attention
remained focused on the problems of establishing the CrimTrac
national database in the face of enduring interjurisdictional issues
(Dearne 2003). Such concerns were addressed in the *Essentially Yours*
inquiry. After considering over 300 written submissions from the general community, experts and interest groups (Opeskin 2002), the
inquiry made 144 wide-ranging recommendations.[6] This chapter focuses on its recommendations concerning forensic DNA profiling and
databasing.

---

[6] A full list of the recommendations pertaining to forensic DNA technologies is
available from: http://www.austlii.edu.au/au/other/alrc/publications/reports/96/
_6_List_of_Recommendations.doc.html (accessed 15 March 2008).

Essentially Yours and beyond

First, with regard to *law enforcement*, key recommendations emphasised developing nationally consistent rules governing the collection, use, storage, destruction and index matching of forensic materials and the DNA profiles created from such material; the inclusion of independent members on CrimTrac's board (e.g. legal academics and ethicists); the periodic public auditing of database systems; and parliamentary reportage of sample collection and storage. With regard to *criminal proceedings*, key recommendations concerned gaining better understandings of DNA evidence, standards for DNA analysis and court presentation, and long-term retention of forensic DNA evidence to allow review of convicted people claiming innocence or wrongful conviction (so-called 'innocence cases'; Innocence Project 2009). Finally, with regard to *criminal investigation*, recommendations focused on DNA sampling by consent or judicial order, guidelines for mass screening programmes, accreditation of laboratories for DNA analysis, coherency of disclosure and use of database information, and harmonisation of cross border legislative and administrative arrangements (Hindmarsh 2008a).

An all-encompassing recommendation was to create an independent statutory body – a Human Genetics Commission of Australia (similar to that of the UK) – to provide advice to the Australian Government about social, legal, ethical and scientific issues and aspects of human genetics, including impacts on human rights. Instead, the recommendation was implemented by establishing a Human Genetics Advisory Committee as a committee of the National Health and Medical Research Council. This body lacks community representation and is dominated by health, medical and genetics experts, with no representation for forensics DNA issues (Hindmarsh and Abu-Bakar 2007).

Returning to media coverage, the post-2004 rise in biocivic concerns appeared to reflect rising dissatisfaction about slow governmental response to the inquiry's recommendations. For example, Stranger *et al.* (2005) acclaimed, 'the country now stands at a crossroads. It either addresses the recommendations of the inquiry ... or continues to ignore the recommendations, fostering cynicism regarding the consultation process and an even greater level of distrust ...'. Reinforcing that contention, a consistent media narrative focused on the failure of most Australian states to join CrimTrac because of the enduring differences in legislation (Cox 2006; Lebihan 2006). Moreover, a community attitudes study conducted in May 2004 for the OFPC showed that while 68% of respondents supported government departments sharing information for policing initiatives, 24% were strongly concerned about their privacy (Curtis 2004). Soon thereafter, in September 2005, the Australian Senate again urged the Government to legislate to protect citizens' genetic privacy (Anon.2005a).

Function creep was also a growing key issue, with the South Australian Attorney General, for example, wanting to expand DNA collection to include suspects of 'less-serious' crimes including assaults, property damage, vandalism and stalking (Wiese Bockmann 2006). A shock for biocivic interests was when the Queensland Government suggested that details gathered from motorists tested for 'drug driving' (using mouth swabs) should be entered into the national DNA database. Both civil libertarians and opposition politicians responded by pushing the need for informed community debate (Finnila 2006). Much attention was also paid to the call of a North Queensland MP, 'for DNA samples to be taken from all Australians and entered into a national database' (Australian Associated Press 2005). One adverse implication posed was that as 'the DNA pool increased, so did the chances that an innocent person would be convicted through a coincidental match' (Walsh cited in the Courier-Mail (Anon. 2005b)). Perhaps of more immediate concern was that the South Australian police taskforce regularly broke laws controlling the state's DNA database, with DNA samples taken from suspects cleared of crimes illegally retained (James 2006).

In July 2007, the NSW Council of Civil Liberties expressed concern that the NSW Government was also planning to introduce 'invasive' DNA testing for a wide range of offences that even included jaywalking, traffic infringements and littering. Civil libertarians worried this might lead to a population-wide database and the state monitoring or surveillance of all citizens. The media reported that the New South Wales Cabinet had approved this development as part of a legislative package focused largely on anti-terrorism measures (ABC [Australian Broadcasting Corporation] News 2007). The state's Police Commissioner simplistically retorted, '[w]e have to be concerned about the civil liberty of being able to leave your car on the street without it being stolen' (Tadros and Smith 2007: 1).

Biosurveillance expansion and continuing biocivic concerns

In this broader context of public distrust, in August 2007, seven of Australia's eight state and territory governments finally struck agreement to share their forensic DNA data through CrimTrac's NCIDD (Johnston 2007), for the storing of DNA profiles from crime scenes, from serious offenders, suspects, objects belonging to missing persons and from unknown deceased persons. The remaining state, NSW, joined soon after and in early 2008 passed amendments to the Crimes (Forensic Procedures) Act 2000 (NSW) to facilitate interjurisdictional DNA data sharing and matching. Police were quick to claim the utility and promote the promise of the NCIDD. Such claims, though, were not of universal appeal. For example, in 2002, Anna Johnston, the Acting Commonwealth Privacy Commissioner, in warning against function creep (cited in Gesche 2006: 85), stated: '[A]rguments based on utility or cost-effectiveness are frequently cited as a basis for expanding the range of possible uses [of DNA databases] including research. These proposals seriously undervalue the importance of trust in relation to the way our society processes personal information such as that derived from DNA.'

Building on that argument, and returning to the emergent imaging of 'panoptic' biosurveillance over space and time in relation to expanding DNA databases, the New South Wales Law Reform Commission (2001: 1) stressed the need for striking an appropriate balance between modern law enforcement techniques and public concerns. The Commission pointed out that the threshold problem of surveillance is the intrusive act of 'being watched or otherwise monitored', where 'everything in range is captured, whether relevant to the purpose or not'. A most important point that appears relevant to expanding DNA databases was that: 'This could have a dissuasive effect on citizens wishing to participate actively in a democratic society' (New South Wales Law Reform Commission 2001: 5).

By May 2008, a CrimTrac media release reported that the national DNA database had made significant progress with the 'system now holding about 400,000 DNA profiles', with arrests being made and hundreds of matches to previously unsolved crimes since interjurisdictional matching processes had commenced (CrimTrac 2008a: 1).[7] Database expansion was also occurring at the state and territory level. In August 2008, South Australia reported a doubling of DNA profiles over the previous year, from 22 411 (June 2007) to 41 161 (August 2008), with some 1800 DNA profiles of offenders and suspects being added

---

[7] For details on how the NCIDD works, see CrimTrac nd. NCIDD – protection of privacy. Available from: http://www.crimtrac.gov.au/systems_projects/ PrivacyandLegalSafeguards.html (accessed 15 March 2009).

each month (Robertson 2008) (or even 1600, as claimed by another journalist) under the state's new Criminal Law Forensic Procedures Act (enacted May 2007)). That expansion saw the loading of 20 000 suspect profiles and 8400 crime scene profiles onto the NCIDD since South Australia had joined the NCIDD in August 2007. Owen (2007) reported that, by November 2007, '1300 link reports' had resulted, with some 513 being followed up. Another journalist wrote: 'That resulted in spectacular successes by police, including the arrest of a man … tracked down last week in Tasmania' (Robertson 2008). Bolstering the narrative of DNA as a language of truth, a police spokesperson explained: 'It's the application of modern science which put beyond doubt a lot of questions about offenders' identity' (Robertson 2008); also, 'the real power of DNA comes from being able to identify a suspect'.

On its website, CrimTrac promotes such power with a rotating scroll of media stories citing retroactive successes of DNA profiling through the NCIDD, especially of crimes against the person. However, that power was questioned in July 2008, with contaminated samples of DNA evidence surfacing in both Victoria and Western Australia in two court cases involving suspected murderers. In the Victorian case, police had to withdraw two murder charges because DNA evidence was found to have been contaminated in a laboratory nine years previously. Opposition police spokesman Andrew McIntosh charged: 'This is a disaster that undermines potentially the entire criminal justice system in the state' (ABC News 2008a). In the Western Australian case, the defence lawyer cited knowledge of three other cases with similar problems; the judge ruled that the contamination was unacceptable (ABC News 2008b). The Director of Public Prosecutions denied that contamination was a major issue (ABC News 2008c); police said contamination happened in all forensic laboratories worldwide, and the President of the Criminal Lawyers Association agreed with the police: 'In any system where you've got humans operating sometimes human error will happen …' (ABC News 2008d). As this was going on, perhaps ironically, Western Australian police expressed their desire to expand the state's DNA database and take DNA from anyone arrested (O'Connell 2008).

This again raises the enduring issue of all suspect profiles being included in Australian DNA databases, the issue pointed up as a global issue in late 2008 with the ruling of the European Court of Human Rights in the case of S and Marper v. the United Kingdom (2008) that the UK's policy of gathering and storing fingerprints and DNA of all criminal suspects was a violation of human rights. In January 2009, ABC Radio National aired the issue as it is the case in Australia in most jurisdictions that, even where a person arrested for a crime is cleared by their DNA sample, the sample can stay in the database for at least a year and during that time gets matched against every other sample (Jeremy Gans of the Melbourne Law School; cited by ABC Radio National 2009: 3). Although a link might be detected between the person and an unresolved crime, on the downside, the person might be exposed to the risk of an erroneous match. Gans continued (cited by ABC Radio National 2009: 3–4):

> The other [issue] is that the parliaments of all these countries are operating on the idea there's something that distinguishes people who've been merely suspected for [sic] a crime, from the rest of us who don't have our DNA in the database because we've never been suspected of a crime ... Once they've started on that track, you start to ask questions about whether there's something wrong about the way they've framed their database.

Such reservations build on strong criticisms of the growing UK DNA database by prominent figures in the UK, including the inventor of 'DNA fingerprinting' Sir Alec Jeffreys (cited by News-Medical.Net 2008), and the then UK shadow Home Secretary David Davis. In response to a call by a senior police officer for a mandatory national DNA register, Davis responded, 'Do you want to turn Britain into a nation of suspects?' and 'We have always had this presumption in our law that the basis of our freedom is that we don't presume somebody is guilty until they have been proven guilty ... [this] in effect [would] change the relationship between the ordinary citizen and the state' (ABC News 2008e; Hofmann 2006: 129). With regard to Australia, University of Queensland Professor David Hamer, a specialist in criminal evidence law, said: 'I think England has gone too far in some respects with its strong surveillance system. Australia is nowhere near as strong but I wouldn't be surprised if we did increase surveillance in future in the effort to solving crime' (Dyer 2008). So, given such trends, what might be the appropriate balance between DNA forensic technologies and biocivic concerns about (expanding) DNA databases and biosurveillance?

IMPLICATIONS FOR GOOD GOVERNANCE

Clearly, little opportunity has been provided for inclusive civic participation with respect to policy formation about the establishment, operation and expansion of forensic DNA databases in Australia. Instead, decision making in this area – except for the broad consultative practices of the ALRC–AHEC inquiry, which, however, reflected the limited public involvement approach of 'passive consultation' (Ankeny and Dodds 2008) – represents more a technocratic policy approach (Fischer 1990). Indeed, the expert top-down decision-making approach seems endemic in Australian life sciences governance (Ankeny and Dodds 2008; Hindmarsh and Du Plessis 2008; Salleh 2008; Lawson and Hindmarsh 2009). A constant companion of technocratic systems is public distrust, as the new participatory and deliberative governance literature emphasises. As Ankeny and Dodds (2008) highlighted with regard to life sciences governance, transparency and accountability are needed, 'otherwise public trust will not be achieved or maintained'. More specifically (Hindmarsh 2008b: 53):

> At the centre of the new civic policy approach, or what I call, in relation to biotechnology, a 'biocivic' policy approach, is the idea that people participate in a meaningful way in their democracy through deliberative governance. This involves contemplative informed discussion by all stakeholders in a shift from top-down, or hierarchical coercive, administrative and legislative approaches to less formal, more inclusive and flexible ones.

Balanced media coverage is also missing. Biocivic concerns about important issues appear largely buried by the many propagandistic narratives promoting forensic DNA testing and databasing, in alignment with narratives of DNA as a language of truth and reconstructive power, which uncritically support and facilitate the continuing expansion of the CrimTrac national DNA database (408 665 DNA profiles as of June 2008, of which 119 965 were from serious offenders and 147 838 were from suspects; CrimTrac 2008b: 32). Such developments point to the need for robust public debate in which civic-minded journalists questioning obviously one-sided 'positive' media narratives play a key and constructive role (Salleh 2008).

But how can biocivic interests be better represented in governance through the new civic policy approach? An important avenue is to revive the idea of an independent statutory body in Australia – a Human Genetics Commission of Australia, so to speak – to provide balanced advice to the Australian Government about the social, legal, ethical and scientific issues and aspects of human genetic information, drawing on both expert and inclusive public participation in the deliberation of issues and the pathways to resolve them. The societal benefit of the UK Human Genetics Commission was clearly seen in the partnership role that it had in initiating and managing the independent 2008 UK Citizen's Inquiry into key social and ethical issues in the forensic use of DNA centred on the UK National DNA Database (NDNAD), especially with respect to the issue of potential encroachments upon civil liberties through extended police powers to take and retain DNA and build large DNA databases (UK Human Genetics Commission 2008).

In July 2008, the UK Citizen's Inquiry recommended, amongst other things, that: 'The police National DNA Database should be placed under the control of an independent statutory authority. And there should be a vigorous nationwide information campaign to explain why DNA samples are taken, how they are used and why they are retained.' Another majority recommendation that pertains to the major policy issue of suspect samples being retained on databases was recommendation 14: 'A majority of participants concluded that samples should be destroyed and profiles removed from the NDNAD when a suspect is not proceeded against or an accused person is not convicted at the conclusion of criminal proceedings' (UK Human Genetics Commission 2008). Such recommendations echoed the conclusion of Williams and Johnson (2006: 248) that, 'the need for expanded deliberative involvement in determining the future directions and use of the NDNAD is essential'.

Such findings also resonate with persistent biocivic calls in Australia for reform and with the recommendations of the *Essentially Yours* report. They also resonate with the call of Hindmarsh and Abu-Bakar (2007: 505) for the development of a 'community engagement biotrust model':[8] one that can potentially mark a transition of forensic DNA profiling and databasing from top-down governance to partnership governance (see also Jones and Salter 2003; Levitt and Waldon 2005: 320). In this arrangement, following the transition management model (e.g. Loorbach 2002), government is the primary actor in the formation and continuation of participatory 'transition arenas', involving, for example, key stakeholders in the criminal justice system, non-governmental organisations, citizens and research institutions. The questionable 'DNA as a language of truth' narrative might then

---

[8] The UNESCO *Universal Declaration on the Human Genome and Human Rights* provides some guidance about the core values and principles that might inform such a model. Available from: http://portal.unesco.org/en/ev.php-URL_ID=13177&URL_DO=DO_TOPIC&URL_SECTION=201.html (accessed 23 March 2009).

find good opportunity to reflect more a consensus-building 'DNA as a language of trust' narrative, which seems appropriate as a way to temper both excesses of the biosurveillance approach and worries about dissuasion for citizens wishing to participate more *actively* in an engaged democratic society around issues of DNA forensic technologies that are of profound importance to them and society.

ACKNOWLEDGEMENTS

The author wishes to acknowledge the support of the GEN-AU program (www.gen.au) of the Austrian Federal Ministry of Science and Research in the context of the *Genes Without Borders? Towards Global Genomic Governance* project, in which research on the topic of this paper was initially situated; also the support of the Centre for Governance and Public Policy, Griffith University, Brisbane, Australia; the research contributions of A'edah Abu-Bakar and Nicole Shepherd; and comments on the manuscript by Barbara Prainsack, Robin Williams and Charles Lawson.

REFERENCES

ABC News (2007). *NSW rights group slams plan for extended DNA sampling*, 22 July http://www.abc.net.au/news/stories/2007/07/22/1984941.htm (accessed 15 March 2009).

ABC News (2008a). *Vic police criticised over botched DNA case*, 7 August http://www.abc.net.au/news/stories/2008/08/07/2326617.htm (accessed 16 March 2009).

ABC News (2008b). *Contaminated evidence delays murder trial*, 16 July http://www.abc.net.au/news/stories/2008/07/16/2305431.htm (accessed 15 March 2009).

ABC News (2008c). *DNA contamination not widespread: DPP*, 16 July http://www.abc.net.au/news/stories/2008/07/16/2305829.htm (accessed 15 March 2009).

ABC News (2008d). *Criminal Lawyers Association defends police DNA bungle*, 17 July http://www.abc.net.au/news/stories/2008/07/17/2306091.htm (accessed 16 March 2009).

ABC News (2008e). *Opponents raise doubts over British DNA register*, 25 February http://www.abc.net.au/news/stories/2008/02/25/2171849.htm (accessed 16 March 2009).

ABC Radio National (2009). *Rear vision: The history of forensic DNA*, 25 January http://www.abc.net.au/rn/rearvision/stories/2009/2436631.htm (accessed 17 March 2009).

Ankeny, R. A., and Dodds, S. (2008). Hearing community voices: public engagement in Australian human embryo research policy, 2005–2007. *New Genetics and Society*, 27, 217–232.

Anon. (2005a). Damned by your own DNA. *The Australian*, 12 September, 10.

Anon. (2005b). DNA fails foolproof test, says scientist, *Courier-Mail*, 20 May, 13.

Australian Associated Press (2005). Fed: Cornelia case shows need for DNA database: MP, 6 February.

Australian Law Reform Commission (2001). Public consultation a priority on genetic information inquiry, media release. Sydney: Australian Law Reform Commission http://www.alrc.gov.au/media/2001/mr0207.htm (accessed January 2008).

Australian Law Reform Commission–Australian Health Ethics Committee (2003). *Essentially Yours: The Protection of Human Genetic Information in Australia*, Vol 1. Canberra: Commonwealth of Australia.

Barlow-Stewart, K. (2009). Genetic discrimination: Australian experiences and policies. *GeneWatch* (USA), 22, 15–17.

Barlow-Stewart, K. and Keays, D. (2001). Genetic discrimination in Australia. *Journal of Law and Medicine*, 8, 250–263.

Biotechnology Australia (2001). *Biotechnology Public Awareness Survey Final Report*. Canberra: Biotechnology Australia http://www.biotechnology.gov.au/index. cfm?event=object.showContent&objectID=FC7E82A5-BCD6-81AC-131E0FB 742A5806E (accessed 15 March 2008).

Conrad, P. and Gabe, J. (1999). Sociological perspectives on the new genetics: An overview. *Sociology of Health and Illness*, 21, 505–516.

Corns, C. (1990). DNA is watching. *Arena*, 92, 24–28.

Corns, C. (1992). The science of justice and the justice in science, *Law Context*, 10, 7–28.

Cox, N. (2006). Missing DNA link mars crime fight. *Sunday Mail*, 15 January, 30.

CrimTrac (2008a). *Crimtrac's National DNA Database Expands to Fight Crime*. [Media release, 6 May.] Canberra: Commonwealth of Australia.

CrimTrac (2008b). *CrimTrac Annual Report 2007–08*. Canberra: Commonwealth of Australia http://www.crimtrac.gov.au/documents/AnnualReport0708-FullReport.pdf (accessed 15 March 2009).

Curtis, K. [Federal Privacy Commissioner] (2004). Privacy in a hi-tech world: technology, policing and identity management. In *Second International Policing Conference*, Adelaide, 3 November.

Dearne, K. (2003). Crook's DNA data held up. *The Australian*, 20 May, 25.

Dyer, R. (2008). A landmark ruling from the European Court of Human Rights has brought the issue . . . , *Gold Coast Bulletin*, 6 December, 44.

Easteal, P. (1990). *Forensic DNA Profiling: The Need for Australian Databases*. [Report for the Attorney General's Department and the Australian Police Ministers Council.] Canberra: Commonwealth of Australia.

Easteal, P. and Easteal, S. (1990). The forensic use of DNA profiling. *Trends & Issues in Crime and Criminal Justice*, 26, 1–7.

Finnila, R. (2006). DNA may be held: fears over database of drug test results. *Courier-Mail*, 18 September, 5.

Fischer, F. (1990). *Technocracy and the Politics of Expertise*. London: Sage.

Fischer, F. (2006). Participatory governance as deliberative empowerment. *American Review of Public Administration*, 36, 19–40.

Foucault, M. (1990). *The Will To Knowledge: The History of Sexuality*, Vol. 1. London: Penguin. [French version 1976].

Gans, J. (2002). The quiet devolution: how the Model Criminal Code Officers' Committee botched New South Wales's DNA law. *Current Issues in Criminal Justice*, 14, 210–223.

Gerbner, G. (1972). *Mass media and human communication theory*. In Sociology of Mass Communications: Selected Readings, ed. D. McQuail. London: Penguin, pp. 35–58.

Gesche, A. (2006). *Genetic testing and human genetic databases*. In The Moral, Social and Commercial Imperatives of Genetic Screening and Testing: The Australian Case, ed. M. Betta. Dordrecht: Springer, pp. 71-94.

Giles, D. (2003). National DNA in crime data, *Sunday Mail*, 18 May, 18.

Goodman-Delahunty, J. and Tait, D. (2006). DNA and the changing face of justice, *Australian Journal of Forensic Sciences*, 38, 97-106.

Hil, R. and Hindmarsh, R. (2006). *Body talk: genetic screening as a device of crime regulation*. In The Moral, Social and Commercial Imperatives of Genetic Screening and Testing: The Australian Case, ed. M. Betta. Dordrecht: Springer, pp. 55-70.

Hindmarsh, R. (2008a). Investigating Australian biocivic concerns and governance of forensic DNA technologies: confronting technocracy. *New Genetics and Society*, 27, 267-284.

Hindmarsh, R. (2008b). *Edging Towards BioUtopia: A New Politics of Reordering Life & the Democratic Challenge*. Crawley: University of Western Australia Press.

Hindmarsh, R. and Abu-Bakar, A. (2007). Balancing benefits of human genetic research against civic concerns: *Essentially Yours* and beyond - the case of Australia. *Personalized Medicine*, 4, 497-505.

Hindmarsh, R. and Du Plessis, R. (2008). The new civic geography of life sciences governance: perspectives from Australia and New Zealand. *New Genetics and Society*, 27, 175-180.

Hofmann, B. (2006). Forensic uses and misuses of DNA: A case report from Norway. *Genomics, Society and Policy*, 2, 129-131.

Hornig Priest, S. (2001). Cloning: a study in news production. *Public Understanding of Science*, 10, 59-69.

Innocence Project (2009). *Website*. www.innocenceproject.org/ (accessed 28 May 2009).

James, C. (2006). Officers acted illegally, judge rules police broke the law. *The Advertiser*, 29 May, 1.

Jeffreys, A., Wilson, V. and Thein, S. (1985a). Hypervariable 'minisatellite' regions in human DNA. *Nature*, 314, 67-73.

Jeffreys, A., Wilson, V. and S. Thein (1985b). Individual-specific 'fingerprints' of human DNA. *Nature*, 316, 76-79.

Jobling, M. and Gill, P. (2004). Encoded evidence: DNA in forensic analysis. *Nature Reviews Genetics*, 5, 739-751.

Johnston, D. (2007). *Fight Against Crime Gains New Weapon*. [Media release, 15 August.] Canberra: Minister for Justice and Customs http://www.crimtrac. gov.au/files/file/media/dj040-07_fight_against_crime_gains_new_weapon.pdf (accessed 3 November 2007).

Jones, M. and Salter, B. (2003). The governance of human genetics: policy discourse and constructions of public trust, *New Genetics and Society*, 22, 21-41.

Justice Action (2000). Submission, Senate Legal and Constitutional Legislation Committee Inquiry into the Crimes Amendment (Forensic Procedures) Bill 2000. Sydney: Justice Action.

Kellie, D. L. (2001). Justice in the age of technology: DNA and the criminal trial. *Alternative Law Journal* 26, 173-176.

Lappeman, S. (2003). DNA data a boon for the police force. *Gold Coast Bulletin*, 18 January, 25.

Lawson, C. (2008). Newborn screening in Victoria: a case study of tissue banking regulation. *Journal of Law and Medicine*, 16, 523-544.

Lawson, C. and Hindmarsh, R. 2009. Legitimising regulatory decision making about genetically modified organisms under the Gene Technology Act 2000

(Cth). In *The Nexus of Law and Biology: New Ethical Challenges*, ed. B. Hocking. Farnham, UK: Ashgate, pp. 115–173.

Lebihan, R. (2006). DNA database not a perfect match. *Financial Review*, 3 April, 47.

Levitt, M. and Weldon, S. (2005). A well placed trust? Public perceptions of the governance of DNA databases. *Critical Public Health*, 15, 311–321.

Lombroso, C. (1913). *Crime: Its Causes and Remedies*. Boston: Little, Brown.

Loorbach, D. 2002. *Transition Management: Governance for Sustainability*. Maastricht, the Netherlands: International Centre for Integrative Studies.

Lyon, D. (2001). *Surveillance Society: Monitoring Everyday Life*. Buckingham, UK: Open University Press.

Lynch, M., McNally, R., Cole, S. A. *et al.* (2008). *Truth Machine: The Contentious History of DNA Fingerprinting*. Chicago, IL: University of Chicago Press.

Marks, L., Kalaitzandonakes, N., Wilkins, L. *et al.* (2007). Mass media framing of biotechnology news. *Public Understanding of Science*, 16, 183–203.

Martin, P., Schmitter, H. and Schneider, P. (2001). A brief history of the formation of DNA databases in forensic science within Europe. *Forensic Science International*, 119, 225–231.

Model Criminal Code Officers Committee (2000). *Model Forensic Procedures Bill and the Proposed National DNA Database*. Canberra: Model Criminal Code Officers Committee.

Moor, K. (1999). DNA's first strike. *Herald-Sun*, 12 October, 14.

New South Wales Law Reform Commission (2001). *Report 98 (2001)–Surveillance: An Interim Report*. Sydney: Lawlink.

New South Wales Ombudsman (2001). *The Forensic DNA sampling of Serious Indictable Offenders (Under Part 7 of the Crimes (Forensic Procedures) Act 2000.* [Discussion paper.] Sydney: New South Wales Ombudsman.

News-Medical.Net (2008). Real concerns over the ethics of a DNA database. http://www.news-medical.net?id=34184 (accessed 17 March 2009).

O'Connell, R. (2008). Police want to take DNA of all suspects. *West Australian*, 7 July, 1.

Office of the Federal Privacy Commissioner (1996). *The Privacy Implications of Genetic Testing*. Sydney: Office of the Federal Privacy Commissioner.

Office of the Federal Privacy Commissioner (2000). *Senate Legal and Constitutional Legislation Committee Inquiry into the Crimes Amendment (Forensic Procedures) Bill 2000*. [Submission, November.] Sydney: Office of the Federal Privacy Commissioner.

Opeskin, B. (2002). Engaging the public: community participation in the genetic information inquiry. *Reform*, 80, 53–58, 73.

Owen, M. (2007). DNA net tightens on 500 crims. *The Advertiser*, 5 November http://www.news.com.au/adelaidenow/story/0,22606,22701755-910,00.html (accessed 15 March 2009).

Pearce, Y. (2003). DNA database points to old sexual assault. *West Australian*, 30 January.

Peters, G. (2000). Governance and comparative politics. In *Debating Governance: Authority, Steering, and Democracy*, ed. J. Pierre. Oxford: Oxford University Press, pp. 36–53.

Robertson, D. (2008). South Australia's DNA data base has almost doubled in a year. *Australian Associated Press*, 28 August http://www.mako.org.au/dna.html (accessed 1 March 2010).

Rose, N. (2007). *The Politics of Life Itself: Biomedicine, Power, and Subjectivity in the Twenty-first Century*. Princeton, NJ: Princeton University Press.

Salleh, A. (2008). The fourth estate and the fifth branch: The news media, GM risk, and democracy in Australia. [In Life Sciences Governance: Civic

Transitions and Trajectories, eds. R. Hindmarsh, R. Du Plessis.] *New Genetics and Society*, 27(Special Issue), 251–266.

Sanders, J. (2000). *Forensic Casebook of Crime*. London: True Crime Library/Forum Press.

Saul, B. (2001). Genetic policing: forensic DNA testing in New South Wales. *Current Issues in Criminal Justice*, 13, 74–108.

Stranger, M., Chalmers, D. and Nicol, D. (2005). Capital, trust and consultation: databanks and regulation in Australia. *Critical Public Health*, 15, 349–358.

Stringer, P. 2002. *Forensic Sampling and DNA Databases*. [Background/issues paper.] Melbourne: Victorian Parliament Law Reform Committee.

Thompson, W. (1997). A sociological perspective on the science of forensic DNA testing. *University of California Davis Law Review*, 30, 1113–1136.

Thompson, W. (2007). The potential for error in forensic DNA testing. *GeneWatch* (USA), 21, 1–43.

Trados, E. and Smith, A. (2007). Moroney's new recruit: Big brother. *Sydney Morning Herald*, 23 July, 1.

UK Human Genetics Commission (2008). *A Citizen's Inquiry into the Forensic use of DNA and the National DNA Database*. Blackburn, UK: Vis-à-Vis Research Consultancy http://www.genomicsnetwork.ac.uk/media/citizens %27_inquiry_mainfindings.pdf (accessed 6 October 2008).

Victorian Parliament Law Reform Committee (2004). *Report on Forensic Sampling and DNA Databases in Criminal Investigations*. Melbourne: Victorian Parliament Law Reform Committee http://www.parliament.vic.gov.au/lawreform/ inquiries/Forensics/Final%20Report.pdf (accessed 15 March 2009).

Walsh, S. (2005). Legal perceptions of forensic DNA profiling: A review of the legal literature. (part I). *Forensic Science International*, 155, 51–60.

Walsh, S., Ribaux, O., Buckleton, J. *et al.* (2004). DNA profiling and criminal justice: a contribution to a changing debate. *Australian Journal of Forensic Sciences*, 36, 34–43.

Weisbrot D. (2003). The Australian joint inquiry into the protection of human genetic information. *New Genetics and Society*, 22, 89–113.

Wiese Bockmann, M. (2006). Satellite trace on child molesters, *The Australian*, 10 March, 7.

Williamson, R. and Duncan, R. (2002). DNA testing for all. *Nature*, 418, 585–586.

Williams, R. and Johnson, P. (2004). Circuits of surveillance. *Surveillance Society*, 2 (1), 1–14.

Williams, R. and Johnson, P. (2006). Inclusiveness, effectiveness and intrusiveness: issues in the developing uses of DNA profiling in support of criminal investigations. *Journal of Law, Medicine and Ethics*, 34, 234–247.

CASE

S and Marper v. *the United Kingdom* (2008). A summary of the judgment is available from http://cmiskp.echr.coe.int/tkp197/view.asp?action=html&documentId= 843937&portal=hbkm&source=externalbydocnumber&table=F69A27FD8 FB86142BF01C1166DEA398649 (accessed January 2009).

JOHANNA S. VETH AND GERALD MIDGLEY

14

Finding the balance: forensic DNA profiling in New Zealand

INTRODUCTION

Since the mid 1980s, DNA identification techniques have become a routine element of criminal investigations and court proceedings. New Zealand was an early adopter of forensic DNA technologies, developing a close relationship with the Forensic Science Service in the UK where the techniques were pioneered. Forensic DNA laboratories were established in New Zealand in 1988, and DNA profiling immediately proved to be an invaluable tool for solving crime. As a geographically small and isolated nation with a modest population size of less than four million, and with a single legal jurisdiction, New Zealand perhaps was an ideal environment in which to successfully introduce DNA profiling technologies.

The use of DNA testing was recognised as an important and valuable investigative tool and by the early 1990s it was frequently incorporated into serious crime investigations. Yet, there was rising discomfort regarding police powers to obtain samples from suspects in the absence of any legislation. That discomfort was reflected in key judicial decisions, which considered whether or not police actions in obtaining biological samples infringed on the donor's rights. Subsequently, in 1995, New Zealand became the second nation after the UK to legislate for the establishment of a national DNA databank (Harbison *et al.* 1999).

The aim of this chapter is to trace the evolution of DNA profiling and review the legislative aspects of governance in New Zealand. Attention is first directed to forensic DNA analysis prior to legislation, with discussion of two key judicial decisions that illustrate the legal debate surrounding this technology. The provisions within the Criminal Investigations (Blood Samples) legislation that focus on

Genetic Suspects: Global Governance of Forensic DNA Profiling and Databasing, ed. Richard Hindmarsh and Barbara Prainsack. Published by Cambridge University Press. Copyright © Cambridge University Press 2010.

whose samples may be obtained and under what circumstances are examined before summarising the main themes and concerns that emerged in the submissions made to the Parliamentary Select Committee considering the 1994 Criminal Investigations (Blood Samples) Bill and the subsequent 2002 Amendment Bill. Our summary of the submissions emphasises cultural and ethical concerns, as these signal possible unintended consequences that a purely technical evaluation of the technology's effectiveness would not anticipate. The New Zealand public appeared to be untroubled by the governance and practice of forensic DNA profiling, and an exploration of the social context and print media reporting at the time provides some explanation as to why this may have been so. Very little research surveying public perceptions in New Zealand of forensic DNA profiling exists. However, our own research exploring social and ethical concerns regarding forensic DNA testing reveals areas both of contention and acceptance (Baker *et al.* 2006). Finally, the chapter reviews present-day operations of New Zealand's DNA Profile Databank, focusing on profile numbers and link rates, and briefly discusses a new amendment bill announced in February 2009.

FORENSIC DNA ANALYSIS IN A LEGAL WILDERNESS

The discovery of hypervariable minisatellite regions in human DNA (Jeffreys *et al.* 1985) and their application in a forensic context (Gill *et al.* 1985) were first described in the mid 1980s. In New Zealand at that time, the Chemistry Division[1] of the government-funded Department of Scientific and Industrial Research was providing forensic services to the police, including the biological analysis of crime and reference samples using conventional serology methods such as ABO blood grouping and various blood cell isoenzyme-typing methods. The new DNA-based technique offered greater discrimination and in 1988, with funding from the New Zealand Police, the Chemistry Division established DNA laboratories and sent five scientists to the UK for training at the Home Office forensic laboratory (Cordiner *et al.* 2003). Following the successes of the technique in the UK, and with the adoption of a similar model in New Zealand, DNA profiling was being incorporated into criminal investigations by the end of the decade.

---

[1] This was changed to the Institute of Environmental Science and Research in 1992. Following organisational change; it continues to provide forensic science services.

However, there was no legislation in New Zealand governing the use of DNA profiling. A Private Member's Bill (a law change proposed by a Member of Parliament) allowing the police to take bodily samples for investigative purposes was proposed in 1988. Opposing the move, the incumbent Justice Minister, Mr Bill Jeffries, publicly stated that he did not believe statutory guidelines or safeguards were needed (Lowe and Barber 1989). Without government support the Bill lapsed in 1991. However, lawyers and police were predicting that without legal safeguards the value of DNA evidence might be undermined in court (Lowe and Barber 1989). The predictions were well founded. When the first DNA cases were tried, concerns were raised about the circumstances under which samples from individuals were obtained, from whom samples could be obtained and what the appropriate limits of police powers were. Two judicial decisions considering the admissibility of DNA evidence highlight the key issues.

In 1990, the first trial involving DNA evidence began in the Auckland High Court, and the defendant was subsequently found guilty of murder (R v. Pengelly 1990). Justice Thorp instructed the jury to exercise caution when considering the DNA evidence because it was a novel technique in New Zealand. However, he also stressed that the huge potential of DNA for identification purposes was internationally accepted and that it would be 'irresponsible for the New Zealand authorities not to try and take advantage of this new knowledge' (Anon. 1990). The Court of Appeal later heard the case (Pengelly v. R 1991). The appellant argued that the DNA evidence should be dismissed because when he consented to provide a blood sample he was not aware that it would be tested using a DNA profiling technique. The Appeal Court declined the appeal, stating that informed consent did not require a specific knowledge of the testing technique to be used.

In an investigation involving the sexual violation of an 11-year-old boy, the police became aware that the defendant had sought an AIDS test and subsequently seized the defendant's blood sample. A DNA profile was obtained and found to correspond with the foreign DNA recovered from the victim. During the trial in 1991, Justice Williamson ruled the DNA evidence inadmissible, citing that the defendant had not given consent for his blood sample to be used in this way (R v. Montella 1992). Nonetheless, recognising the potential value of DNA evidence, Justice Williamson stated: 'A statute is required that sets out the position both of the police and the accused when DNA testing is a possibility . . . '.

LEGISLATION FOR DNA PROFILING

In 1994, the Criminal Investigations (Blood Samples) Bill was considered by the Justice and Law Reform Select Committee and subsequently enacted in 1995. This Act governed the taking and storage of blood samples from individuals for the purpose of DNA testing in criminal investigations, and also legislated for the introduction of a national DNA profile databank, which began operation in August 1996.

The legislation had two main areas of focus. The first (Part II of the Act) was concerned with specific criminal investigations and allowed police to request blood samples, known as reference samples, from persons suspected of committing any indictable offence. The DNA profiles obtained from reference samples could only be compared with the specific crime or crimes the individual was suspected of committing and were not placed onto any database. Suspects were given the opportunity to volunteer a reference sample, but, depending on the severity of the offence, police could apply for a court order compelling a sample if consent was refused. That a suspect had refused to provide a reference sample which was subject to a compulsion order under Part II of the Act could be given as evidence, allowing the court or jury to draw inference from the refusal. The list of offences for which police might apply for a compulsion order was quite long, but covered the broad areas of murder, manslaughter, serious sexual and violent offences, abduction, kidnapping and robbery.

The second focus (Part III) of the Act was the provision for the establishment of a DNA Profile Databank (hereafter referred to as the Databank). Databank samples could be compared with unidentified DNA profiles obtained from crime scene samples and any matches would be reported to the police for further investigation. The legislation allowed for samples to be obtained from individuals either voluntarily, at any time, or by compulsion after conviction for a relevant offence. Anyone at all could be asked to volunteer a sample for the Databank.

In practice, however, police ask only 'persons of interest' to volunteer a Databank sample, and often make the request in conjunction with a request for a suspect sample under Part II of the Act. When making the request, police must provide information outlining what will happen to the sample and how the profile will be used. They must also inform the donor of his or her rights with regard to refusing to provide a sample or, if consent is given, how to later withdraw that consent and have the profile removed from the Databank.

The police cannot *compel* a sample for the Databank unless the person has been convicted of a relevant offence. A schedule in the Act listed the relevant offences and included all the offences listed in Part II as well as burglary and entering with intent. If an individual refuses to provide a sample, which is subject to either a suspect compulsion order under Part II of the Act or a Databank compulsion order under Part III of the Act, the police may use reasonable force to obtain the sample.

Individuals who have volunteered a Databank sample may request for the DNA *profile* to be removed from the Databank at any time, unless convicted of a relevant offence, in which case the profile is stored indefinitely. The physical *samples* obtained from individuals under Part II or Part III of the Act are destroyed once a DNA profile has been obtained. The samples cannot be used for any other purposes such as biomedical research.

In New Zealand, young people (legally defined as aged between 14 and 16 years) and children (legally defined as aged between 10 and 13 years) who are suspected of breaking the law are accorded protections under the legislation that are not available to adults. Primarily, police cannot request a Databank sample from anyone under the age of 17 years unless the individual has been convicted of a relevant offence and has been compelled to provide a sample under a juvenile compulsion order. Furthermore, if the police require a suspect sample from a child or young person thought to have committed a crime, then special consent provisions apply that are stipulated in Part II of the Act.

The New Zealand Police have principal responsibility for the Databank. However, as the Institute of Environmental Science and Research (ESR; from 1992) is the analytical laboratory that undertakes forensic DNA profiling work in New Zealand, the ESR manages the Databank on behalf of the police. With this arrangement, ESR and the police operate as independent and separate organisations. The police do not have direct access to the information held on the Databank or to the analytical laboratory that processes the samples and houses the Databank computer infrastructure. It is not clear whether this arrangement, which places the custodianship of the Databank with a scientific organisation rather than within the criminal justice system, was struck for purely practical reasons or whether it was to address governance concerns related to the extension of police or state powers.

Although not expressly stated in the Act, the establishment of the Databank infers that the purpose of collecting DNA profiles from individuals is to compare these with crime scene profiles. Therefore, a crime sample database was established to house unidentified DNA profiles obtained from crime scene samples. Samples collected from crime scenes and the DNA products resulting from their analysis may be stored indefinitely.

The Criminal Investigations (Bodily Samples) Amendment Act in 2003 introduced significant changes to this original legislation. One change, requiring the title of the Act to be modified from (Blood Samples) to (Bodily Samples), allowed for buccal samples (mouth swabs) to be obtained from individuals. Mouth swabbing is a less invasive sampling technique, and this change meant that medical staff were no longer required to be present to collect samples as individuals could do this themselves. Furthermore, the Amendment Act expanded the list of relevant offences under Part II of the Act to include burglary and entering with intent. A retrospective clause was introduced that allowed for the compulsory sampling of prisoners who were convicted of relevant offences prior to the 1995 Act coming into force.

CONCERNS AND THEMES ARISING
FROM THE LEGISLATION

Prior to enactment, details of the 1995 Bill and the 2003 Amendment Bill were made public so that organisations and private citizens could consider the proposed legislation and make submissions. The following section, which draws heavily on a review conducted by Morris and Cullen (2004), summarises the tenor and content of these submissions. This is followed by a description of three high profile cases and the impact these had on shaping media reporting and public perceptions of DNA profiling.

Submissions made to the proposed legislation

Morris and Cullen (2004) retrospectively reviewed the submissions made in relation to the proposed DNA legislation and its subsequent amendments in order to gauge public and institutional opinion about the legislation and also to determine if there was any noticeable change in perspectives between 1994 and 2002. In 1994, the Justice and Law Reform Select Committee considered 23 submissions originating from five legal, three medical, three women's organisations, three civil rights organisations, the New Zealand Police Association, the ESR and seven individuals, including two scientists and two lawyers.[2] The submissions were categorised in terms of the degree of support given to the legislation as a whole. The submissions from ESR, the Police Association, the Medical Council and two private citizens were described as being generally supportive; 10 gave qualified support, and eight submissions from mainly civil rights and legal organisations opposed the Bill outright. Some submissions raised issues relating to the length and drafting of the Bill, while others were directed at specific laboratory practices. Of particular interest are the submissions that included considerations of ethical, cultural and societal aspects, as these often reflect core values or reveal potential unintended effects.

Two such cultural issues were the Treaty of Waitangi and the reliability of DNA when testing certain ethnic groups. The Treaty of Waitangi (Tiriti o Waitangi) is Aotearoa/New Zealand's founding document. Representatives of Māori tribes and the Crown signed the document on 6 February 1840. The Treaty secured British sovereignty over Aotearoa/New Zealand through negotiation with Māori and is a statement of principles on which to found a nation state and to build a government. While there are some key differences between the English and Māori versions of the text, these principles acknowledge Māori ownership rights to their lands, forests, fisheries and other possessions, and that Māori should enjoy the same privileges as British subjects. The Treaty of Waitangi is not referred to at all in the Bill and this was raised as a concern in submissions from women's groups, as the collection and storage of DNA could have implications for Māori.[3] However, the Select Committee deemed that the matters dealt with by the Treaty of Waitangi did not overlap those covered by the Bill and no changes to the Bill were made (Morris and Cullen 2004: 9–10).

The second cultural issue, although perhaps better described as a technical issue, was raised by a scientist and concerned the possibility that Pacific Island populations may demonstrate less genetic diversity than populations of European descent. If this were the case then there would be significant implications for calculating match probabilities,[4]

---

[2] In their report, Morris and Cullen (2004) provide full details of which organisations made submissions in relation to the 1994 and 2002 Bills.

[3] For insights into Māori perspectives on genetics, DNA testing and cultural risk, see Gillett and McKergow (2007), and Satterfield and Roberts (2008).

[4] Match probabilities are used to calculate the strength of DNA evidence where the profile obtained from the crime scene corresponds with the suspect's profile. In essence, match probability calculations answer the question, 'What is the probability that someone other than, and unrelated to, the suspect, has this same DNA

and New Zealand, where many Polynesian groups have settled, would be a less than ideal environment for forensic DNA profiling. At the time, ESR countered this viewpoint by arguing that Pacific Island and Māori population data demonstrated appropriate levels of diversity at the DNA sites used in forensic profiling (Morris and Cullen 2004: 10).[5]

Protecting basic rights, protecting the public against crime and the extension of police powers were common themes in the ethical issues raised by civil rights, medical, legal and women's organisations and by private citizens. Some submissions argued that the proposed legislation diminished the right against self-incrimination, the right to silence, the right not to be detained by police unless under arrest and the right to the presumption of innocence. Others took the view that the legislation would be the beginning of a general erosion of basic civil liberties and 'thus encouraging the Authorities [sic] disrespect for such rights' (private citizen quoted in Morris and Cullen 2004: 11). The Medical Council, ESR and the Law Commission disagreed that rights were at risk, arguing inter alia that the legislation struck an appropriate balance between public protection and individual rights, and that the legislation protected citizens by clearly stating their rights with respect to the taking and storage of DNA samples (Morris and Cullen 2004: 11–13).

The provision that allowed the use of reasonable force to obtain samples from individuals refusing to provide a sample that is subject to a compulsion order was referred to in 17 submissions. All but one of the submissions argued that this was an indefensible extension of state power. For example, medical practitioner groups held that taking samples by force when there was no medical reason to do so was a corruption of the role of the medical professional (Morris and Cullen 2004: 14). The sole submission expressing support for the reasonable force provision came from the Police Association, which equated this provision with that which allows the use of reasonable force to obtain fingerprints and photographs (Morris and Cullen 2004: 13–15).

profile?' Population databases containing allele frequencies are used to calculate the match probability. If an ethnic group (subpopulation) within the population displays less genetic variation than the general population, then the probability of two people within that ethnic group having the same DNA profile may be significantly greater than a general population estimate would predict.

[5] After meticulous study of subpopulation allele frequencies in the New Zealand population database, ESR refined its match calculation methods to a more sophisticated model that incorporated subpopulation effects.

Twenty submissions specifically addressed the proposed introduction of a national DNA profile databank. Eight submissions from private citizens, ESR, Doctors for Sexual Abuse Care, the Police Association and the Law Commission were strongly supportive of, or gave qualified support to, this aspect of the Bill. Their arguments included that the Databank would help police to identify recidivist offenders and that this would have a deterrent effect, that innocent people would have nothing to fear and that profiling would help to guard against mistaken identity and to eliminate innocent people from enquiries. Those opposed to the proposed Databank included civil rights, legal and women's organisations. They argued that there was no evidence that a DNA databank would have a deterrent effect, that the proposed legislation theoretically allowed the police the power to test whole populations, that by being on such a database innocent people might be falsely accused and that the number of innocent people on the Databank could quickly outnumber offenders (Morris and Cullen 2004: 15–17).

With regard to the 2002 Amendment Bill, the Law and Order Parliamentary Select Committee considered 11 submissions. These submissions came from three civil or human rights organisations, two legal organisations, the ESR, the Police Association and two private citizens. Submissions from two pressure groups, one concerned about the rights of offenders, the other concerned about the rights of victims, were also received. Seven submissions supported or provided qualified support for the Amendment Bill. Three submissions, from the offenders' pressure group, a legal organisation and a civil rights group, were against the amendments (Morris and Cullen 2004: 19–20). Submissions from mainly civil and human rights organisations raised concerns relating to the protection of individual rights, arguing that the legislation was not an appropriate balance between crime control and the protection of human rights as defined in the New Zealand Bill of Rights Act (Morris and Cullen 2004: 22–23). A private citizen and the victims' pressure group took a contrary view, contending that too much attention was being paid to the rights of criminals (Morris and Cullen 2004: 24). The amendment clause, allowing samples to be obtained by compulsion order from offenders convicted of relevant offences prior to the enactment of the 1995 Act, was viewed by some legal and civil rights groups as being a breach of the principle against retrospective legislation (Morris and Cullen 2004: 24–25).

These submissions provide some insights into the responses to the Criminal Investigations (Blood Samples) Bill and Amendment Bill.

While most were generally supportive of the use of DNA profiling and databanking to enhance criminal investigations, many concerns were raised with respect to specific provisions within the Bills and more generally in relation to cultural perspectives and human rights. For the most part, these concerns were given little consideration by the government and no significant changes were made to the Bills prior to enactment.

Signal crimes

In New Zealand, the past two decades have seen a rise in rhetoric regarding intelligence-led policing (Ratcliffe 2005) and public support for populist penal initiatives such as tougher sentencing (Pratt 2008). Forensic DNA profiling was portrayed as a progressive, 'hi-tech' and entirely appropriate method for the police and the government to get tough on crime. Three New Zealand crime stories were widely reported in the media during the period leading up to the introduction of the Act in 1995: Operation Park, Operation Harvey and Teresa Cormack. These crimes and the media attention they attracted provide a framework within which to consider the social environment during the drafting of the DNA legislation. More importantly, they also reflect the optimism felt by the general public regarding the promise of DNA profiling.

Martin Innes (2003) describes 'signal' crimes as those high-profile cases that can have an effect on a population as a whole. Signal crimes can bring to the surface fears that there is something fundamentally wrong with society, resulting in adaptations in behaviour or belief systems, and often these crimes linger in the public consciousness for years to come (Innes 2003). Three widely reported police investigations serve as examples of signal crimes that may have influenced New Zealanders to accept DNA profiling as a method of crime control. In two of the investigations, DNA profiling would play a role in identifying two of New Zealand's most prolific sex offenders and the third, the 1987 sexual violation and murder of a six-year-old girl, would not be solved until 2001, when advances in DNA profiling finally resulted in the conviction of the offender.

Operation Park, the first of the two serial rape investigations, investigated a series of rapes of women and young girls that occurred in the early 1990s in South Auckland. Police theorised that these crimes were the work of a single serial rapist owing to their similarities and that in at least two cases the offender returned to the scene weeks later to violate the victim a second time.

Blood grouping analysis and single locus probe (SLP) DNA profiling provided confirmation that one offender had committed 15 of the crimes under investigation. In response, the police set up the Criminal Profiling Unit. Using information gained from the scenes and the victims, a behavioural profile of the offender was determined and approximately 3000 men were identified as 'persons of interest'. Using a combination of blood grouping and SLP, a mass screen of these men was conducted. Eventually, only Joseph Thompson was identified as a possible source of the male DNA recovered from the 15 crime scenes linked by DNA. Once apprehended, Joseph Thompson pled guilty to 129 charges, including 61 charges of sexual violation, relating to 50 cases that occurred between 1983 and 1994.

The second investigation, *Operation Harvey*, ultimately linked a homicide and a second series of rapes that occurred in South and central Auckland in the late 1980s to the mid 1990s. Susan Burdett was murdered in 1992 and semen left in the victim's body was tested using SLP. A mass screen of 500 men resulted in no matches. By 1995, SLP and a new three locus short tandem repeat (triplex STR) DNA test linked seven rape cases in the South Auckland area, one rape case from Rotorua and the 1992 homicide. A mass screen of approximately 5000 men using the faster triplex STR system was begun. Meanwhile, a similar pattern of offending was now taking place in central Auckland. The DNA profiling results were obtained from just one of the central Auckland cases but the profile confirmed that this case at least was linked to the earlier crimes. In early 1996, Malcolm Rewa became a suspect but could not be located for questioning. As Malcolm Rewa was not available to provide a DNA sample, samples were taken from his close family members instead. Their DNA profiling results were similar to those obtained from the crime scenes, thus confirming that the offender could be a member of the Rewa family. In May 1996, Malcolm Rewa was apprehended after assaulting a young woman and DNA testing confirmed that he could be the source of the DNA found at the crime scenes where incriminating biological evidence had been collected. He pled guilty to the rape cases in which he was implicated by DNA evidence and was found guilty of rape or attempted rape of an additional 18 women.

Operation Park and Operation Harvey were lengthy and costly investigations involving countless police officers exploring many investigative leads. Compared with the overall scale of both investigations, DNA profiling played a small but crucial role by linking crimes within each series and identifying a suspect. The success of DNA profiling in identifying and ultimately convicting Joseph Thompson and Malcolm Rewa was widely reported in the press and provided hope for the third case: the murder of six-year-old Teresa Cormack. This case shocked the nation and fundamentally changed the perception that small-town New Zealand was a safe place for children.

In June 1987, on the day after her sixth birthday, Teresa Cormack had walked toward her Napier primary school as usual but never arrived. A week later, her body was found in a shallow grave on a beach 16 km north of Napier. New Zealanders were profoundly affected by this case. As Napier city councillor Dave Pipe explained later: 'It was a naivety that we had. We were naive that we thought we could send our children off to school and that they would be safe. You just looked at the world through different eyes after that.' He went on to say: 'It was like an atom bomb was dropped here and the shock waves went right out around the country' (Brown 2002).

The postmortem examination revealed that Teresa had been subjected to sexual assault and had died of suffocation. Samples collected from her body and the gravesite were tested by ESR and the UK's Forensic Science Service. However, the less sensitive techniques available then produced no informative results. Nevertheless, it was believed that future advances in DNA research would eventually result in the identification of the offender (Anon.2002). During the early 1990s when the DNA legislation was being considered in New Zealand, this case remained unsolved, a painful example of the hopes and desires that underscored the promise of forensic DNA profiling. It was not until 2001, when a more sensitive DNA profiling technique became available that a male DNA profile was obtained from a slide containing semen collected from Teresa's body. This profile was found to correspond with the DNA profile of Jules Mikus, an individual identified as a suspect early in the investigation but whose alibi had appeared to clear him. Jules Mikus was subsequently convicted for the sexual violation and murder of Teresa Cormack.

Media coverage

The above three cases and the associated media coverage provide a useful context in which to examine the general acceptance in New Zealand of the use of DNA in criminal investigations and the associated legislation. Positive media coverage of emerging DNA profiling techniques began to appear in the early 1990s. In a 1992 newspaper article titled 'Our forensic heroes', DNA profiling techniques were highlighted as providing a 'quantum leap' in associating individuals with crime scenes (Sarney 1992). A 1994 article devoted to DNA profiling explained in some depth how DNA could be used to solve open homicide cases, including that of Teresa Cormack (Matthews 1994). In contrast, early print media reports of the 1994 Bill tended to be negative in tone and brief, generally reporting the concerns raised in the Parliamentary Select Committee submissions. Many of these reports focused primarily on concerns related to the perceived widening of police powers, specifically the provision for police to use reasonable force to take samples. The latter issue also featured in letters to the editor in major New Zealand newspapers. For example, in a letter published in *The Dominion*, the author begins by congratulating the police for capturing Joseph Thompson but then describes the police power to use force to take a DNA sample as 'obnoxious' (Lacey 1995). However, even though specific issues arising from the proposed legislation tended to be reported negatively, the *idea* of forensic DNA profiling as a useful investigative tool was accepted. For example, an article in *The Dominion* reporting the Auckland District Law Society's criticisms of the Bill also stated: 'DNA has become an important method of detecting those responsible for sex crimes and of eliminating suspects ...' (Anon. 1995a). Contentious issues arose primarily around finding within the law an appropriate balance between protecting individual rights and autonomy while allowing DNA profiling in the interests of public safety.

As details emerged about the importance of DNA in identifying Joseph Thompson as the South Auckland rapist, media reporting began to focus on the power of DNA as an investigative tool, and the concerns that were raised in response to the proposed legislation became secondary. Earlier media reports had expressed the extension of police powers as potentially violating the rights of citizens. Now, the rhetoric shifted from protecting the rights and freedoms of the citizenry as a whole towards assigning people into separate classes – offenders and victims – each entitled to different rights. As a result, some media editorials and features not only concentrated on the public safety benefits of DNA profiling but were also critical of those who raised human rights concerns. For example, in an editorial published in *The Dominion* entitled 'Hand-wringers wrong on DNA' (Anon.1995b), the author began with: 'The most surprising thing about the passage into law of the measure allowing the police to take by force a pinprick blood sample for DNA testing were the cries of outrage from civil libertarians, who lost all sense of perspective along the way'. The piece closed with: 'Those who carp against [the Act] risk being seen as more concerned with protecting offenders than in delivering justice to their past victims and protection to their future targets'.

PROFESSIONAL AND PUBLIC PERSPECTIVES

Until recently, when we began exploring professional and public perceptions of DNA profiling,[6] there had been no research in New Zealand that directly captured public opinions of DNA profiling. Besides our own research, a pilot study using a telephone survey designed to gauge public knowledge about the Databank was conducted in 2003–2004 (Curtis 2009) but to our knowledge was never developed into a fully fledged research project. The results of the pilot survey have been made available and offer some (though limited) insight into public perceptions (Curtis and Thomas 2004). One hundred randomly selected Auckland residents were telephoned and asked a number of questions about their understanding of DNA profiling and the role of the Databank. The questionnaire addressed a number of possible areas of concern. Sixty-three respondents agreed or strongly agreed that they had concerns about privacy issues, 60 that DNA might be used for other purposes, 48 had ethical concerns, 30 had cultural concerns, 54 were worried about DNA evidence being planted at crime scenes by police or others, and 59 were concerned about mistakes being made (Curtis 2009). In addition, despite acknowledging the above concerns, 79 respondents agreed or strongly agreed with the statement: 'I have no concerns about the use of DNA as a crime-fighting tool' (Curtis 2009: 321).

Another study identified factors involved in shaping public perceptions of biotechnology in New Zealand (Cook *et al.* 2004). While not a specific aim of the research, the study contrasted concerns about crime and the use of forensic DNA profiling with other social concerns and biotechnologies. A survey was posted to 2000 randomly selected

---

[6] Our current research, along with our co-researchers Professor Victoria Grace of Canterbury University and Annabel Ahuriri-Driscoll of ESR, is called '*CSI* New Zealand: The Meanings of DNA Evidence'. The main purpose of the research is to investigate whether there is a gap between professional and lay understandings of DNA as it is used in a forensic context, and if this gap exists, what its nature is. Although there are no results yet in the public domain the authors can be contacted directly for further information.

homes in New Zealand and 691 responses were used in the final analysis. The respondents were provided with a list of concerns related to biotechnology, genetically modified organisms, terrorism, pollution, unemployment, crime and violence, and public healthcare. Crime and violence attracted the highest degree of anxiety, with 93.5% of the respondents indicating that this was a concern to them (Cook *et al.* 2004: 26). When asked to consider the acceptability of various biotechnologies such as modifying DNA to avoid cancer, cloning endangered species, using stem cells to treat disease or using DNA to catch criminals, forensic DNA profiling was ranked as the highest acceptable use of biotechnology, with 92.3% of respondents indicating acceptance (Cook *et al.* 2004: 29). The authors posit that this approval of forensic DNA profiling is likely reflecting the importance of crime control as a social issue, the fact that it is already routinely used and the fact that it does not involve genetic modification (Cook *et al.* 2004: 29).[7]

A recent New Zealand study brought together various professionals from within the criminal justice system to explore the social impacts of forensic DNA testing (Baker *et al.* 2006).[8] As one key stakeholder group involved, police officers focused primarily on the operational aspects of DNA profiling, frequently expressing a desire to make as much use of the technology as possible in order to find closure for victims. This desire was often articulated alongside a related desire to widen police discretionary powers to sample suspects, although there was no consensus on what the extent of that expansion should be (Baker *et al.* 2006: 10).

Other stakeholders (e.g. lawyers, scientists and academics) acknowledged the usefulness of DNA profiling in a forensic context but were inclined to consider the issues in much broader terms related to fairness and appropriate safeguards. Most of these stakeholders were comfortable with the status quo with regard to who may be asked to provide a sample for the Databank, although some articulated a preference for sampling solely upon conviction. Some of the stakeholders felt that police could be accused of targeting specific populations or

---

[7] The issue of genetically modified foods had recently been the subject of high-profile public demonstrations and direct action protests.

[8] In this study, three workshops were held. The first involved mainly police officers representing a range of ranks, roles and experience. Representatives from a variety of legal, academic, science research, policy, lobbyist and non-governmental organisations attended the second workshop. Four Māori working in the science, health ethics and police sectors participated in the third workshop.

groups under the current legislation, and they suggested that sampling the entire population was inherently fairer (Baker *et al.* 2006: 30). When the suggestion of sampling the entire population was raised with the police officers, they were reluctant to even consider it, citing that the public would never accept such a proposal and, therefore, it was not worth discussing (Baker *et al.* 2006: 21). Issues of cost were also mentioned as a barrier to implementation (Baker *et al.* 2006: vi).

However, a different perspective was raised by Māori academics. For Māori, DNA is a physical manifestation of *whakapapa*, that is, it contains knowledge about cultural identity, connecting ancestors with those living today and generations yet to be born (Satterfield and Roberts 2008). As a consequence, Māori hold a very inclusive view: that criminals are considered to be part of, rather than separate from, family and society and are, therefore, the responsibility of the family unit and the wider community. During the workshop, this was articulated as a desire to 'humanise' the science around the collection and storage of DNA samples so that proper guidance and care could be offered to those represented on the Databank (Baker *et al.* 2006: 21). *Whakapapa* understands DNA to be an expression of collective, rather than individual, identity (Gillett and McKergow 2007). The implications of this in terms of providing individual consent in a forensic DNA profiling context have not so far been discussed in the wider community. However, the Māori participants argued that it was necessary to consider procedures for obtaining collective Māori consent when using DNA from Māori (Baker *et al.* 2006: 21).

All participants agreed that the samples collected for forensic purposes, and the information derived from those samples, should not be used for any other purpose (Baker *et al.* 2006: 24). The participants felt that it was time for a reassessment of the governance of and safeguards around forensic DNA profiling. They agreed that careful consideration was needed of the longer-term impacts of new and potential DNA technologies. It was felt that without such a review, function creep or incremental decision making might have a detrimental effect on the value or positive perception of DNA profiling (Baker *et al.* 2006: 21).

DNA PROFILING TODAY

Following the enactment of the legislation in 1995, the DNA Profile Databank became operational on 12 August 1996. Now, in its thirteenth year, the Databank contains more than 90 000 profiles and the crime sample database contains approximately 21 000 profiles. On average, 900 profiles from individuals are added to the Databank each month.

The crime sample database and the DNA Profile Databank are routinely compared to detect crime-to-person links. These correspondences are reported to the police for further investigation but may not be used in court as evidence against the accused. The police must first obtain a suspect reference sample from the individual to confirm the link. There are sound reasons for this requirement. First, the individual may have used an alias when providing the Databank sample and this may not be revealed until the DNA match is investigated further. Second, and perhaps more importantly, that an individual has submitted a Databank sample may be viewed prejudicially by jurors in a subsequent trial. Therefore, only the link between the suspect reference sample and the crime scene sample forms evidence in a prosecution; the initial link to the Databank is never referred to before a jury.

The current link rate (or 'hit rate') between the Databank and the crime sample database is 61%. This means that of the crime sample profiles added to the crime sample database, approximately 61% will correspond with profiles held on the Databank. These matches are subsequently reported to the police for further investigation. When the Databank was first being considered, its promise was most often articulated in terms of solving violent crime. However, it soon became apparent that the Databank provided an effective way to solve less-serious crimes where there was no suspect. Approximately three-quarters of the Databank links relate to property crime.

In many violent crime cases, comparison with the Databank is not required as suspects are identified by other means during the initial phases of the investigation. In these cases, DNA profiling serves to confirm (or disprove) the presence of DNA from a suspect by comparison with reference samples. Nevertheless, the Databank has been instrumental in identifying suspects in a number of serious and high-profile cases. Historic cases have also been revisited and solved as a result of Databank links.

It is anticipated that the DNA legislation will be dramatically amended in 2010. New Zealanders marked the end of 2008 with a change of government. When the National Party formed the new government in early December 2008, it announced a 100 day action plan, which included a proposal to significantly amend the DNA legislation. The 2009 Amendment Bill, if passed into law in its present form, will ultimately allow police to collect a Databank sample, without requiring consent, from any person they arrest or intend to charge for any imprisonable offence.[9] No prior judicial approval is required and the decision to sample will be at the police officer's discretion. All biological samples must be destroyed once a DNA profile is obtained, and the profiles may be removed from the Databank if charges are withdrawn or the defendant is acquitted. Profiles may also be removed after a specified period of non-offending. Special provisions apply to young people aged between 14 and 16 years, primarily that DNA may be taken if the offence carries a maximum imprisonment term of seven or more years or if the person concerned has had at least one previous conviction. It is anticipated that these proposed changes will result in an increase of submissions to the Databank from the current yearly rate of approximately 11 000 profiles to at least 24 000 profiles.

The New Zealand Government considers mandatory DNA sampling (i.e. a person arrested for an imprisonable offence must provide a Databank sample if asked) as a key component of its 'getting tough on crime' campaign, and this is illustrated in the following comments made by the Minster of Justice and 2009 Amendment Bill sponsor, Simon Power: 'The Government regards this legislation as critical in the fight against crime' (LawTalk 2009) and '[mandatory testing is] going to be a critical tool for the police to be able to conduct the work they need to do to make New Zealand a safer place' (New Zealand Press Association 2009a).

While Simon Power acknowledged that the proposed amendments would provoke strong opposition, particularly from groups concerned with civil liberties (New Zealand Press Association 2009a, 2009b), the following comments show that the New Zealand Government and police spokespeople presented mandatory DNA testing to the New Zealand public as 'hi-tech' ('[DNA is the] modern-day fingerprint' (Prime Minister John Key, cited in Gower 2009)), benign, at least for non-offenders ('Law-abiding citizens have nothing to fear' (Detective Sergeant Mike Ryder, cited in Caygill 2009)), efficient ('DNA testing has been highly effective in matching people with crimes they haven't otherwise been convicted for' (Prime Minister John Key, cited in New

---

[9] If the legislation is implemented as proposed, a phased approach is anticipated. The first phase, to be implemented by July 2010, will allow police to take DNA from any person whom they intend to charge with a relevant offence (this will be a slightly expanded list of offences than already contained in the current Act). The second phase is expected to be implemented by December 2011 and will allow police to take Databank samples from anyone they arrest or intend to charge with *any* imprisonable offence.

Zealand Press Association 2009c)) and a deterrent ('It's the Government's hope it'll have a serious impact on those giving thought to committing crime' (Simon Power, cited in New Zealand Press Association 2009c)).

In a report to the government, Attorney-General Christopher Finlayson warned that the proposed legislation conflicted with the aims of the Bill of Rights Act (Finlayson 2009). In particular, the fact that police will not require judicial or other independent review before taking DNA from anyone arrested for an imprisonable offence is inconsistent with the right protecting against unreasonable search and seizure. The Justice and Electoral Select Committee will consider these implications when it debates the 2009 Amendment Bill. While these finding may seem like a major impediment to the enactment of the proposed legislation, there have been numerous examples of crime control values being prioritised over the values protected in the Bill of Rights (reviewed by Optican (2007) and Schwartz (1998)).

With the support of both the governing party and major opposition parties, the 2009 Criminal Investigations (Bodily Samples) Amendment Bill passed its first reading, under urgency, in Parliament on 12 February 2009 by 108 votes to 13. The Bill will now go before the Justice and Electoral Select Committee for consideration, and public submissions are being sought. While vigorous debate is expected, the proposed legislation is likely to be enacted. (At the time of publication, the legislation had been passed into law but was not yet operational.)

CONCLUSIONS

This chapter has documented the history of DNA profiling and its governance in New Zealand. During the late 1980s and early 1990s, DNA profiling was exemplified as a modern crime-fighting tool and successes were widely reported in the media. However, with no legislative framework, the potential value of DNA profiling was soon being undermined in New Zealand courts. As a result, a comprehensive statute was enacted and New Zealand became the second country, after the UK, to establish a nationwide DNA databank. Ethical concerns did (and still do) exist, as evidenced by many of the submissions made in advance of the DNA legislation and its subsequent amendments. However, these were often set aside in favour of crime control values. The widespread reporting of so-called 'signal' crimes further promoted crime control concerns and articulated DNA profiling as a technology that only offenders need fear. One consequence of this pervasive concern with crime control, in conjunction with the public fascination with forensic DNA technologies, is that the governance structures surrounding DNA profiling in New Zealand have not been subjected to critical evaluation. While New Zealand research exploring public perceptions and understandings of forensic DNA profiling and databanking is limited, what studies have been done allows us to conclude that New Zealanders are concerned about crime and perceive forensic DNA profiling as a pragmatic and acceptable use of DNA-based technologies. Overall, in the last few years, advanced DNA profiling technologies and applications such as low copy number, Y-chromosome profiling and familial searching (see Chapter 2) have all been introduced in New Zealand, with little judicial challenge. It also appears that this will be true for the proposed new legislation. There is wide public support for forensic DNA technologies and practices owing to successes in detecting criminals and securing convictions. This is well summed up in the claim by Harbison *et al.* (2008) that 'the New Zealand DNA databank [has] already proven to be a remarkable crime-fighting tool'.

REFERENCES

Anon. (1990). Man jailed for life in DNA murder trial. *The Dominion*, 7 April, 6.
Anon. (1995a). Police reject lawyers' criticism of DNA bill. *The Dominion*, 21 February, 2.
Anon. (1995b). Hand-wringers wrong on DNA. *The Dominion*, 23 October, 10.
Anon. (2002). Advances in DNA testing bring Cormack Case to Court. *NZCity* http://www.crime.co.nz/c-files.aspx?ID=24864 (accessed 1 March 2010).
Baker, V., Gregory, W., Midgley, G. *et al.* (2006). *The Ethics of Forensic DNA Technology: Summary Report.* Wellington, New Zealand: Institute of Environmental Science and Research.
Brown, J.-M. (2002). Napier: the town that lost its innocence. *New Zealand Herald*, 2 March www.nzherald.co.nz/nz/news/article.cfm?c_id=1&objectid=1090414 (accessed 23 July 2009).
Caygill, L. (2009). Police keen on DNA testing. *Oamaru Mail*, 12 February, 3.
Cook, A., Fairweather, J., Satterfield, T. *et al.* (2004). *New Zealand Public Acceptance of Biotechnology.* Christchurch, New Zealand: Lincoln University.
Cordiner, S., Harbison, S., Vintiner, S. *et al.* (2003). A brief history of forensic DNA testing in New Zealand. *New Zealand Science Review*, 60, 39–42.
Curtis, C. (2009). Public perceptions and expectations of the forensic use of DNA: results of a pilot survey. *Bulletin of Science, Technology and Society*, 29, 313–324.
Finlayson, C. (2009). *Report of the Attorney-General under the New Zealand Bill of Rights Act 1990 on the Criminal Investigations (Bodily Samples) Amendment Bill.* Wellington, New Zealand: Office of the Attorney-General.
Gill, P., Jeffreys, A. and Werrett, D. (1985). Forensic application of DNA fingerprints. *Nature*, 316, 76–79.
Gillett, G. and Mckergow, F. (2007). Genes, ownership, and indigenous reality. *Social Science and Medicine*, 65, 2003–2104.
Gower, P. (2009). DNA seizure law triggers rights alert. *New Zealand Herald*, 11 February, 6.

Harbison, S., Fallow, M. and Bushell, D. (2008). An analysis of the success rate of 908 trace DNA samples submitted to the Crime Sample Database Unit in New Zealand. *Australian Journal of Forensic Sciences*, 40, 49–53.

Harbison, S., Hamilton, J. and Walsh, S. (1999). The New Zealand DNA databank. In *First International Conference on Forensic Human Identification in the Millennium*. London: Forensic Science Service.

Innes, M. (2003). 'Signal crimes': detective work, mass media and constructing collective memory. In *Criminal Visions: Media Representations of Crime and Justice*, ed. P. Mason. Cullompton, UK: Willan, pp. 51–69.

Jeffreys, A., Wilson, V. and Thien, S. (1985). Hypervariable minisatellite regions in human DNA. *Nature*, 314, 67–73.

Lacey, M. (1995). Forced DNA tests. [Letter to the editor.] *The Dominion*, 21 August, 6.

Lawtalk (2009). DNA bill may breach BORA. *LawTalk* Magazine, 2 March, 14.

Lowe, P. and Barber, J. (1989). DNA test legislation leaves Jeffries cold. *The Dominion Sunday Times*, 15 October, 5.

Matthews, L. (1994). Strands of hair could unlock mysteries. *Sunday Star Times*, 11 September, 7.

Morris, A. and Cullen, J. (2004). *A Review of the Submissions to the Criminal Investigations (Blood Samples) Bill 1994 and the Criminal Investigations (Bodily Samples) Amendment Bill 2002*. Wellington, New Zealand: Crime and Justice Research Centre, Victoria University.

New Zealand Press Association (2009a). Push for mandatory DNA testing in criminal investigations. www.3news.co.nz/News/PoliticalNews/Push-for-mandatory-DNA-testing-in-criminal-investigations/tabid/419/articleID/90496/cat/67/Default.aspx (accessed 10 February 2009).

New Zealand Press Association (2009b). DNA bill likely to trigger fiery debate. *Marlborough Express*, 11 February, 2.

New Zealand Press Association (2009c). Forced DNA testing can help innocent. *Timaru Herald*, 11 February, 4.

Optican, S. (2007). 'Front End/Back End'Adjudication (Rights Versus Remedies) Under s 21 of the Bill of Rights (Search and Seizure). In *The New Zealand Legal Method Conference Series: Rights and Freedoms in New Zealand: The Bill of Rights Act Comes of Age*, Auckland, New Zealand.

Pratt, J. (2008). When penal populism stops: legitimacy, scandal and the power to punish in New Zealand. *Australian and New Zealand Journal of Criminology*, 41, 364–383.

Ratcliffe, J. (2005). The effectiveness of police intelligence management: a New Zealand case study. *Police Practice and Research*, 6, 435–451.

Sarney, E. (1992). Our forensic heroes. *New Zealand Herald*, 13 May, 3.

Satterfield, T. and Roberts, M. (2008). Incommensurate risks and the regulator's dilemma: considering culture in the governance of genetically modified organisms. *New Genetics and Society*, 27, 210–216.

Schwartz, H. (1998). The short and happy life and tragic death of the New Zealand Bill of Rights Act. *New Zealand Law Review*, Part II, 259–312.

CASES

Pengelly v. *R* (1991). 7 CRNZ 333 (CA).
R v. *Montella* (1992). 1 NZLR 63 (HC).
R v. *Pengelly* (1990). 5 CRNZ 674 (HC).

MARIA CORAZON DE UNGRIA AND
JOSE MANGUERA JOSE

15

Forensic DNA profiling and databasing: the Philippine experience

INTRODUCTION

The Philippines, an archipelagic country of over 92 million people (USAID 2008), is a member state of the Association of Southeast Asian Nations. Unlike Thailand, Malaysia, Singapore and Indonesia, the Philippines does not yet have a national DNA database or any legislation that would facilitate its establishment. The Philippine Supreme Court has recognised the admissibility of DNA evidence in court and has provided judicial guidelines for the collection, handling and storage of biological samples in the *Rule on DNA Evidence*, which it promulgated in 2007 (Supreme Court of the Philippines 2007). Forensic DNA testing in criminal investigations is performed by the National Bureau of Investigation (NBI), an agency under the Department of Justice, and the Philippine National Police (PNP), an agency under the Department of Interior and Local Government. A third DNA laboratory, which is based at the University of the Philippines (UP), not only performs forensic DNA testing in both criminal and civil cases but also validates DNA testing procedures and conducts research on the genetics of the Philippine population. The three DNA laboratories operate independently and are under the control and supervision of the institutions to which they are attached.

This chapter provides a historical overview of the development of forensic DNA technology and its use in Philippine courts, as well as a discussion on legislative issues in the establishment of a national DNA database. The implications for governance are discussed and, finally, suggestions are made for effective policing combined with the minimisation of infringements of individual rights.

Genetic Suspects: Global Governance of Forensic DNA Profiling and Databasing, ed. Richard Hindmarsh and Barbara Prainsack. Published by Cambridge University Press. Copyright © Cambridge University Press 2010.

DNA IN THE COURTROOM

The courts before the era of DNA

Eyewitness testimony was the principal, and at times, the sole evidence used at trial to link the accused to the crime for many years. Physical evidence was seldom available and hardly used. Because of insufficient equipment and material and the lack of training of investigators, crime scene evidence was also often improperly collected and stored (Center for Public Resource Management 2003; Office of the Coordinator for Counterterrorism 2008), leaving the court no recourse but to rely on testimonies during trial. In sexual assault cases, for example, the sole testimony of a woman was sufficient to convict a man and have him incarcerated for life (*People* v. *Corpuz* 1993), or even sentenced to death (*People* v. *Echegaray* 1996). This view was upheld despite evidence that testimonies might not be that reliable, since they can be influenced or modified by various psycho-social considerations (Saks and Koehler 2005).

Testimonial evidence also made litigation more time consuming since the testimony of one party was pitted against the testimony of another. The huge caseload of government lawyers and limited funds available to conduct pre-trial investigations added to the delayed resolution of cases (Center for Public Resource Management 2003). A suspect could be detained for three years while the case was being tried if the suspect was unable to post bail or if the offence charged was non-bailable (Bureau of Democracy Human Rights and Labor 2007). In death penalty cases, the Supreme Court recognized that in many instances the reliance on testimonies as the sole primary piece of evidence was not enough to prove the guilt of the accused 'beyond reasonable doubt'. In fact, when the Supreme Court reviewed death penalty cases from 1993 to June 2004 (*People* v. *Mateo* 2004), it discovered that,

> the cases where the judgment of death has either been modified or
> vacated consist of an astounding 71.77% of the total of death penalty
> cases directly elevated before the Court [Supreme Court] on automatic
> review that translates to a total of six hundred fifty-one (651) out of nine
> hundred seven (907) appellants saved from lethal injection.

Given this scenario, the use of DNA evidence offered the Philippine criminal justice system a powerful tool for human identification that could be used to accelerate the fair administration of justice.

The use of DNA evidence in court

The use of forensic DNA typing in Philippine courts started with the case of *People* v. *Paras* (1999), a criminal paternity case involving an allegation of rape resulting in the birth of a child. The DNA test results showed that the accused could not have fathered the child. Four months after the DNA test, the accused was acquitted. However, the accused had to suffer more than six years of incarceration while his case was being tried.

The Supreme Court first formally recognised the utility of DNA evidence in the case of *Tijing* v. *Court of Appeals* (2001), a Solomonic case involving two women claiming to be the mother of the same child. Although no actual DNA testing was conducted, the Supreme Court took judicial notice of the importance of recognising new scientific developments that could assist the judicial system, such as DNA technology. A year later, the Supreme Court admitted DNA as corroborative evidence to convict the accused of raping and killing a nine-year-old girl (*People* v. *Vallejo* 2002). Aside from the victim's DNA, which was detected on the clothing that the accused had been wearing on the day of the child's disappearance, eyewitness testimonies positively identified him as the last person who was seen with the victim. The Supreme Court upheld the death sentence meted out by the trial court because of the ability of the crime scene investigators to reconstruct the events leading to the crime, the confession of the accused and the DNA evidence. The Supreme Court acknowledged that DNA evidence was admissible in court and laid down standards to be considered by courts in assessing its probative value. Two years later, the Supreme Court upheld the conviction of an accused (*People* v. *Yatar* 2004) in another rape–homicide case. In this case, the trial court admitted DNA evidence from the vaginal swabs collected from the 16-year-old victim, which matched the DNA profile of the blood sample collected from the accused. Upon review, the Supreme Court ruled that the accused could be compelled to submit to DNA testing and that the conduct of DNA tests did not violate either the right of the accused to remain silent or his right against self-incrimination under the Philippine Constitution, since no testimonial compulsion was involved in the collection of a blood sample for DNA testing (*People* v. *Yatar* 2004). Subsequently, in the case of *Herrera* v. *Alba* (2005), the Supreme Court upheld the trial court's decision to order the putative father to submit to DNA testing in a case for child support, declaring that DNA analysis may be admitted as evidence to prove paternity.

The Supreme Court's response to the use of DNA evidence in court had not always been positive. In death penalty cases, where any error in sentencing is irreversible once the execution has taken place, the Supreme Court denied the request of three convicted offenders for post-conviction DNA testing as 'unnecessary' or 'forgotten evidence too late to consider now' since they had been properly and duly identified by the prosecution witness during trial (*Andal v. People* 1999). In another case, the accused was convicted of raping his 13-year-old niece once. The niece testified that the rape resulted in the birth of a child. In this case, the Supreme Court denied the accused's petitions for the conduct of post-conviction DNA paternity tests with finality (*People v. de Villa* 2001). Through the initiative of his family, a DNA paternity test was conducted out of court and established that Reynaldo de Villa was not the child's father. On 14 July 2003, de Villa sought to re-open his case by filing a petition for the issuance of a writ of *habeas corpus* (*In Re: Writ of habeas corpus for Reynaldo de Villa* 2004). At this time, the *Rules of Court* did not yet have any provision for post-conviction appeals (Supreme Court of the Philippines 1989). *Habeas corpus* – from a medieval Latin phrase that means 'produce the body' – is a writ whose function is to release a party from unlawful official restraint or detention. In this instance, Reynaldo de Villa questioned the legality of his continued imprisonment since DNA testing had established that he could not have fathered the child born from a single incident of forcible sexual intercourse as testified to by the victim in open court. However, the Supreme Court denied the *habeas corpus* petition because Reynaldo de Villa failed to allege that his constitutional rights had been violated, and the re-examination of the weight and sufficiency of evidence of a final judgment was already outside the scope of a *habeas corpus* petition. The Supreme Court further ruled that the pregnancy of the victim was not an element of rape (consequently, the prosecution does not have to establish that the victim became pregnant to convict the accused of rape). Consequently, the Supreme Court refused to overturn the conviction, which was based on the convincing testimonial evidence of the victim already affirmed on appeal, notwithstanding the DNA evidence showing non-paternity (*In Re: Writ of habeas corpus for Reynaldo de Villa* 2004). Fortunately for Reynaldo de Villa, he was granted executive clemency by President Gloria Macapagal-Arroyo in 2005 and eventually released, albeit only after having been imprisoned for more than 10 years (De Ungria *et al.* 2008).

From 1999 until 2007, the use of DNA evidence in court was governed by the *Revised Rules on Evidence* (Supreme Court of the

Philippines 1989). Because of the novelty of DNA technology, trial court judges and lawyers needed guidance in the use of DNA evidence. Hence, there was need to formulate specific guidelines on the use of DNA evidence at trial and during post-conviction appeals to make sure that the evidence collected was properly utilised and that misuse of the technology was prevented (Ochave 2003). The Supreme Court recognised that the indiscriminate and wholesale acceptance of DNA evidence, which is inherently influential and compelling, could result in even greater injustice (Supreme Court of the Philippines 2007).

The Rule on DNA Evidence

The *Rule on DNA Evidence* was promulgated by the Philippine Supreme Court on 15 October 2007 to provide guidelines on the proper use of DNA evidence in court, and to address issues such as the probative value of DNA evidence and the impact of DNA testing on the right to privacy and post-conviction DNA testing (Supreme Court of the Philippines 2007). The *Rule on DNA Evidence* did not limit the use of DNA evidence to specific cases and to specific technologies but rather opened the doors to the use of DNA evidence for any case – criminal or civil, past, present or future.

The *Rule on DNA Evidence* allows a court to issue DNA testing orders at any time, provided that a biological sample exists that is relevant to the case, that the DNA testing uses a scientifically valid technique and that the test has the scientific potential to produce new information relevant to the proper resolution of the case. The *Rule on DNA Evidence* makes DNA testing orders immediately executory and non-appealable (Supreme Court of the Philippines 2007), thus preventing the use of delaying tactics by counsel of adversarial parties. The *Rule on DNA Evidence* also permits DNA testing to be conducted at the behest of any party, including law enforcement agencies, even before a trial commences. The *Rule on DNA Evidence* allows law enforcement agencies to collect biological samples from a person suspected of committing any crime, with or without a court order. In examining the evidence, however, the trial court is directed by the *Rule on DNA Evidence* to consider (1) the collection and handling of all samples, (2) the DNA testing methodology, (3) the accreditation of the forensic DNA laboratory and the expertise of DNA analysts, and (4) the reliability of the DNA testing results.

To safeguard a person's right to privacy, the *Rule on DNA Evidence* makes DNA profiles and all results obtained from DNA testing confidential and limits access to this information to (1) the person from whom the sample was obtained, (2) counsel representing the parties, (3) duly authorised law enforcement agencies, and (4) other persons authorised by the court. Whoever discloses or uses any information concerning a DNA profile without a court order can be held liable (Supreme Court of the Philippines 2007). In addition, the *Rule on DNA Evidence* requires the preservation of the biological samples and DNA profiles until such time as the individuals from whom samples were collected might submit written requests to the court for the destruction of these samples and the removal of the genetic information from all records. In the absence of any written request, DNA profiles and biological samples can be retained indefinitely.

The *Rule on DNA Evidence* includes for the very first time a provision for post-conviction DNA testing, provided that a biological sample exists that is relevant to the case and that DNA testing of the sample would likely result in the modification of the judgment of conviction (Supreme Court of the Philippines 2007). This provision on post-conviction DNA testing represented a paradigm shift in the mindset of the Supreme Court, which never ordered post-conviction DNA testing during the 13-year period when the death penalty existed in the Philippines (1993–2006).

In the 1987 Constitution, the Philippines was the first country in Asia to abolish the death penalty, following the 'People Power' revolution that toppled the Marcos dictatorship in 1986. However, by the early 1990s, after several well-publicised crimes had led to calls for the re-imposition of the death penalty, the Death Penalty Law was enacted on 13 December 1993 (Republic Act 7659). The number of mandatory death-penalty offences (30) and death-penalty eligible offences (22) in the Philippines was the highest of any country that still had the death penalty at that time (Hood 2002).[1] From 1993 to 2006, seven convicts were executed by lethal injection (Orendain 2008).

The second abolition of the death penalty in June 2006 (Republic Act No. 9346) was greeted with widespread acclaim by the international community (European Commission 2007) and by mixed

---

[1] In mandatory death-penalty offences, the judge has no discretion but to impose the death penalty once the guilt of the accused is established beyond reasonable doubt. In offences eligible for the death penalty, the judge has the discretion to impose the death penalty depending on the circumstances.

reactions locally. While human rights and church groups lauded the death penalty's abolition, anti-crime and victims' rights advocates criticised its removal. In 2008, renewed calls were issued to implement the death penalty once again because of the perceived shortcomings of the national penal system and the continuous occurrences of heinous crimes, many of which remain unsolved.

LEGISLATING ON DNA DATABASES

With the government's heightened interest in DNA technology as an instrument to strengthen law enforcement and combat terrorism in the Philippines, the need to establish a national DNA database became of paramount importance. In 2008, the PNP acknowledged that its database had expanded to include genetic information from known members of the al Qaeda-linked Islamic militant network *Jemaah Islamiah* (Bartolome 2008). Meanwhile, the NBI had vowed to actively investigate all human rights abuse cases using new scientific technologies, such as DNA, to accelerate the process of identifying victims of extrajudicial killings and the persons responsible for these crimes (San Juan 2007).[2]

While several bills have already been filed in the legislature, a law establishing a national DNA database is still pending in the Philippines. With documented reports of abuse by those in power, a history of martial rule, economic instability and jurisdictional issues involving law enforcement agencies, one can presume that any or all of these factors could have influenced the slow progression of any bill on DNA databasing in the legislature.

Effective law enforcement versus individual rights

The DNA profiling and storage of DNA profiles in a database for the purpose of effective law enforcement raises the issue of rights of a state versus the rights of the individual. The Bill of Rights contained in the 1987 Constitution was explicit about the right of individuals to be protected against unwarranted intrusion in the form of unreasonable

---

[2] Extrajudicial killings refer to illegal liquidations, unlawful or felonious killings and forced disappearances in the Philippines. These are forms of extrajudicial punishment commonly attributed either to the state government or state authorities such as the armed forces and police.

searches. Law enforcement agents must provide 'probable cause' before a judge, justifying the search of a person, property or personal papers and effects, in order to secure authorisation in the form of a warrant. The issue of 'unreasonable searches' was raised in relation to DNA searches of a national DNA database (Agabin 2003). Should law enforcement agents secure a warrant before performing any search of the national DNA database? Could any court issue an order for a data-base search? In the *Rule on DNA Evidence*, the Supreme Court allowed a trial court to issue a DNA testing order at any time, provided that the requirements for the issuance of an order were met (Supreme Court of the Philippines 2007).

The right against unreasonable searches is closely linked with the right to privacy, also guaranteed in the 1987 Constitution. The right to privacy is the right of an individual to be free from unwarranted (i.e. without any legal basis such as a court order or arrest warrant) publicity and the right of a person to live without public interference in 'private' matters (de Leon 2002). The constitutional provision on the right to privacy complements the security of a person against unreasonable searches. In fact, the right to privacy is fundamental in maintaining a person's security in their own 'personhood' without the threat of unwarranted public disclosure of 'private matters' (de Leon 2002).

In the Philippines, the issue of effective governance through government access to 'personal matters' versus a citizen's right to privacy was well debated when President Fidel Ramos (in office from 1992 to 1998) issued Administrative Order (A.O.) No. 308 on 12 December 1996 instituting the National Identification System. The rationale behind the order was to increase the efficiency of government agencies in providing basic services to the people through a computerised system of identification that would link information from agencies that handle social security claims, applications for driver's licences, taxes, criminal records, medical records, welfare services, employment records or voter's identification. The scheme was aimed at combating corruption, fraudulent transactions and misrepresentations.

On 24 January 1997, Senator Blas Ople filed a petition with the Supreme Court questioning the constitutionality of A.O. No. 308 against then Executive Secretary Ruben Torres[3] and the heads of the

---

[3] The Executive Secretary belongs to the executive branch of the government and is the head and highest ranking official in the Cabinet. The holder of this office possesses much power since the holder is the 'alter ego' of the President and can issue orders in the name of the President.

government agencies who were charged with its implementation (*Ople v. Torres* 1998). Human rights advocates condemned the issuance of A.O. No. 308 and threatened to launch a civil disobedience campaign to demonstrate their opposition. Groups such as the Philippine Alliance for Human Rights, May 1 Worker's Movement (*Kilusang Mayo Uno*), League of Filipino Students, Student Christian Movement, College Editors of the Philippines, National Federation of Labor, New National Alliance (*Bagong Alyansang Makabayan*) and the Public Interest Law Center expressed strong opposition to A.O. No. 308 (Solidarity Philippines-Australia Network 1997).

On 23 July 1998, the Supreme Court declared A.O. No. 308 unconstitutional (*Ople v. Torres* 1998). The Supreme Court ruling, written by Justice Reynato S. Puno, stated:

> Unlike the dissenters, we prescind from the premise that the right to privacy is a fundamental right guaranteed by the Constitution hence, it is the burden of government to show that A.O. No. 308 is justified by some compelling state interest and that it is narrowly drawn. A.O. No. 308 is predicated on two considerations: (1) the need to provide our citizens and foreigners with the facility to conveniently transact business with basic service and social security providers and other government instrumentalities and (2) the need to reduce, if not totally eradicate, fraudulent transactions and misrepresentations by persons seeking basic services. It is debatable whether these interests are compelling enough to warrant the issuance of A.O. No. 308. But what is not arguable is the broadness, the vagueness, the overbreadth of A.O. No. 308, which if implemented will put our people's right to privacy in clear and present danger.

Further, the Supreme Court recognised that (1) the establishment of a single national identification system must be via the passage of a law, not an order from the executive branch of government; and (2) there is an absence of safeguards in A.O. No. 308 that would protect the privacy of individuals. The Supreme Court in this judgment (*Ople v. Torres* 1998) was emphatic about the importance of a person's right to privacy. It stated:

> The right to privacy is one of the most threatened rights of man living in a mass society. The threats emanate from various sources – governments, journalists, employers, social scientists, etc. In the case at bar, the threat comes from the executive branch of government, which by issuing A.O. No. 308 pressures the people to surrender their privacy

by giving information about themselves on the pretext that it will
facilitate delivery of basic services.

However, the Supreme Court did not permanently close the doors to
ordinances which mandate a government agency to compile information on citizens for a defined purpose, such as to improve law enforcement, to accelerate the delivery of public service or to better manage
financial activities of the government.

Since the Supreme Court ruling's in *Ople* v. *Torres* (1998) various
senators have introduced bills proposing to establish a national identification system. However, none of the bills have been enacted. In
rejecting the idea of a national identification system, three reasons
have been identified: (1) the danger of 'function creep', wherein information would be used for purposes other than the original intent for
gathering the information (see Chapter 2); (2) the potential for misuse
resulting from identity fraud; and (3) privacy issues (Encinas-Franco
2005). The same issues are likely to be raised once the Philippine
Congress starts public hearings on a law establishing a national DNA
database.

Because of the inability of Congress to enact legislation to create
a national identification system and cognisant of the Supreme Court
ruling on A.O. No. 308, President Gloria Macapagal-Arroyo issued
Executive Order (E.O.) No. 420 in 2005 requiring all government agencies and government-owned corporations to streamline their identification systems for the purpose of consolidating and improving existing
government identity systems.

Congressional critics filed House Resolution No. 611 on 22
February 2005, reiterating their opposition to the national identifica-
tion system and questioning the constitutionality of E.O. No. 420.
Congressmen Francis Escudero, Eduardo Zialcita and Lorenzo Tanada
III, members of human rights groups, namely Nation (*Bayan*), Human
Rights (*Karapatan*), Courage, Science (*Agham*), the National Council of
Churches and the Association of Major Religious Superiors of the
Philippines, petitioned the Supreme Court to declare E.O. No. 420
unconstitutional (*Kilusang Mayo Uno v. The National Economic Development
Authority and the Department of Budget and Economic Development* 2006).
However, this time the Supreme Court upheld the constitutionality of
E.O. No. 420. In its ruling, the Supreme Court recognised the authority
of the executive to issue orders for the purpose of reducing govern-
ment costs while enhancing the ability of government to provide
service to the people. The Supreme Court did not consider this order

as a usurpation of the legislative power of Congress by the President. The ruling stated:

> On its face, E.O. No. 420 shows no constitutional infirmity because it even narrowly limits the data that can be collected, recorded and shown compared to the existing ID systems of government entities. E.O. No. 420 further provides strict safeguards to protect the confidentiality of the data collected, in contrast to the prior ID systems which are bereft of strict administrative safeguards. The right to privacy does not bar the adoption of reasonable ID systems by government entities. In the present case, E.O. No. 420 does not establish a national ID system but makes the existing sectoral card systems of government entities less costly, more efficient, reliable and user-friendly to the public.

Based on these Supreme Court rulings, two considerations must be addressed with regard to the establishment of a unified national identification system: (1) the need for a law with the necessary budget appropriation, and (2) the necessity of placing appropriate safeguards in the law to protect the citizen's right to privacy and the security of information contained in the database. The same considerations must also be addressed in the establishment of a national DNA database.

Custodian of the national DNA database

Another important issue raised is the identification of the appropriate agency to establish and maintain a national DNA database. Should the DNA database be entrusted to law enforcement agencies tasked with investigating crimes or to an independent institution that is not involved in the investigation? If the task is assigned to law enforcement, which agency, for example the NBI or the PNP, should be the database custodian? How would that affect the investigations made by the other agency? In the absence of an actual DNA database law, this issue is best examined by studying the various bills already filed in the legislature, for an understanding of the influence in this matter of the complexities of Philippine history and politics.

Moves to create a national DNA database began in 1998 with Senate Bill (S.B.) No. 914, filed by Senator Robert Barbers, which proposed The National Bureau of Investigations Reorganization and Modernization Act of 1998. In the bill, the control and supervision of the database was assigned to a reorganised NBI charged with functioning (1) as a national clearing house for all criminal records and (2) as a repository for identification records such as fingerprints and dental records. The bill assigned to the NBI the task of establishing and maintaining modern forensic science laboratories nationwide and of promoting the application of scientific knowledge in criminal investigation. The bill, if enacted, would have given the NBI director the power to 'establish policies and standards for effective and economical operation of the office in accordance with the programs of government'.

Another bill, S.B. No. 1765, known as the DNA Act of 2001, was filed by Senator Gregorio Honasan for the creation and establishment of a centralised, nationwide DNA database. The agencies identified to generate and maintain the database were the NBI, the PNP and the UP Natural Sciences Research Institute (UP-NSRI) DNA Analysis Laboratory. The bill identified the Department of Health as the agency that would coordinate with these laboratories in obtaining database information needed for health research. The bill restricted the collection of samples from an individual to only when (1) the person provided consent in writing, (2) the person had been convicted of an offence or crime, or (3) the person was detained and there was reasonable ground to believe that s/he was involved in another crime. Under S.B. No. 1765, only the DNA laboratories of PNP, NBI and UP, the accused and courts wherein the case was filed have access to the genetic information relevant to the case.

In 2001 and 2002, in similar but separate Senate bills, Senators Renato Cayetano (S.B. No. 1172) and Manuel Villar, Jr (S.B. No. 2300) questioned the objectivity of laboratory results conducted by law enforcement laboratories, given their involvement in the prosecution of crimes. Both bills recognised the need to regulate access of law enforcement to the national DNA database.

In 2006, Senator Edgardo Angara filed S.B. No. 2245, otherwise known as the DNA Analysis Enhancement Act of 2006. This bill proposed the establishment of a comprehensive national DNA database; the enhancement of the DNA testing capacities of the DNA laboratories of NBI, PNP and UP-NSRI; the support of research into new DNA testing technologies; the provision of training programmes for the appropriate collection, handling and storage of DNA evidence; and the provision of funds for post-conviction DNA testing. The overall goal was to improve the criminal justice system. The bill also pushed for the creation of a DNA Advisory Board that would include scientists working in public and private forensic laboratories as well as those not affiliated with a forensic laboratory. Notably, the status of this Board was merely recommendatory. The power to determine the quality assurance protocols required of a forensic laboratory, to establish the requirements for the performance of DNA analysis and to formulate standards for testing the proficiency of laboratories and analysts would be given to the NBI director. A parallel bill, House Bill No. 136, which was identical to S.B. No. 2245, was also filed in the House of Representatives by Senator Edgardo Angara's son, Congressman Juan Edgardo Angara.

Unlike the bills filed by the two Angaras, House Resolution No. 659, filed by Congressman Lorenzo Tanada III on 13 March 2005, called on Congress 'to institutionalize and maximize the value of forensic DNA technology in forensic and crime scene investigations in the country' by establishing a DNA technology center at UP. This House resolution proposed entrusting the management of the DNA database solely to a DNA technology center that would be established by a DNA technology law. The proposed assignment of the custody and management of the DNA database to an institute other than to law enforcement agencies was intended to improve the protection of the population against unreasonable searches and misuse that might occur if law enforcement laboratories had uncontrolled access to the information derived from a national database (L. Tanada, personal communication).

In a 2006 meeting of the Subcommittee on Research and Development and Technology Transfer of the Congressional Committee on Science and Technology, Maria Socorro Diokno, Secretary General of the Free Legal Assistance Group (FLAG; an organisation of lawyers committed to the protection of human rights), spoke about the importance of restricting data access as a safeguard against possible misuse of the genetic information contained in the database and of the excess biological samples that were stored after the DNA profiles were generated. The compilation of important personal as well as genetic information in a single database might tempt the unscrupulous to make use of the database for personal gain. Diokno cited the 'perception that the government agencies performing DNA testing [were] corrupt, inefficient and abusive' (Committee on Science and Technology 2006). The potential for the misuse of information in general crime databases was earlier recognised by Senator Miriam Defensor-Santiago in 1999 (S.B. No. 1636) and Congressman Gerry Salappudin in 2001 (House Bill No. 3452), who proposed legislative action against law enforcement personnel involved in lawless undertakings. Diokno reiterated the need to identify the appropriate agency to establish and maintain the national DNA database in order to assure the public of the integrity of any investigation which might make use of the database (Committee on Science and Technology 2006).

This resistance against unrestricted government access to personal information voiced by Maria Socorro Diokno on behalf of FLAG may have originated from the experience of her father, the late Senator Jose Diokno, and her knowledge of FLAG cases. Senator Diokno had been detained by the military from 1972 to 1974 without being charged in a civilian court. After his release in 1974, Senator Diokno had founded FLAG with lawyers and with other political opponents of the Marcos Government (Bailen 1998). Since its foundation, FLAG has taken the lead role in prosecuting erring military personnel and policemen (Free Legal Assistance Group 2006).

When Martial Law was declared on 21 September 1972 by Proclamation No. 1081, President Marcos assumed absolute powers and suspended many civil rights on the pretext of national security. Marcos assumed the legislative responsibilities of Congress, and he directly controlled the judiciary by the creation of military commissions to try civilians, even if civilian courts continued to exist in a semi-independent state (Civil Liberties Union of the Philippines 1975). Persons were arrested on an Arrest, Search and Seizure Order (ASSO) issued by President Marcos or the Secretary of National Defense. Arrests were made by members of the Armed Forces of the Philippines, the Integrated National Police, which later became the PNP, and other law enforcement agencies. (General Order No. 60, 1977, pp. 1–2; Letter of Instruction No. 621, 1977). Some 70 000 people are believed to have been detained, tortured and/or killed during the 14-year Marcos regime, which was overthrown in 1986 by a People Power Revolution (Task Force Detainees of the Philippines 1998).

After the overthrow of the Marcos regime, there was renewed hope for a fresh beginning of a more democratic society. However, even to the present day, although the number of persons killed or missing differs between government agency estimates and those provided by human rights groups, the occurrence of extrajudicial killings and arrests of civilians without a warrant continue (Puno 2007). As of 2007, the estimated number of victims varies from 100 to more than 2000 civilians (Sevilla 2007). Groups such as Families of Involuntary Disappearance, Human Rights (Karapatan) and the Asian Human Rights Commission consider members of the military and police to be responsible.

To address the painful problem of the growing number of victims, President Macapagal-Arroyo ordered the formation of an

independent group headed by retired Supreme Court Justice Jose Melo
and known as the Melo Commission to conduct public hearings to
determine the veracity of claims made in each case of extrajudicial
killing or forced disappearance (Melo Commission 2007). In a report
released on 22 January 2007, the Melo Commission concluded that
'there is no direct evidence linking some elements in the military to
the killings; there is no official or sanctioned policy on the part of the
military or its civilian superiors to resort to what others euphemisti-
cally call alternative procedures – meaning 'illegal liquidations' (Melo
Commission, 2007). General Hermogenes Esperon, Jr, Armed Forces of
the Philippines Chief of Staff, categorically denied the accusations of
military involvement made by human rights groups (Ebdane 2007).
In the same meeting, General Oscar Calderon, PNP Chief, identified
sources of the problems in resolving cases of extrajudicial killings,
which include the reluctance of witnesses to cooperate in investigations out of fear of reprisals, limited forensic equipment and the
absence of a centralised database on missing persons and recovered
cadavers (Calderon 2007). He proposed solutions focused on the need to
strengthen the government's witness protection programme, to
upgrade investigation facilities and equipment, to conduct training
on crime scene investigation and to develop the missing persons database (Calderon 2007).

Another noticeable issue is the duplication of functions of the
PNP and the NBI, which are both mandated to investigate crime and
other offences against Philippine laws (Republic Acts No. 157 and
6975). A study commissioned by the Philippine Supreme Court
identified the problems arising from a lack of demarcation of functions between these two agencies. They included potential operational conflicts and conflicts between institutions; contradictory
results from investigations, which might lead to problems for the
prosecution; an erosion of accountability; and inefficiency and was-
tage of human, financial and material resources (Center for Public
Resource Management 2003). To strengthen the law enforcement
capabilities of both agencies, the study recommended that the functions of each agency must be properly delineated (Center for Public
Resource Management 2003). It remains to be seen whether the two
agencies, with defined functions, can work together to establish a
single and larger national DNA database, rather than establishing
separate DNA databases. It is not cost-effective to maintain more
than one national DNA database, each containing some but not all
DNA profiles. For example, the DNA profiles of repeat offenders

may be included in one or both databases, thereby creating problems in annotating and matching information from several crime scenes or victims.

A question of economics

Given the high cost of infrastructure, equipment, chemicals and salaries for technical personnel needed for DNA profiling, analysis and databasing, the establishment of a national DNA database would entail a political decision to allocate a significant measure of resources from an already limited resource pool. For developing countries such as the Philippines, this would imply a choice between spending for basic healthcare, education and poverty amelioration programmes with perceptible immediate short-term effects or spending on a database that could provide benefits for the longer term. For example, the cost of creating a DNA database of all the convicted offenders already detained in seven national penitentiaries by 2002, consisting of 25 002 samples (Center for Public Resource Management 2003), could be conservatively estimated to cost at least one billion Philippine pesos (PhP). This amount, which does not include the DNA analysis of crime scene samples and infrastructure costs, represents more than 1.5% of the PhP 65.2 billion allocated for defence in a national budget for 2009 of PhP 1.415 trillion (Ambat 2008). This cost would rise considerably if the national database included samples from persons detained in 79 provincial, 25 subprovincial, 135 district, 1003 municipal and 85 city jails (Center for Public Resource Management 2003). At the time of writing, the actual number of detainees in these jails was not available.

Hence, the cost of establishing and maintaining a national DNA database must be balanced against the need to improve national security and crime prevention. With the current global economic crisis, which led to a double-digit inflation rate in the Philippines in 2008, the country faces a serious challenge in improving the lives of its citizens (Ambat 2008). The Asian Development Bank identified conflict and security issues, particularly the continued problem of terrorism in Mindanao, Southern Philippines, as being amongst the five major causes of poverty in the Philippines (Asian Development Bank 2005). When national security is not maintained, poverty can result from failure of government to maintain the rule of law. Advocates who are pushing for the enactment of laws to decrease criminality argue that such laws are necessary to increase investor confidence and contribute to economic growth. Hence, the Arroyo administration declared its commitment to fighting for peace and good governance with the aim of alleviating poverty (Ambat 2008).

DE FACTO FORENSIC DNA DATABASES IN THE PHILIPPINES

To date, there are three de facto forensic DNA databases in the Philippines. These databases are maintained at the DNA laboratories of the NBI, PNP and UP-NSRI. In the absence of any legislation on forensic databases, biological samples related to all offence types have been entered into these databases. The NBI DNA laboratory became operational in 1999. In 2001, the PNP also established its DNA laboratory. These two laboratories now handle the majority of criminal cases. Because they receive their material from crime scene investigators, the forensic databases of NBI and PNP are much larger than that of the UP-NSRI. Based on current developments, one can presume that the data contained in the de facto DNA databases maintained by NBI and PNP have significantly increased since the promulgation of the *Rule on DNA Evidence*, as more parties take advantage of DNA testing.

Meanwhile, the forensic database maintained at the UP-NSRI consists of two separate, non-overlapping sections: the reference population genetic database and the casework database (including data pertaining to both civil and criminal cases). The UP-NSRI reference population genetic database comprises samples from persons who have volunteered a biological sample for population genetic research and for inclusion in a reference database (Lessig *et al.* 2003; De Ungria *et al.* 2005; Salvador *et al.* 2007). This database consists of over 2000 DNA profiles and is in an electronically searchable format, with all personal individual information removed from the profiles. All biological samples left after the DNA profiles have been generated are stored indefinitely. In contrast, the casework database is in a non-searchable format and comprises over 3000 DNA profiles.

In criminal casework, the UP-NSRI laboratory has conducted DNA tests in 12 criminal paternity cases (De Ungria *et al.* 2008) and 156 sexual assault cases (Delfin *et al.* 2005), including the case of *People v. Yatar* (2004). Of the 12 criminal paternity cases, nine accused were found not to have fathered the victim's child and five had already been acquitted by their respective courts. In one case, *People v. de Villa* (2001),

DNA test results were used to obtain a pardon for de Villa, who had been initially sentenced to death, with the sentence later commuted to life imprisonment because of his advanced age (De Ungria *et al.* 2008). The convicted offenders in two other cases, both serving life sentences, are using their DNA test results and the new *Rule on DNA Evidence* to attempt to have their convictions overturned. In cases involving children, the UP-NSRI laboratory has been able to conduct DNA tests on samples collected from child victims and submitted by the network of child protection units (CPU-Net) to the laboratory. Many of these cases have not proceeded to court because of the economic costs and for social reasons such as pressure from family members. The suspected perpetrators of the abuse are commonly a person known to the child, such as the child's father, stepfather, uncle, brother or grandfather (B. Madrid, personal communication). In these cases, reference samples have not been collected from suspects for DNA testing.

As a matter of policy, genetic information obtained from one case may be used in another case only if the concerned parties have given their consent. To date, the UP-NSRI laboratory has not performed any search on DNA profiles stored in its casework database. Biological samples are routinely stored for five years by agreement between the laboratory and the parties concerned. Biological samples associated with criminal investigations are stored indefinitely or until the court orders the physical destruction of said samples. To prevent unauthorised access to sensitive information, records of genetic information in both sections of the UP-NSRI database are archived in computers not accessible via the Internet to ensure protection against unwanted or inappropriate intrusions.

CONCLUSIONS

What then are the implications for governance? The Philippine justice system has adopted forensic DNA technology as an instrument to expedite the administration of justice. To fully utilise the new technology, the Philippines needs to take the next step by establishing a national DNA database. The *Rule on DNA Evidence* has already opened the doors to a DNA database by providing guidelines for the proper collection, handling and storage of biological samples. The viability of a national DNA database is dependent on congressional acceptance of government access to an individual's private affairs in exchange for security. The use of forensic databases to fight crime and terrorism and in aid of identification of missing persons and disaster victims is widely accepted (Asplen 2008). However, not only is the establishment of such databases for criminal investigation and security expensive but in addition the databases are open to abuse if the appropriate safeguards and controls are not put in place (Thompson 2007). In developing countries such as the Philippines, where scarce resources have to be allocated among different pressing needs – including basic ones such as shelter, food and access to clean water – and where the temptation for personal gain by misuse of the database may be strong, it is imperative for the legislature to find creative solutions to protect the credibility of the institution and the objectivity of the technology at a reasonable cost.

REFERENCES

Agabin, P. (2003). DNA as evidence. Lecture at the *Conference on Bioscience and Biotechnology: Science and the Law*, Philippine Judicial Academy Development Center, Tagaytay City.

Ambat, G. (2008). *Conquering Poverty: Funding Requirements, Issues and Challenges.* Manila: Senate Economic Planning Office.

Asian Development Bank (2005). *Poverty in the Philippines: Income, Assets and Access.* Manila: Asian Development Bank.

Asplen, C. (2008). The DNA connection: international updates. *Forensic Magazine*, April/May. www.forensicmag.com/ (accessed 29 June 2009).

Bailen, A. (1998). *The Odyssey of Lorenzo M. Tañada.* Quezon City: University of the Philippines Press.

Bartolome, N. (2008). *PNP Gets Interpol Support to Develop Counter Cyber Crime and Bioterrorism Capability.* [News Release No. 08-0306.] Quezon City: Philippine National Police.

Bureau of Democracy Human Rights and Labor (2007). *Country Report on Human Rights Practices, 2006.* Washington, DC: US Department of State.

Calderon, O. (2007). Report from the Philippine National Police. In *National Consultative Summit on Extrajudicial Killings and Forced Disappearances: Searching for Solutions.* Manila: Supreme Court.

Center for Public Resource Management (2003). *Strengthening the Other Pillars of Justice through Reforms in the Pillars of Justice.* Manila: Supreme Court.

Civil Liberties Union of the Philippines (1975). *The State of the Nation after Three Years of Martial Law.* Manila: Civil Liberties Union of the Philippines.

Committee on Science and Technology (2006). *Prospects of DNA Technology.* [Summary of the Meeting of theSubcommittee.] Quezon City: House of Representatives Committee News.

de Leon, H. (2002). *Textbook on the Philippine Constitution.* Quezon City: Rex Printing.

Delfin, F., Madrid, B., Tan, M. *et al.* (2005). Y-STR analysis for detection and objective confirmation of child sexual abuse. *International Journal of Legal Medicine*, 119, 158–163.

De Ungria, M., Roby, R., Tabbada, K. *et al.* (2005). Allele frequencies of 19 STR loci in a Philippine population generated using *AmpFlSTR* multiplex and ALF singleplex systems. *Forensic Science International*, 152, 281–284.

De Ungria, M., Sagum, M., Calacal, G. *et al.* (2008). Forensic DNA evidence and the death penalty in the Philippines. *Forensic Science International: Genetics*, 2, 329–332.

Ebdane, H. (2007). Report from the Armed Forces of the Philippines. In *National Consultative Summit on Extrajudicial Killings and Forced Disappearances: Searching for Solutions*. Manila: Supreme Court.

Encinas-Franco, J. (2005). *National Identification System: Do we Need One?* Manila: Senate Economic Planning Service.

European Commission (2007). *Furthering Human Rights and Democracy Across the Globe*. Brussels: European Commission.

Free Legal Assistance Group (2006). *Newsletter*. Manila: Free Legal Assistance Group http://flagfaqs. blogspot.com/ (accessed 10 April 2008).

Hood, R. (2002). *The Death Penalty: A Worldwide Perspective*. Oxford: Oxford University Press.

Lessig, R., Willuweit, S., Krawczak, M. *et al*. (2003). Asian online Y-STR haplotype reference database. *Legal Medicine (Tokyo)*, 5(Suppl. 1), 160–163.

Melo Commission (2007). *Report of the Independent Commission to Address Media and Activist Killings*. Manila: Melo Commission.

Ochave, J. (2003). The double helix in chambers: forensic DNA evidence in criminal investigation and prosecution. *Integrated Bar of the Philippines*, 29, 88–115.

Office of the Coordinator for Counterterrorism (2008). *Country Reports: East Asia and Pacific Overview*. Washington, DC: Department of State.

Orendain, J. (2008). *Not in our name*. Quezon City: Free Legal Assistance Group.

Puno, R. (2007). Keynote address. In *National Consultative Summit on Extrajudicial Killings and Forced Disappearances: Searching for Solutions*. Manila: Supreme Court.

Saks, M. and Koehler, J. (2005). The coming paradigm shift in forensic identification science. *Science*, 309, 892–895.

Salvador, J., Calacal, G., Villamor, L. *et al.*(2007). Allele frequencies for two pentanucleotide STR loci Penta D and Penta E in a Philippine population. *Legal Medicine (Tokyo)*, 9, 282–283.

San Juan, J. (2007). Expedite human rights cases, prosecutors told. *Business Mirror*, 2, 1–2.

Sevilla, N. (2007). Report from the families of involuntary disappearance. In *National Consultative Summit on Extrajudicial Killings and Forced Disappearances: Searching for Solutions*. Manila: Supreme Court.

Solidarity Philippines-Australia Network (1997). Ramos orders ID card. *Kasama*, 11, 1–5.

Supreme Court of the Philippines (1989). *Revised Rules on Evidence*. Manila: Supreme Court.

Supreme Court of the Philippines (2007). *Rule on DNA Evidence*. Manila: Supreme Court.

Task Force Detainees of the Philippines (1998). Philippines – political prisoners: the forgotten heroes. *Human Rights Solidarity*, 8, 1626.

Thompson, W. (2007). The potential for error in forensic DNA testing. *GeneWatch* (USA), 21, 1–43.

USAID (2008). *Country Health Statistical Report: Philippines*. Washington, DC: US Agency for International Development.

CASES AND LEGISLATION

Administrative Order No. 308, 1996.

Andal v. *People* (1999). Supreme Court General Register No. 138268 and 138269-69. En Banc.

Executive Order No. 420, 2005.
General Order No. 60, 1977.
Herrera v. *Alba* (2005). Supreme Court (First Division) General Register No. 148220 (Justice Antonio T. Carpio).
House Bill No. 3452, 2001.
House Bill No. 136, 2006.
House Resolution No. 611, 2005.
House Resolution No. 659, 2005.
In Re: Writ of habeas corpus for Reynaldo de Villa (2004). Supreme Court General Register No. 158802. En Banc.
Kilusang Mayo Uno v. *The National Economic Development Authority and the Department of Budget and Economic Development* (2006). Supreme Court General Register No. 167798. En Banc.
Letter of Instruction No. 621, 1977.
Ople v. *Torres* (1998). Supreme Court General Register No. 1276 85. En Banc.
People v. *Corpuz* (1993). Supreme Court General Register No. 101005. En Banc.
People v. *de Villa* (2001). Supreme Court General Register No. 124639. En Banc.
People v. *Echegaray* (1996). Supreme Court General Register No. 117472. En Banc.
People v. *Mateo* (2004). Supreme Court General Register No. 147678-87. En Banc.
People v. *Paras* (1999). Criminal case no. 85974–85978, Pasig City, Branch 163.
People v. *Vallejo* (2002). Supreme Court General Register No. 144656. En Banc.
People v. *Yatar* (2004). Supreme Court General Register No. 150224. En Banc.
Philippine Constitution, 1987.
Proclamation No. 1081, 1972.
Republic Act No. 157, 1957.
Republic Act No. 9346, 2006.
Senate Bill No. 914, 1998.
Senate Bill No. 1636, 1999.
Senate Bill No. 1172, 2001.
Senate Bill No. 1765, 2001.
Senate Bill No. 2300, 2001.
Senate Bill No. 2245, 2006.
Tijing v. *Court of Appeals* (2001). Supreme Court (Second Division). General Register No. 125901. March 8, 2001 (Justice Leonardo A. Quisumbing).

Section 3 Conclusions

16

Beyond borders: trends and challenges in global forensic profiling and databasing

This book did not set out to evaluate or assess the phenomenon of forensic DNA databasing in any normative way. As editors, we did not approach this book project with a particular message in mind that we sought to get across. Neither did we encourage our contributors to present linear and causal interpretations of the emergence and governance of forensic DNA databases. Instead, we encouraged them to provide 'thick descriptions' (Geertz 1973) of how forensic DNA profiling and databases operate in different criminal justice systems, and in different societal/political systems, more broadly.

In this context, the analyses were carried out in the tradition of analysing technology-in-practice (Timmermans and Berg 2003). This means that we sought to avoid technological determinism, namely the analysis of policy-making and societal developments in the field of forensic DNA databasing as resulting from, or being driven by, the technology that underpins practices in the field. We also sought to avoid the opposite approach, namely to regard forensic DNA databases merely as a technology through which political and societal values are enforced and realised (Timmermans and Berg 2003). In other words, our contributors set out to provide a better understanding of how forensic DNA databases are enacted, discussed and regulated, and what the implications of these insights are for governance.

To allow for a strong comparative component, we invited contributions from authors at the forefront of discussions on regulatory, ethical and operational governance aspects of forensic DNA databases in a range of countries. Subsequently, the contributions to this volume span a wide range of cases in terms of geography, stakeholder perspectives and the extent of development of forensic DNA databasing.

Genetic Suspects: Global Governance of Forensic DNA Profiling and Databasing, ed. Richard Hindmarsh and Barbara Prainsack. Published by Cambridge University Press. Copyright © Cambridge University Press 2010.

Despite the different focal points across the chapters, there are several themes that most contributions have identified as crucial and particularly challenging for the governance of forensic DNA databasing in the contemporary era of the biopolitics of the life sciences.

FUNCTION CREEP

The term function creep refers to a widening of the scope of purposes for which DNA profiling and databasing are used (Dahl and Rudinow 2009). We are currently witnessing a push, perhaps trend, to expand databases in the majority of countries, with centralised DNA databases for police and forensic use. These database expansions take different forms, from including DNA profiles from a wider range of suspects and arrestees to broadening the kinds of information that can legally be obtained from the analysis of DNA samples (see below). As Hindmarsh argues (Chapter 13), governmental authorities often justify such database expansions by cost-effectiveness or other utility considerations, so that critical voices can be dismissed as unreasonable. Williams (Chapter 7) sees function creep as intimately linked with a prevalent techno-centric policy approach when he argues that government investment in DNA profiling and databasing is stimulated by a narrow understanding of the uses of bioinformation that highlights DNA at the cost of a focus on other aspects of policing. In this sense, the expansion of forensic DNA databases becomes a self-perpetuating 'machine': the more effective forensic DNA technologies promise or prove to be in identifying and convicting offenders, the more the criminal justice system relies on the technology. In turn, the more reliance on the technology, the more important it becomes to improve its efficacy.

COST CONSIDERATIONS

Cost considerations were raised by authors mainly when they criticised actors using the cost argument strategically to justify database expansion, for example, by arguing that the inclusion of a broader range of profiles in the database from suspects to those arrested will eventually bring down investigation costs. However, Machado and Silva (Portugal; Chapter 11) and De Ungaria and José (the Philippines; Chapter 15) draw attention to the struggle of some countries to afford centralised DNA databases. Machado and Silva qualified this as a 'question of the proportionality between the possible benefits of the forensic DNA database and its economic costs in a developing country [Portugal] with a relatively low level of serious crime (in the European context)'. De Ungaria and José contrasted the costs of the establishment of a central-ised forensic DNA database in the Philippines, with the costs of press-ing societal needs – including basic ones such as shelter, food and access to clean water. But countries with large and expensive DNA databases for police and forensic uses are also affected by cost issues. The most pressing implication of scarce financial resources is the problem of backlogs in the analysis of DNA samples, which regularly slow down the identification of suspects. It can be expected that infra-structural problems caused by limited financial resources will increase as a result of the global economic crisis. Private companies have already started to make stronger inroads into criminal justice systems by closing the gaps that lack of public funding has opened. A likely consequence of an increasing number of private companies offering their services to countries with considerable backlogs in the analysis of DNA samples, such as the Netherlands (Chapter 9), might well be a stronger reliance on corporate or private providers of DNA analysis in the criminal justice field in many countries.

DEPARTING FROM THE INFALLIBILITY
ASSUMPTION OF FORENSIC DNA SCIENCE

Although any increasing reliance of the criminal justice system on corporate or private service providers for DNA analysis is one circum-stance in which the governance of DNA databasing might not follow the trajectory of other forensic technologies, it still is important to note, as Cole and Lynch (Chapter 6) remind us, that in many respects, 'procedures for governing DNA data have not been discovered anew but have followed patterns established by earlier forensic and biomet-ric technologies, particularly fingerprinting'. This insight gives rise to the hope that the infallibility assumption – the idea that forensic DNA technologies are the pinnacle of 'sound forensic science' (Lynch et al. 2008; Prainsack and Kitzberger 2009) – will be replaced by more sober assessments of both the strengths and the shortcomings of forensic DNA technologies. So far, the most apocalyptic predictions about for-ensic DNA profiling and databasing, for example that governments will establish universal DNA registries to exploit their access to the genetic 'blueprints' of all citizens, have not materialised. Neither, however, has DNA profiling and databasing fulfilled the expectation of providing ultimate and infallibly truthful answers to questions of guilt. Very accurately in this context, Hindmarsh (Chapter 13) places expectations of forensic DNA technologies in a longer history of using the body – and particularly the current molecular version of it – as a source of truth (see also Lynch *et al.* 2008).

MORE TRANSPARENCY AND BROADER
PUBLIC DEBATES

More sober expectations of forensic DNA technologies necessitate increasing transparency and broader public debates about forensic DNA profiling and databasing. While there are extensive public debates on the governance of medical data collections in virtually all countries covered in this book, discussions about the scientific soundness, governance and oversight of forensic DNA technologies have been limited to professional experts in most countries (with the exception of the UK and, to some extent, the USA). However, as Aronson (Chapter 12) argues, those who decide what counts as 'good science' and 'good governance' in connection with DNA databases should not have vested interests in their use. Indeed, most contributors to this volume argue that civic engagement in decision making about the scope and use of forensic DNA databasing and profiling is important. As Williams (Chapter 7) concludes: 'There is also a pressing need to document and consider the varied ways in which efforts to govern the uses of these technologies reflect and reconstitute understandings of the social and political identities of the subjects whose bio-identities they seek to capture and profile' (Williams and Johnson 2006). Furthermore, as Machado and Silva (Chapter 11) assert, giving voice to public perspectives and their representation is vital for public confidence and good governance.

One concept deserving broader and deeper public discussions is the notion of the 'volunteer' in forensic DNA databasing (Chapters 3, 4 and 11). The volunteer concept in the medical realm, where it typically signifies a person who donates bodily material (e.g. DNA or tissue), to a database or repository for altruistic reasons (usually to foster disease research), cannot be easily transposed to the criminal justice field, where the stakes, risks and benefits are very different. These differences, however, are insufficiently conceptualised at present, and more research needs to be carried out in the field (Prainsack and Gurwitz 2007). Similarly, the notion of 'consent' in connection with volunteering/donating/providing a DNA sample to a DNA database for police and forensic use carries quite different meanings and different weight than it typically does in the medical field. Also in this regard, deeper discussions including perspectives from a broader range of disciplines beyond criminal justice and law enforcement are called for. In this context, Tutton and Levitt's contribution (Chapter 5) provides a useful basis for thinking about similarities and differences between the functional fields of medical and forensic DNA databasing, and about how some concepts can travel between them, while others cannot.

Turning to the potent issue of the reification of 'race', and the proliferation of ethnic biases inherent in the criminal justice system via the use of forensic DNA profiling and databasing, Washington also concludes that 'the public must be informed and invited into the conversations and policy dialogue about race, security and genetic science' (Chapter 4). For similar reasons, authors in this volume call for more education about forensic DNA technologies for jurors (Chapter 4) and defence lawyers (Chapter 10). In addition, Hindmarsh (Chapter 13) deems necessary more robust and balanced media coverage of forensic DNA technologies in contributing to the debate about their scope and use. Although Prainsack (Chapter 8) does not invite more education for prisoners on DNA technologies, she shows how the scientific authority associated with DNA profiling has increased the power gap between those who use the technology and those upon whom it is used. This issue relates to questions about the equitable governance of forensic DNA technologies, which is, in turn, intimately linked to the issue of public trust in their 'correct' use.

PUBLIC TRUST

Deeper and broader public discussions will also enhance public trust with regard to forensic DNA technologies and especially database expansion. Writing from the perspective of law enforcement in Israel, Zadok et al., (Chapter 3) also make a number of recommendations about the use of intelligence-led mass DNA screenings, or DNA dragnets, when all other means have failed, most importantly that they should be used only in conjunction with other intelligence and investigative tools. In line with Prainsack (Chapter 2) and Williams (Chapter 7), Zadok et al. caution against the reflex-like priorisation of DNA technologies in the context of investments (financial and otherwise) and training at the cost of other aspects of police work, such as human experience, observation and judgment. Finally, in Chapter 13, in finding the 'DNA as a language of truth' narrative questionable in view of the many 'biocivic' issues raised about it, Hindmarsh calls for a new consensus-building 'DNA as a language of trust' narrative in the context of a governance shift to participatory governance.

FORENSIC DNA DATABASES: ANALYSING THE EMERGENCE AND GOVERNANCE OF NEW TECHNOLOGIES

The insights of this book shed light on many implications for the analysis of the emergence and governance of new forensic technologies locally and more broadly. There is great variety in the ways in which forensic DNA databases are set up, managed and monitored, for example in the criteria for inclusion (and later deletion) of DNA profiles in the database and in the policies and practices pertaining to the storage and the location of DNA samples. There is also considerable variety among countries in the public and regulatory discourses that preceded and accompanied the establishment of national forensic DNA databases – a process that is still ongoing in some countries – or the discourses that relate to new development. For example, the current trend towards an expansion of the scopes and uses of DNA databases for police and forensic purposes is now attracting increasing discussion and can meet different levels of public concern in different countries. In contrast to the findings of most contributors, Veth and Midgley (Chapter 14) found that expansion of the forensic DNA database in New Zealand was not accompanied by any public controversies; instead, the public supported it as a further step to provide an effective tool for fighting crime. Even within those countries where such concerns led to greater public debate, such as the USA (Chapter 12), the focus and sometimes even the direction of the concerns were different. For example, as Machado and Silva point out in Chapter 11, low levels of public confidence in the efficiency of forensic DNA databasing in Portugal are mirrored by low levels of trust 'in public institutions in general, and in the justice system in particular, which is generally seen as corrupt'.

Therefore, it is apparent that obtaining and maintaining public trust in forensic DNA databases depends on a much wider range of factors than the design and the oversight model of the database itself (Gottweis and Petersen (2008) outline similar insights on the governance databases in the medical realm). The variety in both regulatory approaches and public debates pertaining to forensic DNA databasing results from a variety of different operational and political traditions – that is, of different established practices and understandings of 'how things are done', for example in relation to granting access to the database, appointing monitoring authorities, and so on. Such variety also results from differences in what Gottweis calls political metanarratives, which 'describe general concepts and values of the social order' (Gottweis 1998: 33). An example of the productive force of such metanarratives would be the high level of trust that the public in Norway or Austria have in forensic science institutions that are run by, or affiliated with, the national government (Chapters 8 and 10). As Dahl diagnoses, '[r]etaining DNA analysis within the realm of government-owned institutions is perceived as helping to ensure the rule of the law'. In many other countries, such intimate links between the production of science and governmental authority would arguably be grounds for caution rather than increased trust (Pinkerton 2009). Issues related to surveillance (e.g. Lyon 2006) are addressed by Hindmarsh (Chapter 13) in the Australian context and by Williams (Chapter 7) in the context of England and Wales.

Here, again, questions about 'good' governance of forensic DNA databases cannot be answered at a general level. What is 'good' in this context depends on the particular configurations of collective values, norms and narratives in each country – even if we agreed that 'good' would need to meet the requirement of striking a fair balance between the public interest in efficient criminal investigation on the one hand, and individual civil rights and liberties on the other, as most contributors argued.

This difficulty of producing any generalised imperatives for 'good governance' in this field poses challenges for the increasing move towards transnational collaboration in criminal investigation. In the European context, The Prüm Decision was signed by Belgium, Germany, Spain, France, Luxembourg, the Netherlands and Austria in 2005 and is an instructive example of this trend. Three years after the Decision was signed, the European Union (EU) Council adopted it into EU law, which now obliges every EU Member State to render their DNA, motor vehicle data and fingerprint databases searchable on a hit/no-hit basis for authorities in other EU countries by 2011. Those countries that do not yet have centralised databases for forensic and police use will now need to establish them.

Why is this development instructive? Because it remains to be seen what spill-over effects the implementation of the Prüm Decision will have on national legislations and policy practices. For example, it has been argued that by allowing one-locus mismatches between DNA profiles to be treated as 'matches' in the first instance, the Prüm

Decision could allow for familial searching 'through the back door' in countries where the practice is currently prohibited. The Netherlands, for example, has already initiated a change in legislation to legalise familial searching, arguably to preempt such a scenario (Prainsack and Toom, unpublished data). Whether legal change aligns with principles of 'good governance' of DNA databases in a given country depends on whether or not the public and other stakeholders deem these changes compatible with other values and norms that organise their social and political space. For example, acceptance of the possibility that biological relatives can become the subject of police investigations purely on the basis of their biological relatedness to somebody whose DNA profile is stored in a police database depends on cultural and legal traditions and practices. These include negotiating the boundary between individual rights and common interests, common understandings of kinship and how the practice of familial searching – if and when legalised – would be enforced in practice.

SUMMARY

In conclusion, as the contributions to this volume have shown, an analysis that takes into consideration how DNA databases operate in practice, rather than focusing only on the formal and institutional regulatory framework, has facilitated a range of useful insights into the societal, regulatory and operational implications of these practices. These insights also draw our attention to what DNA databases for police and forensic uses 'do' beyond the boundaries of the criminal justice field. We suggest that this exercise contributes well to informing a range of audiences, both professional and non-professional (because anyone at one time or another might be cast as a 'genetic suspect'), and that it will aid individual and collective deliberation and decision making on this very important topic.

REFERENCES

Dahl, J. Y., and Rudinow, S. (2009). 'It all happened so slowly': on controlling function creep in forensic DNA databases. *International Journal of Law, Crime and Justice*, 37, 83–103.
Geertz, C. (1973). Thick description: toward an interpretive theory of culture. *The Interpretation of Cultures: Selected Essays*, Ch. 3. New York: Basic Books, pp. 3–30.
Gottweis, H. (1998). *Governing Molecules: The Discursive Politics of Genetic Engineering in Europe and in the United States*. Cambridge, MA: MIT Press.

Gottweis, H. and Petersen, A. (eds) (2008). *Biobanks: Governance in Comparative Perspective*. London: Routledge.

Lynch, M., Cole, S. A., McNally, R. *et al.* (2008). *Truth Machine: The Contentious History of DNA Fingerprinting*. Chicago, IL: University of Chicago Press.

Lyon, D. (ed) (2006). *Theorizing Surveillance: The Panopticon and Beyond*. Portland, OR: Willan.

Pinkerton, J. (2009). DNA backlog still plagues HPD crime lab. *Houston Chronicle*, 2 October http://www.allbusiness.com/government/government-bodies-offices-regional-local/13102321-1.html (accessed November 2009).

Prainsack, B. and Gurwitz, D. (2007). 'Public fears in private places?' Ethical and regulatory concerns regarding human genomics databases. [Editorial.] *Personalized Medicine Special Focus Issue*, 4, 447–452.

Prainsack, B. and Kitzberger, M. (2009). DNA behind bars: 'other' ways of knowing forensic DNA technologies. *Social Studies of Science*, 39, 51–79.

Timmermans, S. and Berg, M. (2003). The practice of medical technology, *Sociology of Health and Illness*, 25, 97–114.

Williams, R. and Johnson, P. (2006). Inclusiveness, effectiveness and intrusiveness: issues in the developing uses of DNA profiling in support of criminal investigations. *Journal of Law, Medicine and Ethics*, 34, 234–247.

Index

Printed in the United States
by Baker & Taylor Publisher Services